Carbon Accounting for Sustainability and Environmental Management

As the world increasingly focuses on sustainability and climate justice, this book sets the scene by establishing the significance of carbon accounting in today's environmental management practices.

It provides a comprehensive exploration of the subject, with a specific focus on the Chinese context and sheds light on how carbon accounting practices are being integrated into corporate and national strategies. While the book has a strong focus on empirical realities in China, its exploration of carbon accounting and environmental management holds international importance. The book bridges the gap between theory and practice, offering readers in-depth insights into the intricate dynamics of carbon accounting and its implications for sustainable development, drawing on data and case studies to provide practical insights into the effectiveness of various carbon accounting approaches and their impact on environmental sustainability. It focuses on the crucial role of the development of green accounting, its future growth, and its wider impact on climate change. Additionally, the book examines how accurate tracking and reporting of carbon emissions are integral to developing effective environmental strategies and evolving environmental policies. Accompanied by real-world case studies and actionable recommendations, this book is a valuable resource for anyone navigating the intricate landscape of carbon accounting and reporting, offering a road map to informed strategic decision-making and sustainable business practices.

It will be particularly beneficial for professionals in environmental management, policy formulation, and corporate sustainability, as it translates complex carbon accounting concepts into tangible, practical strategies.

Tarek Rana is an associate professor of accounting at RMIT University, Melbourne, Australia.

Md Jahidur Rahman is an assistant professor of accounting at Wenzhou-Kean University in Wenzhou, China.

Peter Öhman is a professor and head of business administration at Mid Sweden University, and director of Centre for research on Economic Relations (CER).

For more information about this series, please visit www.routledge.com/Routledge-Studies-in-Accounting/book-series/SE0715

Carbon Accounting for Sustainability and Environmental Management

Case Studies from China

Edited by Tarek Rana, Md Jahidur Rahman and Peter Öhman

Routledge
Taylor & Francis Group

LONDON AND NEW YORK

First published 2025

by Routledge
4 Park Square, Milton Park, Abingdon, Oxon OX14 4RN

and by Routledge
605 Third Avenue, New York, NY 10158

Routledge is an imprint of the Taylor & Francis Group, an informa business

British Library Cataloguing-in-Publication Data
A catalogue record for this book is available from the British Library

ISBN: 978-1-032-78678-0 (hbk)
ISBN: 978-1-032-78685-8 (pbk)
ISBN: 978-1-003-48896-5 (ebk)

DOI: 10.4324/9781003488965

Typeset in Times New Roman
by Apex CoVantage, LLC

Tarek Rana dedicates this book to his wife, Afsana
Siddique, in recognition of her unconditional love,
boundless inspiration, and continuous support throughout
his career.

Md Jahidur Rahman dedicates this book to his beloved
wife, Rakiba Sultana, with deep thanks for her endless love
and support, which have been a steady source of strength
and inspiration throughout his life.

Peter Öhman dedicates this book to his family and
friends in gratitude for their unwavering love, steadfast
encouragement, and constant support, which have been his
foundation throughout his academic journey and life.

Contents

Figures

Tables

Editors' Biographies

Tarek Rana is an associate professor of accounting at RMIT University in Melbourne, Australia. He holds a PhD from UNSW Australia, an MBA from the University of Saskatchewan, and both a BCom (Hons.) and MCom from the University of Dhaka, Bangladesh. He is a fellow of CPA Australia (FCPA), an associate of the Chartered Institute of Management Accountants (ACMA, UK), and a chartered accountant (CA, Australia & New Zealand). His research interests lie at the intersection of accounting and digitalization, with a focus on risk management, performance measurement, public sector management, and sustainability. Recognized for his academic contributions, Rana has received awards such as the Dean's Award and the College's Award for Research Excellence. In addition to editing several research books and guest-editing special issues in esteemed journals, he has published widely in leading academic journals, including *Accounting, Auditing & Accountability Journal, British Accounting Review, Financial Accountability & Management, Public Money & Management, Business Strategy and the Environment, Intelligent Systems in Accounting, Finance and Management, Meditari Accountancy Research, Qualitative Research in Accounting & Management, Sustainability Accounting, Management and Policy Journal, Journal of Intellectual Capital, Journal of Accounting & Organizational Change*, and *International Journal of Managerial and Financial Accounting*. Beyond academia, he has significant professional experience, having served in partnerships within accounting firms, as well as on boards and governance bodies in executive and non-executive roles.

Md Jahidur Rahman is an assistant professor of accounting at Wenzhou-Kean University, Wenzhou, China. He earned his PhD in accountancy from the City University of Hong Kong, an MBA in business management from Ritsumeikan Asia Pacific University, Japan, and a BBA (Hons.) and MBA in accounting from the University of Dhaka, Bangladesh. He is also a CPA (Certified Practising Accountant, Australia). Rahman's excellence in teaching has been recognized with multiple awards, including the Excellence in Teaching Award, the Most Dedicative Professor Award, and the Advanced Educator Award from the Wenzhou Education Bureau in China. He has edited numerous research volumes and contributed extensively to the field, publishing in prominent journals such as *British Accounting Review,*

Accounting & Finance, International Journal of Auditing, Journal of Contemporary Accounting and Economics, Advances in Accounting, Meditari Accountancy Research, Managerial Auditing Journal, Journal of Family Business Management, Corporate Governance, and *Business Ethics.*

Peter Öhman is a professor and head of business administration at Mid Sweden University, where he also serves as the director of the Centre for Research on Economic Relations (CER). His research centres on behavioural issues in accounting and auditing, with additional interests in the banking and property sectors. He has published extensively in top-tier academic journals, including *Contemporary Accounting Research, Behavioral Research in Accounting, European Accounting Review* and *British Accounting Review,* Professor Öhman is also a prolific author of books, book chapters, and reports, and he has presented at numerous international conferences. He is an active member of several editorial boards and serves as a reviewer for high-impact international journals. He has earned several scientific and educational awards. Additionally, he supervises multiple doctoral research projects.

Contributors

Jiayi Chen graduated from the Department of Accounting at the College of Business and Public Management, Wenzhou-Kean University, a Sino-American university affiliated with Kean University, NJ, USA.

Zhou Dongying graduated from the Department of Accounting at the College of Business and Public Management, Wenzhou-Kean University, a Sino-American university affiliated with Kean University, NJ, USA.

Xinyi Huang graduated from the Department of Accounting at the College of Business and Public Management, Wenzhou-Kean University, a Sino-American university affiliated with Kean University, NJ, USA.

Zhao Jiayi graduated from the Department of Accounting at the College of Business and Public Management, Wenzhou-Kean University, a Sino-American university affiliated with Kean University, NJ, USA.

Sajjad Hossain Khan has over 20 years of experience teaching accounting at the University of Melbourne, Charles Sturt University, La Trobe University, Federation University, Charles Darwin University, and the University of Dhaka. He specializes in auditing, forensic accounting, management accounting, sustainability research, and has strong academic leadership and publication experience.

Tu Qianqian graduated from the Department of Accounting at the College of Business and Public Management, Wenzhou-Kean University, a Sino-American university affiliated with Kean University, NJ, USA.

Zhou Qin graduated from the Department of Accounting at the College of Business and Public Management, Wenzhou-Kean University, a Sino-American university affiliated with Kean University, NJ, USA.

Xia Qinglu graduated from the Department of Accounting at the College of Business and Public Management, Wenzhou-Kean University, a Sino-American university affiliated with Kean University, NJ, USA.

Xue Wenxin Shirley graduated from the Department of Accounting at the College of Business and Public Management, Wenzhou-Kean University, a Sino-American university affiliated with Kean University, NJ, USA.

Zhang Sijie graduated from the Department of Accounting at the College of Business and Public Management, Wenzhou-Kean University, a Sino-American university affiliated with Kean University, NJ, USA.

YE Xiaowei graduated from the Department of Accounting at the College of Business and Public Management, Wenzhou-Kean University, a Sino-American university affiliated with Kean University, NJ, USA.

Zhou Yanyan graduated from the Department of Accounting at the College of Business and Public Management, Wenzhou-Kean University, a Sino-American university affiliated with Kean University, NJ, USA.

Cheng Yibo graduated from the Department of Accounting at the College of Business and Public Management, Wenzhou-Kean University, a Sino-American university affiliated with Kean University, NJ, USA.

Ying Yilu graduated from the Department of Accounting at the College of Business and Public Management, Wenzhou-Kean University, a Sino-American university affiliated with Kean University, NJ, USA.

Cheng Zeyu graduated from the Department of Accounting at the College of Business and Public Management, Wenzhou-Kean University, a Sino-American university affiliated with Kean University, NJ, USA.

Hongtao Zhu is a PhD student in the Department of Accountancy at the City University of Hong Kong. His research has been published in journals including *British Accounting Review, Managerial Auditing Journal, Accounting & Finance, Journal of Contemporary Accounting & Economics, Advances in Accounting,* and *Meditari Accountancy Research.* He is also a reviewer for academic journals in his field.

Shao Zichun graduated from the Department of Accounting at the College of Business and Public Management, Wenzhou-Kean University, a Sino-American university affiliated with Kean University, NJ, USA.

Preface

In an era where sustainability and environmental management are at the forefront of global concerns, carbon accounting has become a critical tool for aligning corporate strategies with ecological responsibility. This book, *Carbon Accounting for Sustainability and Environmental Management: Studies from China*, edited by Tarek Rana, Md Jahidur Rahman, and Peter Öhman, offers a timely and comprehensive exploration of carbon accounting practices within the context of Chinese enterprises. As one of the world's largest economies, China faces unique environmental challenges driven by rapid industrial growth and increasing pressure from regulatory bodies and stakeholders to mitigate carbon emissions. This volume addresses these challenges by presenting empirical insights and rigorous analyses of carbon accounting in publicly listed firms across China. The chapters within this book are organized to cover a wide array of topics related to carbon accounting, including the influence of ownership structures, corporate governance, and management control systems on carbon disclosure and sustainability performance. Each chapter is based on extensive research, offering readers valuable data and interpretations that illuminate the intricate relationship between carbon accounting practices and corporate sustainability. A key strength of this book is its empirical approach, utilizing data from several years of analysis of Chinese manufacturing and industrial companies. This robust dataset enables comprehensive statistical analysis, yielding insights that are academically rigorous and practically relevant. The findings not only contribute to the academic discourse on environmental accounting but also provide actionable guidance for policymakers, investors, corporate managers, and environmental auditors. One of the core messages is the transformative role of carbon accounting in enhancing corporate transparency, accountability, and financial performance, all while reducing environmental impact. The evidence presented here underscores the critical need for integrating carbon metrics into performance management systems, highlighting the positive impact of carbon accounting on mitigating climate risks, improving resource efficiency, and fostering long-term sustainability. These insights are particularly relevant for stakeholders seeking to understand the implications of carbon accounting within the rapidly evolving regulatory landscape in China. The global relevance of this book is undeniable. As climate-related issues increasingly shape business practices worldwide, the lessons learned from China's experience with carbon accounting offer insights for other regions. This book serves as an essential resource for a global audience interested in sustainable environmental management practices and the intersection of carbon accounting and corporate performance. We are deeply grateful to the contributors whose expertise and dedication have made this book possible. Their research enriches our knowledge of carbon accounting in China and contributes to the broader discourse on corporate sustainability. We also extend our appreciation to our academic institutions for their support and to the reviewers for their invaluable feedback. As we face the environmental challenges of the 21st century, integrating carbon accounting principles into corporate strategies will be vital in advancing sustainable development. This book stands as a testament to the importance of carbon accounting in shaping the future of business and offers a comprehensive guide for those committed to sustainability and environmental management.

Tarek Rana, Md Jahidur Rahman, Peter Öhman
September 2024

Part 1

Carbon Accounting Development and Social Impact

1 Carbon Accounting for Sustainability and Environmental Management

Tarek Rana, Md Jahidur Rahman,
and Peter Öhman

1. Introduction

Carbon accounting, an essential facet of environmental accounting, has gained prominence as organizations strive to address climate change and enhance sustainability practices. This book delves into the various dimensions of carbon accounting, including its evolution, methodologies, frameworks, challenges, and its integration into corporate sustainability strategies. The concept of carbon accounting emerged in response to the increasing recognition of climate change as a critical global issue. Initially, carbon accounting was primarily focused on national-level greenhouse gas (GHG) inventories, driven by international agreements such as the Kyoto Protocol and the Paris Agreement (Guenther et al., 2016). Over time, the focus expanded to include corporate-level accounting as businesses began to realize its significant role in mitigating climate change (Bebbington & Larrinaga-González, 2008).

Various methodologies have been developed to standardize the measurement and reporting of carbon emissions. One of the most influential frameworks is the Greenhouse Gas Protocol (GHG Protocol), which provides comprehensive guidelines for organizations to quantify and manage their GHG emissions (WRI & WBCSD, 2004). The GHG Protocol categorizes emissions into three scopes: Scope 1 (direct emissions from owned or controlled sources), Scope 2 (indirect emissions from the generation of purchased electricity), and Scope 3 (all other indirect emissions that occur in the value chain) (Huang et al., 2013).

Carbon accounting has significant strategic implications for organizations. By identifying emission hotspots and inefficiencies, it enables organizations to develop targeted reduction strategies and improve operational efficiency (Liesen et al., 2015). Furthermore, transparent carbon reporting enhances stakeholder trust and can lead to competitive advantages by differentiating companies as sustainability leaders (Kolbel et al., 2020).

Integrating carbon accounting with corporate sustainability strategies is crucial for achieving meaningful environmental impact. Effective integration requires organizations to align their carbon management practices with broader sustainability goals, such as resource efficiency, waste reduction, and corporate social responsibility (CSR) initiatives (Ioannou et al., 2015). This alignment ensures that

DOI: 10.4324/9781003488965-2

carbon accounting is not conducted in isolation but as part of a holistic approach to sustainability.

Regulatory developments will continue to shape carbon accounting practices. Governments around the world are increasingly implementing policies aimed at reducing GHG emissions, such as carbon taxes, emission trading schemes, and mandatory reporting requirements (European Commission, 2020). These regulatory measures provide a strong incentive for organizations to adopt robust carbon accounting practices and align their operations with climate goals.

Technological innovations are poised to transform carbon accounting practices. Blockchain technology, with its decentralized and immutable ledger, offers potential for improving the traceability and verification of carbon emissions (Kumar et al., 2020). Similarly, Internet of Things (IoT) devices can enable real-time monitoring and data collection, enhancing the accuracy and granularity of emission measurements (Zhang et al., 2021). These technologies can address some of the current challenges in carbon accounting, such as data quality and availability.

2. Background and Motivation

Carbon accounting has emerged as a tool in the quest for sustainability and effective environmental management. This book explores the methodologies, challenges, and implications of carbon accounting, emphasizing its role in corporate sustainability strategies and regulatory compliance. A comprehensive review of existing literature and case studies highlight the importance of accurate carbon measurement, reporting, and verification in achieving environmental objectives.

The growing awareness of climate change and its impacts has led to increased scrutiny of GHG emissions. Being a subset of environmental accounting, carbon accounting has become essential in tracking and managing these emissions (Ascui & Lovell, 2011). This book examines the principles and practices of carbon accounting, exploring its significance in sustainability and environmental management.

Carbon accounting involves quantifying and reporting carbon emissions, typically measured in carbon dioxide equivalents (CO_2e) (Bebbington & Larrinaga-González, 2008). The literature identifies two primary approaches: corporate carbon accounting and national carbon accounting. The former focuses on emissions from business activities, while the latter addresses emissions at the country level (Stechemesser & Guenther, 2012).

Regulatory and market drivers play crucial roles in shaping carbon accounting practices. Government regulations, such as carbon pricing mechanisms and emission trading schemes, incentivize organizations to measure and reduce their carbon footprints (European Commission, 2020). Additionally, market pressures from investors, consumers, and other stakeholders are increasingly demanding greater transparency and accountability in carbon management (CDP, 2020).

The European Union Emissions Trading System (EU ETS) is another prominent example of carbon accounting in action. The EU ETS, a cap-and-trade system, sets a limit on the total emissions allowed for covered sectors and enables organizations to trade emission allowances (European Commission, 2020). This market-based approach incentivizes organizations to reduce emissions and invest in cleaner technologies. The success of the EU ETS in driving emissions reductions underscores the potential of regulatory frameworks to enhance carbon accounting practices.

Accurate carbon accounting requires robust methodologies. As mentioned, the GHG Protocol, developed by the World Resources Institute and the World Business Council for Sustainable Development, is a widely recognized standard (WRI & WBCSD, 2004). The GHG Protocol outlines methods for measuring emissions across scopes described below.

Scope 1 emissions are direct emissions from sources that are owned or controlled by the organization, such as fuel combustion in company-owned vehicles or facilities (Huang et al., 2013). Accurate measurement of Scope 1 emissions provides a clear picture of an organization's direct impact on the environment. Techniques such as direct measurement and the use of emission factors are commonly employed to quantify these emissions (Stechemesser & Guenther, 2012).

Scope 2 emissions result from the generation of purchased electricity, heat, or steam consumed by the organization. These emissions occur at the facility where electricity is produced and are considered indirect because the organization does not directly control the production process (WRI & WBCSD, 2004). To account for these emissions, organizations often use market-based or location-based methods, depending on the availability of data on emission factors specific to the electricity grid (Kolk et al., 2008).

Scope 3 emissions encompass all other indirect emissions that occur in the organization's value chain, including upstream and downstream activities. These emissions are often the most challenging to measure due to their broad and diffuse nature (Ioannou et al., 2015). Scope 3 emissions can include emissions from purchased goods and services, business travel, employee commuting, and the use of sold products (Huang et al., 2013). The complexity and breadth of these emissions highlight the need for robust data collection and estimation methods.

Despite its importance, carbon accounting faces several challenges. Measurement accuracy is a significant concern, as errors can undermine the credibility of reported data (Ascui & Lovell, 2011). Additionally, the lack of standardized reporting frameworks complicates comparisons across organizations and sectors (Kolbel et al., 2020), and there is often a disconnect between carbon accounting practices and broader corporate sustainability strategies (Liesen et al., 2015).

The availability and quality of data are fundamental to accurate carbon accounting. In many cases, organizations rely on estimations and proxies due to the lack of direct measurement capabilities. This reliance can introduce uncertainties and affect the reliability of reported emissions (Ascui & Lovell, 2011). To address these issues, organizations are increasingly adopting advanced data management systems and technologies to enhance data collection and verification processes (Liesen et al., 2015).

Corporate case studies further illustrate the practical applications and challenges of carbon accounting. For instance, companies like Unilever and Microsoft have integrated carbon accounting into their sustainability strategies, setting ambitious reduction targets and implementing comprehensive emission management programmes (Unilever, 2020; Microsoft, 2020). This demonstrates how leading organizations leverage carbon accounting to drive sustainability initiatives, enhance operational efficiency, and achieve regulatory compliance.

Examining real-world applications of carbon accounting reveals its practical challenges and benefits. The Carbon Disclosure Project (CDP) collects and disseminates information on corporate carbon emissions, providing insights into the effectiveness of carbon management strategies (CDP, 2020). Another example is the EU ETS, which illustrates how regulatory frameworks can drive improvements in carbon accounting practices (European Commission, 2020).

Carbon accounting plays a pivotal role in sustainability and environmental management. It enables organizations to identify emission hotspots, set reduction targets, and monitor progress (Ioannou et al., 2015). Additionally, transparent carbon reporting enhances stakeholder trust and can lead to competitive advantages (Kolbel et al., 2020). From a policy perspective, robust carbon accounting supports the development and implementation of effective climate policies (Stechemesser & Guenther, 2012).

To this end, the evolution of methodologies and frameworks has provided organizations with tools to measure, manage, and report their carbon emissions. Despite challenges related to data quality, standardization, and integration, carbon accounting offers strategic benefits, including improved operational efficiency, enhanced stakeholder trust, and regulatory compliance. Future advancements in technology and regulatory frameworks are expected to further enhance the accuracy and effectiveness of carbon accounting practices, contributing to global efforts to mitigate climate change.

3.   Overview of the Chapters

Chapter 2 provides a comprehensive examination of the state of green accounting in China, detailing both its achievements and challenges. Over several decades, China has made strides in the development of green accounting, yet significant hurdles remain in theory and practice. While the country's basic understanding of traditional accounting and ecological principles is robust, further research and adaptation are necessary to align green accounting with China's unique socioeconomic context. One of the primary challenges facing the implementation of green accounting is the lack of a comprehensive theoretical framework. Unlike developed countries where green accounting has been steadily evolving, China's progress remains in its infancy. Theoretical research in China is still nascent, leading to a lack of standardized measurement methods and operational challenges. The absence of clear accounting standards and inconsistent methodologies further complicate the implementation of green accounting practices. A crucial aspect of advancing green accounting in China is the cultivation of skilled professionals. Currently, there is a shortage of talent proficient in green accounting, stemming from the interdisciplinary nature of the field. To address this gap, dedicated training programmes and academic courses need to be established in universities. Furthermore, collaboration between accounting and environmental science disciplines is essential to equip professionals with the necessary expertise. Legal and regulatory frameworks play a vital role in shaping the landscape of green accounting. While China has made strides in environmental legislation, there is still a lack of specific regulations governing green accounting practices. The incorporation of green accounting into existing accounting laws and environmental regulations is imperative to provide a legal framework for its implementation. Clear guidelines and enforcement mechanisms are needed to ensure compliance and accountability among enterprises. Enterprise awareness and commitment to environmental stewardship are pivotal in driving the adoption of green accounting. Businesses must recognize the long-term benefits of environmental sustainability and integrate it into their strategic objectives. Governmental support through incentives and penalties can incentivize firms to prioritize environmental accountability. Additionally, transparent reporting and disclosure of environmental accounting information are essential for fostering accountability and public trust. Despite the challenges, there are significant opportunities for the advancement of green accounting in China. Increasing public awareness and concern for environmental issues provide a conducive environment for the adoption of green accounting practices. International collaboration and knowledge exchange can enrich China's understanding of green accounting methodologies and best practices. Technological advancements, particularly in data analytics and artificial intelligence (AI), offer innovative solutions for environmental monitoring and accounting. This chapter illustrates that the development of green accounting in China is a multifaceted endeavour that requires concerted efforts from various stakeholders. While challenges persist in theoretical development, talent cultivation, regulatory frameworks, and enterprise engagement, opportunities abound for progress. By addressing these challenges and capitalizing on opportunities, China can establish itself as a global leader in green accounting, contributing to sustainable development and environmental stewardship on a national and international scale.

Chapter 3 provides a comprehensive investigation into the relationship among green accounting, corporate sustainability, and financial performance (FP) in Chinese companies. The research focuses on non-financial companies listed on the Shanghai and Shenzhen stock exchanges over the period from 2011 to 2020. The study aims to determine whether the implementation of green accounting positively impacts corporate sustainability and FP. To achieve this, the authors employ regression analysis and the Hausman test to decide between fixed- and random-effect models. The findings suggest a significant positive relationship between green accounting practices and both corporate sustainability and FP, aligning with previous studies in the field. The research methodology involves selecting a sample of over 1,500 companies across various industries, excluding financial firms, from the Shanghai and Shenzhen stock exchanges. Data for these companies were obtained from the China Stock Market & Accounting Research (CSMAR) database, covering a ten-year period. The study employs multiple linear regression models to test the hypotheses, with green accounting as the independent variable and corporate sustainability and FP as the dependent variables. The empirical analysis reveals that green accounting positively impacts both corporate sustainability and FP. The descriptive statistics show a substantial variation in corporate sustainability among Chinese companies, while most firms exhibit positive FP. Regression analyses confirm the positive relationship, with the coefficients for green accounting showing statistical significance. Robustness tests, including multicollinearity checks and alternative variable measurements, support the study's findings. Variance inflation factor (VIF) values indicate no multicollinearity issues, and alternative measures of dependent variables yield consistent results. The discussion section interprets the empirical results, comparing them with findings from previous studies. The positive relationship between green accounting and corporate sustainability aligns with global research, although the results for FP show some inconsistencies due to data variability. The chapter concludes that green accounting significantly enhances both corporate sustainability and FP in Chinese companies. This supports the broader adoption of green accounting practices in China and provides a foundation for future research focusing on specific industries and smaller firms. The study's findings also offer valuable insights for policymakers and corporate managers aiming to integrate environmental considerations into financial decision-making. By integrating green accounting, Chinese companies can better balance profitability with environmental responsibility, thereby contributing to sustainable development goals. The chapter underscores the importance of further research to refine green accounting practices and explore their impacts across different sectors. This chapter contributes to the literature by providing empirical evidence from a large and diverse sample of Chinese companies, thereby enhancing the understanding of green accounting's role in promoting corporate sustainability and FP.

Chapter 4 investigates the relationship between environmental accounting and firm performance, focusing on listed Chinese oil companies. Using data from 2015 to 2019, the study employs regression analysis through SPSS and Stata to analyse financial and CSR reports from companies on the Shanghai Stock Exchange (SSE) and Shenzhen Stock Exchange (SZSE). Despite extensive efforts, the study reveals no significant correlation between environmental accounting practices and firm performance. Additionally, it highlights a prevalent lack of environmental information disclosure (EID) in annual reports. The study recommends government-mandated disclosure of environmental costs to improve transparency and accountability. Environmental issues are global challenges that have intensified scrutiny on corporate environmental responsibilities. This has led to a growing emphasis on "green growth", "low-carbon development", and "sustainable development". Firms across various sectors are pressured to reduce their environmental impact, driven by increasing stakeholder awareness and demand. Enhanced environmental performance (EP) can positively influence a firm's business value, as demonstrated by existing research, which indicates that firms adopting environmental strategies may experience improved financial outcomes and sustainable performance. China's rapid economic growth over the past few decades has come at a significant environmental cost. The Chinese government has responded by implementing regulatory policies to address environmental degradation and promote corporate environmental responsibility. Key measures include the SSE's "Guidelines on Environmental Information Disclosure for Listed Companies" and the 2015 Environmental Protection

Law. Despite these regulations, compliance among oil companies remains limited, negatively impacting their corporate image and FP. EID is essential for companies to communicate their environmental responsibilities to the government and public, enhancing transparency and accountability. Studies, such as those by Okafor (2018) on Nigerian oil companies, show that environmental costs can positively impact firm performance. However, in China, many companies are still hesitant to fully engage in environmental reporting, leading to an unclear relationship between EP and FP. This study aims to bridge this gap by focusing on listed Chinese oil companies and assessing the impact of their EP on business value. Environmental accounting includes various definitions and approaches, generally involving the assessment of all environmental costs associated with activities and products. It helps companies identify cost-saving opportunities, enhance overall performance, and make informed management decisions. Research indicates that environmental accounting can reduce production costs and increase profitability, while also serving as a tool for demonstrating a company's commitment to creating a sustainable investment environment. Empirical evidence on the relationship between environmental accounting and firm performance is mixed, with some studies finding positive correlations and others finding none. This chapter's regression analysis indicates an insignificant relationship between environmental costs and most FP indicators, except for a significant relationship with dividend per share (DPS). The findings suggest that environmental accounting disclosure does not significantly impact FP, corroborating some previous studies while contradicting others. The chapter concludes that while environmental accounting is crucial for managing environmental costs and enhancing transparency, it does not significantly impact FP in the Chinese oil industry. Future research should explore other industries and external factors, such as company size and external economic conditions, to provide a more comprehensive understanding of the relationship between environmental accounting and firm performance.

Chapter 5 explores whether improvements in carbon scores and carbon disclosures impact firm value, using data from China's industrial listed businesses between 2013 and 2023. The research uses Tobin's Q as the dependent variable and employs Wind and CSMAR databases. Methodologically, the chapter utilizes a two-stage least squares regression model and the Hausman fixed-effect model to address endogeneity issues related to carbon scores and disclosures. The findings indicate a negative correlation between an enterprise's carbon score and its value, as well as a detrimental effect of carbon disclosure on firm value. The study builds on the work of Xie et al. (2020), which concluded that carbon-neutral commitments positively affect enterprise value and that improved environmental, social, and governance (ESG) performance enhances market response to these commitments. The chapter identifies a gap in the literature, noting that few studies have explored the significance of corporate carbon ratings and their impact on corporate value, particularly in the context of China. This investigation aims to fill that gap by examining whether a firm's carbon-neutral rating and the extent of its carbon disclosure influence its value. The chapter highlights the growing attention to carbon neutrality in China, driven by the government's commitment to peak carbon emissions by 2030 and achieve carbon neutrality by 2060. This policy shift reflects a broader recognition that environmental conservation can drive high-quality economic development. However, the study suggests that the impact of carbon neutrality on enterprises is two-sided: while it supports CSR, it also involves increased costs for environmental protection, which can reduce firm value. Globally, carbon neutrality is a critical issue, with increasing attention from investors and stakeholders. For instance, PricewaterhouseCoopers (PwC) noted a significant rise in interest in carbon emissions over the past decade. The adverse effects of exceeding GHG limits, as outlined in the Paris Climate Agreement, underscore the importance of corporate carbon management. The chapter posits that consumer trust and business transactions are closely linked to an organization's carbon score and its commitment to CSR. The practical relevance of this study is underscored by its focus on China's industrial sector, which is a major contributor to GDP, energy consumption, and pollution. The study argues that while greater disclosure of carbon neutrality can enhance ESG performance and drive value growth, it also entails significant costs that may not yield immediate positive benefits. The chapter provides empirical evidence that both carbon score and carbon disclosure negatively impact enterprise value, while CSR has a positive effect. Theoretically, publication of carbon dioxide data enhances corporate credibility and stakeholder confidence. However, the costs associated with carbon information disclosure can reduce firm value. The chapter concludes that while carbon neutrality commitments can lead to long-term benefits, they also present immediate financial challenges for enterprises. The study's findings offer valuable insights for national policymaking and corporate strategies aimed at balancing environmental responsibilities with economic growth. The chapter contributes to the literature by addressing the gap in carbon information disclosure studies within China's industrial market. It provides empirical evidence that high-quality carbon disclosure improves ESG performance and drives enterprise value growth, albeit not immediately. The results also highlight the need for the Chinese government to formulate low-carbon incentive policies to support green industry development and optimize the energy structure.

Chapter 6 explores the relationship between corporate EP and FP, focusing on publicly listed manufacturing companies in China. The primary objective is to ascertain whether the impact of EP on FP is direct and to investigate the existence of a bidirectional relationship between these variables. The study employs an empirical approach, analysing data from 49 manufacturing companies over a period from 2010 to 2020. EP is evaluated using the possession of environmental certificates, specifically ISO14001 and ISO9001, as objective indicators. A panel model with cross-sectional data is utilized to scrutinize the causal relationship between EP and FP. The findings reveal that investments in EP significantly enhance FP, particularly in terms of market value, as reflected by the price-to-book value (PBV) ratio. This impact is both direct and substantial. However, no significant effect is observed on the accounting index – earnings before interest, taxes, depreciation, and amortization (EBITDA)/total assets (TA). This indicates that while environmental efforts positively influence market perceptions and valuations, they do not immediately translate into improvements in traditional accounting measures.

Furthermore, the study indicates that FP, whether measured by PBV or EBITDA/ TA, does not significantly motivate companies to invest in environmental improvements. This challenges the notion of a bidirectional relationship, suggesting instead a unidirectional influence where EP positively affects FP, but not vice versa. The lack of a reciprocal relationship is critical for understanding the strategic motivations behind environmental investments. The chapter delves into various hypotheses and theoretical frameworks to contextualize these findings. It references the natural resource-based view (NRBV), which posits that environmental initiatives can enhance a company's competitive advantage by promoting sustainability and efficiency. The adoption of environmental management systems like ISO14001 is highlighted as a key driver in achieving superior EP, which in turn boosts financial outcomes by reducing waste, enhancing resource utilization, and fostering innovation in eco-efficient products and services. Additionally, the study addresses the contentious debate over the direct versus indirect impact of EP on FP. While some scholars argue that environmental efforts indirectly benefit financial performance through improved corporate reputation and customer satisfaction, this study provides empirical evidence supporting a direct positive impact on market value. This direct relationship underscores the importance of environmental certifications as a tangible signal of a company's commitment to sustainability, which investors and stakeholders increasingly value. In examining the broader implications, the chapter discusses the role of national policies and regulatory frameworks in shaping corporate environmental strategies. China's unique institutional context, characterized by stringent environmental regulations and a pressing need for sustainable development, provides a compelling backdrop for this analysis. The findings suggest that robust EP not only aligns with regulatory requirements but also offers a strategic advantage in the market, thereby encouraging companies to adopt proactive environmental measures. This chapter contributes to the ongoing discourse on corporate sustainability by elucidating the positive impact of EP on financial outcomes in China's manufacturing sector. It challenges the bidirectional hypothesis, reinforcing the significance of environmental investments in driving market value. These insights are instrumental for policymakers and corporate leaders in formulating strategies that balance economic growth with environmental stewardship. The study underscores the potential for environmental initiatives to serve as a catalyst for financial success, advocating for an integrated approach to corporate sustainability.

Chapter 7 investigates the relationship between ESG performance and corporate FP, with a particular focus on the moderating effect of firm ownership. Utilizing data from Chinese A-share listed companies spanning 2014–2020, the study employs a fixed-effect model to conduct empirical regression analyses addressing these relationships. The FP of firms is quantified using economic value added (EVA), while ESG scores are sourced from the Bloomberg database. The analysis begins by elaborating on the importance of ESG in contemporary corporate governance and performance assessment. ESG measures evaluate the sustainability and societal impact of an enterprise's activities, extending beyond traditional financial metrics. This study underscores the significance of understanding the interplay between ESG performance and financial outcomes, especially in the context of Chinese enterprises, where corporate governance structures and market dynamics differ significantly from their Western counterparts. A critical aspect of the research design is the use of the Heckman two-stage model to address potential endogeneity issues. Endogeneity arises when explanatory variables correlate with the error term, potentially biasing the results. The Heckman model helps mitigate this by, firstly, predicting a selection equation to calculate an inverse Mills ratio (IMR), which is then included in the second-stage regression to correct for sample selection bias. The empirical findings reveal a significant positive relationship between ESG performance and corporate FP. Specifically, the results indicate that firms with higher ESG scores tend to exhibit better financial outcomes as measured by EVA. This positive correlation supports the hypothesis that ESG initiatives can enhance corporate value by improving operational efficiencies, enhancing reputation, and reducing risk. Moreover, the study explores the moderating effect of firm ownership on the ESG–FP relationship. The results show that non-state-owned enterprises (non-SOEs) exhibit a stronger positive relationship between ESG performance and FP than state-owned enterprises (SOEs). This differential impact is attributed to the distinct operational and strategic orientations of SOEs and non-SOEs. Non-SOEs, driven by profit motives and competitive market pressures, may be more incentivized to leverage ESG practices to enhance their market standing and FP. In contrast, SOEs, which often benefit from state support and face different performance expectations, may not experience the same level of financial benefit from ESG activities. The robustness of these findings is confirmed through additional analyses, including a robustness test focusing on the computer, communication, and electronic equipment manufacturing industry. This sector-specific test reinforces the generalizability of the results across different industrial contexts, further validating the positive impact of ESG on FP. The chapter concludes by discussing the theoretical and practical implications of the findings. Theoretically, it contributes to the growing body of literature affirming the positive link between ESG and corporate FP. It also addresses the moderating role of firm ownership, highlighting the nuanced ways in which organizational characteristics can influence the effectiveness of ESG initiatives. Practically, the results provide actionable insights for policymakers and corporate managers. For policymakers, the study suggests the need to tailor regulatory frameworks that encourage both SOEs and non-SOEs to engage in ESG activities. For corporate managers, particularly in non-SOEs, the findings highlight the strategic value of integrating ESG into core business operations to drive FP. This chapter enriches the understanding of the ESG–FP nexus, demonstrating that ESG investments are not only socially and environmentally beneficial but also financially advantageous, especially for non-SOEs in China. This dual benefit underscores the critical role of ESG in fostering sustainable corporate growth and enhancing overall market performance.

Chapter 8 explores the effectiveness of EID policy in China by examining data from 248 companies over the period 2011–2019. The chapter emphasizes the importance of EID as a tool for corporate environmental responsibility, which can be influenced by stakeholder pressure and economic incentives. It employs six evaluation components to quantify EID levels and utilizes a regression model with fixed effects to assess the impact of mandatory disclosure policies. The analysis reveals that mandatory disclosure policies significantly and positively affect EID scores, demonstrating their efficacy in enhancing environmental transparency among companies. However, the presence of state-owned shares reduces the positive impact of mandatory policies on EID levels. This finding suggests that SOEs may engage in rent-seeking behaviour, potentially obscuring adverse environmental information to maintain favourable relations with the government. The book reviews literature on EP and disclosure, highlighting mixed results regarding the relationship between these variables. Stakeholder theory suggests that firms disclose environmental information to manage stakeholder impressions and fulfil accountability. Company characteristics such as size, ownership structure, and economic incentives are critical factors influencing EID. SOEs, due to their political connections, might disclose less environmental information to maintain a favourable image, supporting the rent-seeking theory. Methodologically, the study uses data from the CSMAR database, covering Chinese listed companies from 2011 to 2019. It evaluates EID using six components: ISO environmental system authentication, environmental protection policies, loans related to environmental protection, waste disposal and recycling, environmental improvement projects, and other related information. The analysis employs a regression model and introduces state-owned shares as a moderating variable. Results indicate that mandatory disclosure policies significantly enhance EID scores. However, state-owned shares negatively moderate this relationship, suggesting that higher proportions of state ownership weaken the effectiveness of mandatory policies. This underscores the need for stricter supervision of SOEs to ensure authentic and comprehensive environmental reporting. The chapter concludes with practical recommendations for policymakers. It advocates for a comprehensive system of mandatory EID policies and enhanced external supervision, particularly for SOEs. The study also suggests that future research should include data post-2019 to assess the impact of COVID-19 on EID and explore reasons behind the declining EID scores among firms subject to mandatory disclosure policies since 2017. Chapter 8 provides a thorough empirical analysis of the impact of EID policies in China, highlighting the effectiveness of mandatory disclosure in improving EID scores while identifying challenges associated with state ownership. The findings contribute to the broader discourse on environmental governance and corporate transparency, offering actionable insights for regulators to enhance environmental accountability and sustainability.

Chapter 9 examines the critical factors influencing the degree of sustainability reporting within the Chinese medicine manufacturing industry. The analysis is centred on the relationship among specific company characteristics, board composition, audit firms, and the extent of sustainability disclosure among 64 publicly listed Chinese companies. A primary theme in the chapter is the significant role of sustainability reporting as an essential component of corporate governance. It highlights how sustainability disclosures serve as a testament to a company's commitment to transparency, environmental responsibility, and social accountability, in line with global trends and increasing regulatory demands. Among the company characteristics examined, firm size emerges as a significant determinant of sustainability reporting. Larger firms are found to disclose more comprehensive sustainability reports, likely due to their greater resources and the heightened scrutiny they face from stakeholders and regulators. This drives these firms to adopt more robust disclosure practices. Interestingly, firm age does not show a significant influence on the extent of sustainability reporting, indicating that older firms are not necessarily more inclined to prioritize sustainability disclosures than younger firms. The composition of the board is another critical theme. The study finds no significant correlation between board size or independence and the level of sustainability reporting, suggesting that merely increasing board size or its independence is insufficient to enhance sustainability disclosures without other supportive mechanisms. However, board gender diversity shows a significant positive relationship with the degree of sustainability reporting. Companies with a higher proportion of female directors are more likely to engage in comprehensive sustainability disclosures, reflecting a broader perspective and greater sensitivity to sustainability issues. The involvement of prominent audit firms, particularly the Big Four (Deloitte, PwC, Ernst & Young, and KPMG), is positively associated with higher levels of sustainability reporting. These firms' rigorous standards and reputational concerns drive better disclosure practices among their clients, emphasizing the role of quality audits in enhancing corporate transparency. The findings of the chapter suggest significant room for improvement in sustainability reporting practices within the Chinese medicine manufacturing sector. Companies are encouraged to enhance their governance frameworks and leverage diverse board compositions to drive better sustainability outcomes. Additionally, engaging reputable audit firms can further ensure the credibility and comprehensiveness of sustainability reports. In terms of policy and future research, the chapter calls for stronger regulatory frameworks and industry standards to promote uniform and comprehensive sustainability disclosures. It also identifies areas for future research, such as exploring the impact of other corporate governance mechanisms and external pressures on sustainability reporting.

Chapter 10 examines the significance of EID within the Chinese food industry, particularly against the backdrop of increasing environmental concerns and regulatory pressures. The chapter sets the stage by highlighting China's shift towards sustainable development due to severe environmental pollution issues. The transformation from a resource- and pollution-intensive economy to a green economy necessitates enhanced corporate transparency in EP. EID is crucial as it mitigates information asymmetry, allowing stakeholders to make informed decisions regarding corporate environmental responsibility. Since the implementation of the Measures for the Disclosure of Environment Information (MDEI) in 2008, there has been a steady rise in EID among Chinese companies, although inconsistencies and strategic disclosures remain prevalent. The chapter focuses on understanding the relationship between the quantity (number of words) and quality of EID among 43 listed Chinese food companies in 2019. It addresses whether an increase in the quantity of disclosed information correlates with improved quality. The study employs content analysis, utilizing sector-specific Global Reporting Initiative (GRI) guidelines to evaluate EID. The findings reveal a significant relationship between company size and both the quantity and quality of EID. Larger companies tend to disclose more extensive and higher quality environmental information than smaller firms. The study also identifies a high positive correlation between the quantity and quality of EID, suggesting that more detailed disclosures often equate to higher quality information. A comparative analysis between top-performing and bottom-performing companies in terms of EID quantity and quality supports these findings. The results indicate that top-performing companies, as ranked by market capitalization, quantity, and quality of EID, significantly outperform their lower-ranked counterparts. This reinforces the notion that larger and more resource-rich companies are better equipped to meet stringent disclosure requirements. It emphasizes the need for standardized and uniform guidelines to ensure consistent and high-quality EID across the industry. The role of GRI guidelines is particularly highlighted as a comprehensive framework that can guide Chinese companies towards better EID practices. Chapter 10 underscores the critical role of EID in promoting corporate accountability and transparency in environmental practices. The positive correlation between the quantity and quality of disclosures suggests that enhancing the amount of information disclosed can improve its overall quality. The findings advocate for more robust regulatory frameworks and the adoption of international guidelines like GRI to standardize EID practices in China's food industry.

Chapter 11 examines the factors influencing the quality of environmental accounting information disclosure (EADI) within China's chemical manufacturing sector. This research spans data from 2011 to 2020, sourced from the CSMAR database, and employs a multiple regression model, specifically the ordinary least squares (OLS) method, to analyse these determinants. The study underscores the importance of environmental accounting disclosures for the sustainable development of society, aligning with contemporary concerns for a low-carbon economy. The introduction outlines the global and regional factors influencing environmental accounting disclosure. It highlights the significance of both internal factors, such as organizational stakeholders and innovation strategies, and external factors, such as customer pressure and cultural contexts, on environmental disclosure practices. The literature review explores the mixed results in existing research, noting the lack of focus on developing countries, particularly China, and the chemical manufacturing industry. This gap underscores the need for a deeper understanding of the relationships among FP indicators, stakeholder factors, and EDI within this specific context. The results reveal that firm size positively correlates with EDI quality, suggesting that larger firms tend to disclose higher quality environmental information, consistent with legitimacy theory and stakeholder theory. Conversely, leverage exhibits a negative association with EDI, indicating that companies with higher debt levels are less likely to engage in extensive environmental disclosures due to financial constraints. However, the study finds no significant relationships among earnings per share (EPS), return on asset (ROA), and ownership concentration (LAR) with EDI, contradicting some previous research and highlighting the complexity and context-specific nature of these associations. This chapter highlights the critical role of environmental accounting disclosures in promoting sustainable development and the importance of firm size and financial stability in enhancing EDI quality. The study provides valuable insights for policymakers, investors, and corporate managers in understanding and improving environmental disclosure practices within China's chemical manufacturing industry. Future research directions include examining different types of companies, incorporating external stakeholder factors, and exploring longitudinal trends to further enrich the understanding of EDI determinants.

Chapter 12 presents a comprehensive analysis of EAID within China's steel industry. It establishes an evaluation index system targeting three principal aspects: relevance, reliability, and compliance. The research uses a case study approach, selecting 20 listed steel enterprises and employing 22 evaluation indicators. A projection pursuit model and an accelerated genetic algorithm are utilized to process the multidimensional, nonlinear, and non-normally distributed data. The study finds the overall quality of EAID to be poor, indicating significant gaps in the disclosure practices of these enterprises. The chapter highlights that voluntary disclosure of environmental information is limited, often influenced by governmental pressure rather than corporate initiative. Key indicators such as environmental liabilities, energy consumption, and "three wastes" emissions emerge as critical factors in the assessment. The research underscores the necessity for systematic evaluation to identify weaknesses, enhance credibility, and guide improvements in disclosure practices. Additionally, the chapter discusses the historical and regulatory context of EAID in China, noting the evolving but still insufficient regulatory framework. The methodology section details the sample selection and the construction of the evaluation index system, emphasizing the adaptability and advantages of the projection pursuit model. The findings reveal a disparity in disclosure quality among enterprises, with larger companies not necessarily exhibiting better EAID quality. The chapter concludes with recommendations for policy enhancements to encourage comprehensive and transparent environmental disclosures, thus supporting sustainable development within the industry.

Chapter 13 investigates whether Chinese firms that implemented ESG policies during the COVID-19 pandemic experienced reduced adverse effects on their firm value. The analysis utilizes data from companies listed in the Chinese market between 2011 and 2021. Firm performance, the dependent variable, is measured using the corporate ESG score. To mitigate endogeneity issues, the study employs the instrumental variables method, the Heckman method, propensity score matching (PSM), and replacement fixed effects. The study aims to determine if Chinese firms that adopted ESG policies during the COVID-19 pandemic were less adversely affected than those that did not implement such policies. This examination addresses whether ESG policies can provide resilience against the pandemic's impact on firm value. The study leverages panel data from 2011 to 2021 for Chinese-listed companies. Firm performance, operationalized through corporate ESG scores, serves as the key metric. The research design incorporates advanced econometric techniques, including the instrumental variables method, Heckman method, PSM, and replacement fixed effects, to address potential endogeneity and ensure robust results. Empirical results reveal that firms adopting ESG policies experienced less severe impacts from the COVID-19 pandemic on their firm value. Specifically, these firms showed greater resilience in their stock prices, indicating stronger resistance to the crisis compared to firms that did not implement ESG policies. The consistency of the endogeneity test and robustness test results further corroborates the study's robustness. This research contributes to the existing literature by elucidating the role of ESG policies in mitigating the impact of significant security emergencies, such as the COVID-19 pandemic, on firm market capitalization. It provides empirical evidence supporting the hypothesis that ESG policies enhance corporate resilience during crises. The findings have practical implications for business managers and practitioners, suggesting that ESG adoption can serve as a strategic tool for enhancing firm stability and value during uncertain times. The study's findings suggest that business managers should consider integrating ESG policies to bolster their firms' resilience against future crises. Policymakers may also find the results useful for encouraging broader adoption of ESG practices across industries to enhance overall market stability. The study concludes that ESG policies play a significant role in shielding firms from the adverse impacts of the COVID-19 pandemic. By fostering better relationships with stakeholders and demonstrating a commitment to sustainable practices, firms can achieve greater resilience in their financial performance and stock market stability during crises. Future research could explore the long-term benefits of ESG policies across different market conditions and industries.

Chapter 14 explores the interplay among carbon dioxide emissions, carbon accounting emissions, and temperature fluctuations, focusing on Chinese provinces from 2010 to 2018. Through meticulous analysis of data on carbon emissions and temperature records, the provinces are categorized into seven distinct regions. Employing advanced quantitative methodologies, including curve fitting techniques, the research unveils a noteworthy trend: the escalating proportion of carbon accounting emissions within the broader spectrum of carbon dioxide emissions. Furthermore, it elucidates the affirmative influence of carbon accounting emissions on the amplification of both carbon dioxide emissions and temperature variations. These findings carry significant implications for policy formulation, urging stakeholders across governmental, corporate, and societal spheres to heighten their vigilance towards temperature shifts. The disclosure of the intricate dynamics between temperature alterations and carbon emissions underscores the imperative for robust public oversight and proactive governmental intervention to mitigate the impending challenges posed by climate change.

Chapter 15 investigates the nexus between China's harvested wood products (HWPs) and carbon accounting, recognizing the profound implications for global climate change mitigation efforts. Leveraging a robust dataset sourced from CSMAR and China's forestry spanning the period from 1990 to 2020, the research presents a comprehensive analysis of HWPs' carbon dynamics and its implications for carbon flux. Drawing upon prior research in carbon accounting methodologies and the application of HWPs, this investigation extends our understanding of the intricate relationship between HWP utilization and carbon emissions. It underscores the pressing nature of the issue, particularly in light of China's status as a significant carbon emitter. The findings underscore the pivotal role of HWPs in influencing carbon flux, highlighting the urgency of addressing this challenge. In alignment with China's carbon emission standards and international protocols such as the "Paris Agreement", accurate assessment and reporting of HWP carbon stocks are imperative for effective climate change mitigation strategies. Recognizing the complexity of this task, the study identifies three key methods adopted by the Chinese government to enhance carbon accounting practices related to HWPs. These measures are geared towards aligning with international standards and domestic emissions targets, thereby facilitating progress towards carbon neutrality and fostering global environmental stewardship. By shedding light on the interplay between HWP utilization and carbon emissions, this study not only advances scholarly discourse but also provides actionable insights for policymakers and practitioners tasked with steering climate change mitigation efforts. The imperative to achieve accurate carbon accounting in HWPs underscores the imperative of international collaboration and concerted efforts towards sustainable environmental management.

Chapter 16 examines the impact of ownership structure, corporate governance mechanisms, and external supervision on the quality of environmental accounting information disclosure (EAIDQ) within China's mining sector. Utilizing a dataset comprising 71 Chinese A-share listed mining companies spanning the period from 2012 to 2021, the empirical analysis investigates the relationship between these factors and EAIDQ. The dependent variable, EAIDQ, is dissected into monetary (M-EAIDQ) and non-monetary (NM-EAIDQ) dimensions to capture variations in information expression. Ownership concentration serves as a proxy for ownership structure, while the proportion of independent directors reflects corporate governance. External supervision is gauged through the Pollution Information Transparency Index (PITI) index and media reports. To address potential endogeneity concerns, the study employs a robust multiple regression model and PSM technique. Findings reveal a general inadequacy in EAID in China, albeit with a gradual improvement trend. Notably, ownership structure exhibits a significant correlation with EAIDQ, while corporate governance exerts a substantial influence on NM-EAIDQ. Regarding external supervision, public oversight demonstrates a significant impact on M-EAIDQ, whereas government supervision exhibits a delayed significance on NM-EAIDQ. These results withstand robustness checks and endogeneity assessments, underscoring their reliability. The study's outcomes provide theoretical insights for firms aiming to enhance EAIDQ and offer practical implications for government efforts to strengthen environmental oversight mechanisms. This research contributes to advancing understanding in corporate environmental disclosure practices and underscores the importance of effective governance and oversight mechanisms in fostering sustainability within the mining industry.

Chapter 17 explores the role of carbon accounting within the realms of management control, performance measurement, and climate risk management, highlighting its significance in fostering corporate sustainability. As global environmental challenges intensify, carbon accounting has become an essential tool for organizations to quantify, manage, and reduce GHG emissions. This chapter provides a comprehensive exploration of the theoretical foundations, practical applications, and strategic implications of integrating carbon accounting into management practices. The integration of carbon accounting into management control systems (MCS) is vital for embedding sustainability into corporate strategies. Carbon accounting provides data that support decision-making processes, enabling organizations to set measurable targets, monitor progress, and implement effective strategies for reducing environmental impacts. By incorporating carbon metrics into MCS, companies can align their financial and environmental goals, fostering a holistic approach to performance management. Performance measurement frameworks traditionally focus on financial indicators; however, the inclusion of carbon metrics offers a more comprehensive assessment of organizational performance. Integrating carbon accounting into performance measurement systems allows companies to track their progress toward sustainability goals, ensuring that environmental considerations are embedded in strategic and operational decisions. This holistic approach enhances accountability, transparency, and stakeholder trust. Effective climate risk management requires accurate measurement and reporting of carbon emissions. Carbon accounting supports compliance with regulatory requirements and voluntary reporting standards such as the GRI and the Carbon Disclosure Project (CDP). By integrating carbon accounting into climate risk management frameworks, organizations can enhance their resilience to climate-related risks and capitalize on opportunities arising from the transition to a low-carbon economy. The chapter emphasizes the necessity of clear and comprehensive regulatory frameworks to standardize and institutionalize carbon accounting practices. Regulatory compliance not only enhances transparency and accountability but also builds stakeholder trust and supports broader sustainability goals. The chapter advocates for the formulation of specific laws and regulations tailored to the intricacies of green accounting, encompassing aspects such as environmental asset valuation, liability assessment, and disclosure requirements. Emphasized is also the need for dedicated training programmes to address the shortage of skilled professionals in the field of green accounting. By integrating carbon accounting principles into academic curricula and professional development initiatives, organizations can cultivate a workforce equipped with the knowledge and skills required to navigate the complexities of green accounting. The chapter underscores the importance of interdisciplinary collaboration in designing and implementing these training programmes. The empirical evidence presented demonstrates a positive correlation between robust carbon accounting practices and improved EP and FP. Companies that adopt green accounting practices tend to exhibit superior sustainability outcomes, including improved resource efficiency, reduced emissions, and enhanced stakeholder relationships. These practices not only contribute to environmental sustainability but also drive financial value by fostering innovation, operational efficiencies, and corporate reputation. The potential of technological advancements such as blockchain, AI, and the IoT is highlighted, and these technologies can improve data collection, reporting, and analysis. The chapter argues that the integration of carbon accounting within management control, performance measurement, and climate risk management frameworks is essential. By providing critical data for decision-making, supporting regulatory compliance, and driving financial value, carbon accounting could help organizations achieve their sustainability goals.

4. Conclusion

At an aggregated level, this book delves into the evolving field of carbon accounting and its significance in sustainability and environmental management, particularly within the dynamic context of China. With its empirical foundation and focus on practical applications and policy implications, the book serves as a resource for academics, environmental professionals, and policymakers. It provides a thorough exploration of carbon accounting practices in China, illuminating their impact on environmental sustainability and offering a roadmap for future advancements in this crucial area.

In focus is the role of carbon accounting in sustainability and environmental management, and how accurate tracking and reporting of carbon emissions are integral to developing effective environmental strategies and evolving environmental policies. Also highlighted is the development of green accounting and its future development and wider impact on climate change. By analysing carbon accounting systems and their impact, the book discusses implications for practice and policy. It offers actionable recommendations for stakeholders, including government bodies, corporations, and environmental auditors, contributing to the broader discourse on sustainable environmental practices.

Accompanied by real-world case studies, empirical realities in China, and actionable recommendations, this book is of interest for anyone navigating the intricate landscape of carbon accounting and reporting. Addressing a global audience interested in environmental management and sustainable practices, the content is essential for investors, policymakers, business leaders, managers, and scholars seeking actionable insights into carbon accounting and its role in sustainability and environmental management.

References

Ascui, F., & Lovell, H. (2011). As frames collide: Making sense of carbon accounting. *Accounting, Auditing & Accountability Journal, 24*(8), 978–999.

Bebbington, J., & Larrinaga-González, C. (2008). Carbon trading: Accounting and reporting issues. *European Accounting Review, 17*(4), 697–717.

CDP. (2020). *Carbon disclosure project.* www.cdp.net/en

European Commission. (2020). *EU emissions trading system (EU ETS).* https://ec.europa.eu/clima/policies/ets_en

Guenther, E., Endrikat, J., & Guenther, T. W. (2016). Environmental management control systems: A conceptualization and a review of the empirical evidence. *Journal of Cleaner Production, 136*, 147–171.

Huang, Y., Weber, C. L., & Matthews, H. S. (2013). Categorization of scope 3 emissions for streamlined enterprise carbon footprinting. *Environmental Science & Technology, 47*(21), 12369–12376.

Ioannou, I., Li, S. X., & Serafeim, G. (2015). The effect of target difficulty on target completion: The case of reducing carbon emissions. *The Accounting Review, 90*(5), 1467–1492.

Kolbel, J. F., Busch, T., & Jancso, L. M. (2020). How media coverage of corporate social irresponsibility increases financial risk. *Strategic Management Journal, 41*(2), 194–223.

Kolk, A., Levy, D., & Pinkse, J. (2008). Corporate responses in an emerging climate regime: The institutionalization and commensuration of carbon disclosure. *European Accounting Review, 17*(4), 719–745.

Kumar, S., Teichroew, J., & Singhal, D. (2020). Blockchain-based carbon accounting: An accountability perspective. *Journal of Cleaner Production, 275*, 124160.

Liesen, A., Hoepner, A. G., Patten, D. M., & Figge, F. (2015). Does stakeholder pressure influence corporate GHG emissions reporting? Empirical evidence from Europe. *Accounting, Auditing & Accountability Journal, 28*(7), 1047–1074.

Microsoft. (2020). *Carbon negative by 2030: Our plan to cut carbon emissions.* https://blogs.microsoft.com/blog/2020/01/16/microsoft-will-be-carbon-negative-by-2030/

Stechemesser, K., & Guenther, E. (2012). Carbon accounting: A systematic literature review. *Journal of Cleaner Production, 36*, 17–38.

Unilever. (2020). *Sustainable living: Reducing greenhouse gas emissions.* www.unilever.com/sustainable-living/reducing-ghg/

World Resources Institute (WRI) & World Business Council for Sustainable Development (WBCSD). (2004). *The Greenhouse Gas Protocol: A corporate accounting and reporting standard* (Revised ed.). https://ghgprotocol.org

Xie, X., Wang, T., Yue, X., Li, S., Zhuang, B., & Wang, M. (2020). Effects of atmospheric aerosols on terrestrial carbon fluxes and CO2 concentrations in China. *Atmospheric Research, 237*, 104859.

Zhang, D., Cai, J., & Huo, X. (2021). Real-time carbon footprint tracking: IoT-based intelligent carbon accounting framework for sustainability. *Sustainable Production and Consumption, 27*, 1263–1273.

2 Green Accounting in China

Challenges, Opportunities, and
Future Directions

*Md Jahidur Rahman, Tarek Rana, Hongtao Zhu,
and Cheng Zeyu*

1. Introduction

This chapter examines the determinants driving the adoption of green accounting practices in Chinese enterprises and evaluates their current status. It explores the historical development of green accounting in China and the institutional and cultural factors influencing its adoption. The key focus is on assessing the extent of green accounting implementation among Chinese businesses and identifying the barriers to effective adoption. The study also aims to evaluate the current state of green accounting practices, pinpointing factors that contribute to their growth, and to propose strategies for overcoming obstacles and promoting sustainable development initiatives. Furthermore, it investigates future trajectories for green accounting in China, considering the roles of technological innovation, policy reform, and international collaboration.

This research is motivated by several factors. Green accounting originated in Western contexts, and its initial uptake was limited. However, increasing resource consumption and environmental degradation have prompted a global shift towards environmental accountability, leading to the establishment of environmental accounting frameworks. In China, demographic pressures and resource scarcity, coupled with regional economic disparities, highlight the need for environmental stewardship, fostering a growing awareness of green accounting.

Current production paradigms often prioritize profit maximization over environmental concerns, and while theoretical insights into green accounting exist, empirical studies are lacking. This study aims to fill this gap by providing empirical evidence of green accounting practices in China, enhancing the understanding of their effectiveness and challenges. Additionally, the literature on green accounting in foreign contexts is fragmented, and comprehensive studies on its evolution and application within China are scarce. This research seeks to synthesize the existing literature and analyse financial statements to provide a holistic view of green accounting practices in the Chinese context.

The study will systematically review the domestic literature on green accounting, culminating in a synthesis of prior research. In this study, we will outline the fundamental principles of green accounting and analyse a selected company's

DOI: 10.4324/9781003488965-3

financial reports from 2018 to 2022, creating a green accounting balance sheet and profit statement for empirical analysis.

Ultimately, this study aims to contribute to the literature on green accounting in China by providing empirical insights into its corporate efficacy and offering actionable recommendations aligned with current economic and social realities. It will also examine the post-pandemic landscape of green accounting practices, identifying challenges and proposing solutions, while enhancing transparency and accountability in financial reporting.

The subsequent sections of the chapter will detail the genesis and evolution of green accounting in China, the theoretical framework guiding the research, a comprehensive literature review, the theoretical underpinnings of green accounting, and an analysis of Zhuzhou Smelter Group, culminating in a synthesis of research findings.

2. Background

The evolution of human society has been marked by continuous exploration, with modern advancements reflecting a history of experimentation. A pivotal moment in this journey was the Industrial Revolution of the 19th century, which saw the integration of scientific principles into production processes. This era triggered significant technological innovation, leading to not only increased material comforts but also considerable environmental degradation (Chen et al., 2019). The rapid population growth and resource exploitation revealed a troubling truth: economic growth often comes at the cost of the environment. The prevailing mindset of "development first and governance later" highlighted how economic goals frequently overshadowed environmental concerns. This pattern persists in contemporary China, where economic ambitions often take precedence over environmental stewardship, resulting in various environmental issues.

In contrast, foreign companies have recognized the seriousness of environmental challenges for years. Major international firms like BP and Shell have proactively addressed climate change for over a decade. Meanwhile, China has pursued research and policy initiatives aimed at combating climate change and reducing greenhouse gas emissions (Chen et al., 2019). The experiences of these global enterprises, coupled with China's efforts, carry significant implications for domestic companies navigating climate change and carbon reduction complexities.

Reconciling economic growth with environmental preservation is crucial, particularly for enterprises at the forefront of this challenge. While economic growth and environmental conservation are often viewed as opposing forces, achieving a balance is essential. Green accounting offers a promising approach to align financial objectives with environmental stewardship (Hu et al., 2019). By adopting green accounting practices, companies can pursue ecologically sound production and operations without sacrificing their financial goals, fostering a symbiotic relationship between growth and sustainability. Rahman et al. (2023) investigated the relationship between corporate social responsibility (CSR) and financial distress and the moderating effect of firm characteristics, auditor characteristics, and coronavirus disease 2019 (COVID-19) in China.

Green accounting, or environmental accounting, extends beyond traditional accounting that focuses solely on economic metrics. It promotes a holistic approach that integrates economic and ecological goals, creating synergies between development and environmental conservation (Zhu et al., 2019). Companies are encouraged to incorporate green accounting principles into their financial frameworks, aligning their operations with environmental preservation. Rahman and Wu (2023) explored the effect of mergers and acquisitions (M&As) on corporations' environmental, social, and governance (ESG) performance and values in the Chinese financial market.

The study of green accounting's evolution in Chinese enterprises is particularly relevant in light of growing global demands for sustainability. However, China faces significant challenges in establishing a robust green accounting framework, including gaps in regulations, monitoring systems, and organizational understanding. The contrast between domestic and international research on green accounting reveals China's emerging landscape, highlighting substantial opportunities for growth and development in this area.

3. Theoretical Framework

Sustainable development (SR) serves as the theoretical foundation for green accounting, promoting the integration of economic growth, environmental preservation, and social equity. This approach aims to balance the tensions between economic advancement, environmental stewardship, and social justice, fostering a harmonious relationship between human development and ecological integrity (Sun, 2020). The Industrial Revolution and subsequent economic growth have led to environmental degradation and social inequalities, highlighting the need for sustainable development (Hu et al., 2019). This paradigm stresses the importance of long-term planning and the holistic integration of economic, social, and environmental factors in decision-making (Ma & Ma, 2019). Green accounting (GR) embodies these principles within the accounting field, urging businesses to consider the environmental and social impacts of their activities and to incorporate sustainability into financial reporting and strategic decisions (Wang, 2004; Jiang et al., 2020).

The theoretical basis of green accounting aligns with ecological economics, social responsibility, and stakeholder theory, emphasizing the need to account for the broader impacts of economic activities (Ma & Ma, 2019). This framework encourages a balanced development approach, integrating economic, environmental, and social imperatives into financial practices. Green accounting represents the convergence of accounting, ecological principles, and management strategies, aimed at promoting environmental preservation and mitigating negative externalities. Both green accounting practices and the disclosure of related information require further research. The following discussion will explore existing studies on green accounting and its information disclosure, both domestically and internationally.

4. Literature Review

4.1 Scope and Content of Green Accounting

The scope and content of green accounting have evolved significantly, marked by key milestones that clarify its theoretical foundations and practical applications. A major turning point was the 1998 United Nations meeting in Geneva, where the "Green Accounting and Reporting Position Statement" highlighted the need to integrate environmental considerations into accounting frameworks (Zhang, 2016).

Before China's adoption of green accounting, several European countries and advanced economies like Japan had already championed its principles. Notably, Japan articulated the importance of green accounting in 2000 and issued relevant documents (Zhang, 2016). Concurrently, the United Nations published "Combination of Enterprise Environmental Performance and Financial Performance Indicators – Method for Standardization of Eco-efficiency Indicators", which outlined various aspects of green accounting, including disclosure scopes, definitions of environmental costs and assets, and assessment methodologies for environmental liabilities (Zhang, 2016).

This landmark document represented a shift from theory to practical implementation, advancing the operationalization of green accounting principles. To accurately reflect an entity's green assets, it introduced the "green asset depletion" account, which quantifies the financial implications of utilizing natural resources and their depletion (Zhang, 2016). This account functions as a contra-account to the "green assets" account, allowing for a net valuation of green assets by factoring in resource usage.

Naturally formed resource assets are recorded at their appraised value, while artificially invested resources are recorded at acquisition cost, adjusted annually based on production costs and the asset's inherent value (Zhang, 2016). Resources acquired through alterations are appraised at their purchase value or estimated worth, further illustrating the detailed methodology behind green accounting's asset valuation framework.

4.2 Prior Research on Green Accounting

In the past two decades, China has rapidly developed into the world's second-largest economy, yet it still faces challenges typical of a transitioning economy. The early stage of its green accounting system presents both obstacles and opportunities. Lu (2019) identified four key challenges: the theoretical foundations of green accounting are underdeveloped, lacking the depth needed for effective implementation; the absence of a coherent framework leads to inconsistent reporting practices; weak supervisory mechanisms deter investment in green accounting; and the lack of a dedicated audit system encourages speculative reporting.

Tang (2020) similarly highlighted the limited awareness and institutional support for green accounting in China, noting a shortage of specialized talent and unequal treatment compared to traditional accounting practices. Yang et al. (2021)

pointed out issues in high-polluting industries, where regulatory mandates often result in falsified or incomplete disclosures, undermining stakeholders' access to reliable information.

These insights reveal the complex challenges facing the development of green accounting in China. Overcoming these obstacles requires strengthening institutional frameworks, raising awareness, developing specialized talent, and improving regulatory oversight. Only through such comprehensive efforts can China harness the potential of green accounting for sustainable development.

4.3   Accounting Information Disclosure of Green Accounting

Accounting information disclosure serves as a vital complement to the financial statements of a company, offering stakeholders deeper insights into its operations and performance. Within the ambit of green accounting, information disclosure assumes added significance as it elucidates enterprises' endeavours pertaining to environmental pollution and governance. Such disclosures play a pivotal role in galvanizing the implementation of green accounting initiatives within the country.

4.3.1   Research Quality of Green Accounting Information Disclosure

The assessment of green accounting information disclosure goes beyond financial reporting to include various environmental factors. Li et al. (2021) introduced the concept of environmental accounting information disclosure quality (EAIDQ) as a measure for evaluating the quality of environmental information provided by companies. Interestingly, Liu found that EAIDQ is not significantly correlated with company ownership structures but is strongly linked to environmental regulations, especially in state-owned enterprises.

Similarly, Nguyen (2019) investigated the relationship between EAIDQ and financial performance in Vietnamese companies. His research indicated that companies that disclose environmental information tend to have better financial performance than those that do not. In contrast, Yang and Liang (2017) reported a significant negative correlation between the quality of green accounting information disclosure and company liabilities and profitability, while its relationship with tradable shares and developmental capabilities is unclear.

In the context of social media's rise, Luo (2019) studied how media disclosures affect environmental information disclosure quality. Analysing data from 842 heavily polluting firms, Luo found that positive media coverage has little effect on disclosure quality, while negative coverage adversely impacts it, inversely correlating with corporate debt financing costs. Together, these studies highlight the complex dynamics influencing the quality of green accounting information disclosure. Factors such as regulatory frameworks, financial performance, and media influence interact to shape disclosure quality, providing valuable avenues for further research and improvement. Rahman et al. (2024) investigated whether and how the intensity of social distancing from the COVID-19 pandemic influences the CSR disclosure index.

4.3.2 Research on the Practice of Domestic Green Accounting Information Disclosure

Green accounting has gained recognition among the Chinese public and enterprises in recent years, but many small and medium-sized enterprises (SMEs) still overlook its importance. While the benefits of environmental management accounting (EMA) are acknowledged, SMEs in China and elsewhere often fail to adopt effective EMA practices (Kong et al., 2022). According to Kan (2015), many Chinese companies have only a basic understanding of green accounting and often adopt a passive approach to environmental responsibility, sometimes even shirking their duties. Jin (2017) noted that despite the implementation of the "Circular Economy Promotion Law" in 2009, which mandates 16 heavily polluting industries to disclose environmental information, China's guidelines differ significantly from Japan's, particularly in legislation concerning reuse and recycling. Furthermore, Tang (2020) highlighted that China's disclosure system for green accounting information is lax and lacks comprehensive content, leading to potential confusion for users. Rahman and Zhu (2024) explored the relationship between ex ante expected changes in ESG and future stock returns. This hinders the healthy development of green accounting practices in Chinese enterprises.

5. Research Theory

5.1 The Connotation of Green Accounting

5.1.1 Connotations

Green accounting merges accounting, ecological, and management methodologies to enhance environmental conservation, reduce degradation from developmental activities, and promote sustainable growth alongside corporate advancement. Grounded in the principle of "monetary measurement", green accounting incorporates complex environmental factors and employs various non-monetary measurement techniques, such as physical units or percentages, when necessary. This approach is rooted in societal interests, supported by legal frameworks for environmental protection, and aligned with sustainable development strategies.

Green accounting encompasses financial data related to operational activities, including income and expenditures for environmental development and maintenance. It facilitates the measurement, reporting, analysis, and evaluation of these factors. By examining the relationship among financial gains, environmental performance, and resource consumption, it aids decision-making processes aimed at maximizing economic, environmental, and societal benefits. This enables enterprises to enhance ecological protection while achieving economic objectives, fostering a model of green, low-carbon development.

The key components of green accounting include calculating and analysing environmental pollution costs, assessing natural resource depletion costs, and improving resource utilization efficiency while minimizing environmental costs. By integrating these elements, green accounting aims to ensure enterprise sustainability while upholding environmental preservation.

5.1.2   *Differences from Traditional Accounting*

The disparities in the origins of traditional accounting and green accounting underscore their divergent historical contexts. Traditional accounting traces its roots to 13th-century Italy, a period marked by a relatively developed commodity economy. Luca Pacioli's seminal work in 1494 introduced the widely adopted "Venice accounting method", laying the groundwork for traditional accounting practices. In contrast, the emergence of green accounting stems from escalating environmental resource depletion and critiques of conventional accounting practices. Since the 1870s, escalating population growth and demand have intensified natural resource consumption, leading to energy shortages and escalating environmental degradation. It is within this backdrop that green accounting emerges as a response to contemporary environmental imperatives.

The divergences manifest in the fundamental meanings and assumptions underpinning traditional and green accounting methodologies. Traditional accounting relies on documented evidence and employs specialized procedures to systematically record and report the economic transactions of entities, furnishing managers with vital financial insights. Its focus solely on economic considerations sidelines environmental concerns. Rooted in the notion of maximizing economic benefits, traditional accounting views enterprises as "economic entities" with the primary goal of enhancing financial worth, irrespective of resource utilization and social ramifications. In contrast, green accounting adopts a holistic perspective, conceptualizing enterprises as "social ecological economic entities". This paradigm encourages businesses to factor in social and environmental considerations alongside financial objectives, reflecting a broader understanding of organizational responsibilities.

The scope of accounting objects differs significantly between traditional and green accounting. Traditional accounting primarily focuses on monetary transactions within the production process, encapsulating capital movement within its purview. Conversely, green accounting transcends conventional capital-centric perspectives by considering the broader societal impacts of economic activities. In addition to financial transactions, green accounting encompasses the valuation of natural cycles and societal production and consumption dynamics. This expansive scope necessitates redefining accounting elements to incorporate environmental considerations fully. Under green accounting frameworks, assets extend beyond traditional categories to include "natural assets", encompassing tangible resources like air, water, and forests, and intangible assets such as mining rights, pollutant discharge permits, and environmental patents. By broadening the accounting scope, green accounting provides a comprehensive assessment of environmental impacts and resource utilization, surpassing the confines of traditional accounting practices.

5.2   The Theoretical Basis of Green Accounting

5.2.1   Sustainable Development

Sustainability serves as a fundamental prerequisite for achieving sustainable development, encapsulating the capacity of societal and ecological systems, as well as other evolving systems, to function effectively without depletion. This concept inherently embodies a long-term temporal dimension, emphasizing the ability of systems to endure over extended periods. Typically, sustainability serves as a yardstick for assessing the ecological impact of production activities on environmental resources. On the other hand, sustainable development denotes the enduring progression towards development over time, encompassing the amalgamation of sustainability and developmental pursuits. While prioritizing development, sustainable development mandates the consideration of environmental, social, and other resource-related facets to ensure harmony and equilibrium.

The concept of sustainable development has evolved since the mid-20th century, gradually coalescing into a comprehensive and actionable framework. Central to this theory is the notion of coordinated development across population, environment, resources, and economy. As such, sustainable development constitutes a holistic approach to societal advancement, advocating for balanced growth that safeguards environmental integrity and social equity.

The environment holds a unique status as a shared public resource. In the pursuit of wealth creation, enterprises often grapple with the challenge of reconciling their profit motives with environmental responsibilities. This inherent tension arises from the perception that assuming environmental responsibilities may entail financial sacrifices for businesses. However, within the framework of sustainable development, this apparent conflict is re-evaluated, prompting a re-examination of corporate roles and responsibilities vis-à-vis environmental stewardship. Various theoretical frameworks and practical tools, including the advancement of circular economy principles and the adoption of environmental accounting practices by enterprises, serve as catalysts in navigating and resolving this complex interplay between business imperatives and environmental sustainability.

5.2.2   Opportunity Cost Theory

Opportunity cost theory stands as the cornerstone of environmental accounting, providing a direct theoretical and methodological framework for addressing environmental costs and benefits. This theory facilitates the identification and measurement of environmental benefits, thereby serving as the foundational principle and methodological basis for environmental accounting. Given the presence of economic externalities, there exists a reciprocal feedback mechanism whereby economic activities and natural resources exert mutual influence and interaction. Consequently, every economic endeavour undertaken by an enterprise incurs costs, with the total cost of products stemming from the entirety of natural resources under human intervention.

Environmental accounting delves into the examination of the cyclical processes within material systems, encompassing not only material and labour costs but also environmental costs. These costs encapsulate the broader impact of economic activities on the natural environment. In order to sustain the normal functioning of the human economic system, individuals acknowledge and actively engage in environmental protection endeavours. These activities can be categorized into two primary domains: pollution prevention and control initiatives, such as wastewater and waste gas treatment, and resource restoration efforts, such as the recycling of waste products. Through these concerted actions, society endeavours to mitigate the adverse effects of economic activities on the environment, thereby fostering a more sustainable trajectory of development.

6. Findings

6.1   *The Necessity and Feasibility of Implementing Green Accounting*

From the vantage point of sustainable development, the adoption of green accounting and the advancement of sustainable economic growth form a symbiotic relationship, each complementing the other in an indispensable manner. Their shared objective is to incentivize enterprises to actively pursue both social and economic benefits while adhering to the principles of environmental resource conservation and circular development. By meticulously planning and conserving various natural resources, enterprises strive to address environmental concerns as a primary objective. Given the finite nature of resources crucial for human survival, effective environmental protection becomes imperative for fostering improved living conditions. Green accounting, as a methodological approach, quantifies the loss of natural resources alongside economic compensation, facilitating the judicious utilization of resources and the measurement of various social factors pertinent to sustainable economic and social development in China, thereby propelling sustainable economic growth forward.

From the standpoint of enterprise development, implementation of a green accounting system enables enterprises to assess both the environmental resource protection costs and their profitability accurately. Additionally, it enables the precise depiction of a company's business operations and management outcomes, aiding in the analysis of financial management risks and enabling stakeholders to make informed evaluations of managerial decisions. Conventional enterprise development models often entail high-cost inputs and yield low-efficiency outputs, resulting in excessive waste of environmental and human resources, ecological pollution, inefficiency, and inadequate utilization of environmental resources, thus impeding the enterprise's progress. Given the heightened competitive pressures facing enterprises, ensuring continued survival in increasingly competitive markets becomes paramount. As Sun (2019) asserted, green accounting is imperative for the sustainable development of all enterprises, as it fosters resource rationalization, promotes resource recycling, and encourages the pursuit of new resources and energy sources, thereby enhancing enterprises' competitive edge in the market.

From the perspective of accessing international markets, despite the temporary setbacks stemming from the recent global pandemic, China's full-scale opening up necessitates increased engagement in international trade. Notably, environmental protection has emerged as a prominent international concern. To penetrate international markets and accrue greater economic benefits, Chinese enterprises must offer products that align with stringent environmental standards. Insufficient adherence to green indicators often leads to product rejections in foreign markets. Therefore, comprehensive implementation of green accounting throughout the product lifecycle not only ensures compliance with "green" standards but also serves as a requisite for navigating green trade barriers and enhancing competitiveness in the international arena. As highlighted by Shao (2020), prioritizing the comprehensive implementation of green accounting safeguards the legal rights and economic interests of Chinese enterprises in international trade activities, positioning them for dominance in global market competition.

6.2 Practice of Green Accounting

Zhu Ye Group, situated in the Qingshuitang Industrial Zone of Zhuzhou, Hunan Province, has traversed several decades of challenges and triumphs since its inception. Founded in 1956, the Zhu Ye Smelter underwent reorganization and attained listing on the Shanghai Stock Exchange in 2004 under the stock code 600961. Primarily engaged in the production of lead, zinc, and alloy products and the comprehensive recovery of rare metals such as copper, gold, silver, bismuth, cadmium, indium, tellurium, and sulphuric acid, the company boasts an annual production capacity of 650,000 tonnes of lead and zinc products. Distinguished by its exceptional comprehensive recovery rate of valuable metals, Zhu Ye Smelter Group stands as the pioneer in China's lead and zinc industry to achieve ISO9001 quality system certification, ISO14001 international environmental management system certification, and OHSMS18001 occupational health and safety system certification. The company's developmental trajectory is guided by a commitment to "leading technology" and "green smelting", epitomizing exemplary practices in environmental stewardship, comprehensive recycling, and adaptability to raw materials. Serving as a beacon among resource-based enterprises, Zhu Ye Smelter Group not only furnishes technical expertise to China's lead–zinc smelting technology but also makes substantial contributions to environmental conservation, cementing its position as an industry exemplar.

Embracing the mantra of "developing circular economy and fostering a green Zhu Ye Industry", Zhu Ye Group has adopted a proactive stance towards environmental preservation, focusing on prevention and amalgamating prevention with control. By aligning environmental protection efforts with the enterprise's core policy, the company has achieved an impressive resource utilization rate exceeding 85%. This accomplishment has been underpinned by investments in equipment upgrades, research and development initiatives, the adoption of new technologies, the realization of clean production practices, and the reduction of "three-waste" discharge in lead and zinc smelting processes. Positioning itself at the forefront of lead and zinc smelting technology, Zhu Ye Group's technological prowess rivals that of global leaders.

Demonstrating a sustained commitment to environmental stewardship, Zhu Ye Smelter Group has made substantial investments in scientific research, particularly since 2011, when it embarked on aggressive environmental pollution control measures. Committed to complying with national policies, the company earmarked a staggering 40 million yuan for scientific research to mitigate environmental pollution. Furthermore, in 2012, the company allocated 12 million yuan to procure 2,000 tonnes of sulphur dioxide emission rights from the Hunan Provincial Main Pollutant Discharge Reserve Trading Center, with a discharge period of five years. Subsequently, in 2013, the company purchased 10,515.42 tonnes of sulphur dioxide emission rights and 149.17 tonnes of chemical oxygen demand emission rights from the Zhu Ye Main Pollutant Emission Rights Reserve Exchange, further underscoring its environmental commitment. The year 2016 marked a pivotal juncture in the company's evolution, as it embarked on a transformative journey towards upgrading its operations, achieving significant technological breakthroughs, and contributing substantially to environmental conservation efforts.

Central to the implementation of green accounting are three pivotal concepts: green assets, green liabilities, and green benefits. Green assets represent accounting entities capable of effectively controlling economic activities and projecting potential environmental and resource benefits. Assets such as land, forests, oceans, minerals, and ecological assets are considered green assets if they offer direct or indirect benefits to the enterprise in the future. Green liabilities encompass the costs associated with mitigating the environmental impact of the enterprise's economic activities. Characterized by traceability, joint responsibility, and uncertainty, green liabilities mandate enterprises to assume responsibility for environmental restoration costs, even if the actions causing environmental harm were legal at the time. Green benefits denote the economic gains accruing from the enterprise's environmental initiatives within a specified period. These benefits encompass tax incentives, rewards for environmental protection, enhanced social reputation, and increased stock value.

Drawing insights from Zhu Ye Group's financial reports spanning 2018–2022, a comprehensive examination of the company's green accounting practices emerges. The calculated green assets exhibit fluctuations over the five-year period, attributed primarily to variations in government subsidies impacting natural green assets. Conversely, green liabilities demonstrate a declining trend, indicating a reduction in environmental restoration costs associated with lead and zinc smelting processes. However, it is noteworthy that the reported figures fail to fully capture the company's environmental expenditures, including the significant costs incurred in relocating the smelter and remediating pollution in the Qingshuitang area, estimated at nearly 100 billion yuan. Despite these challenges, the company's unwavering commitment to environmental stewardship and sustainable development underscores its leadership in the realm of green accounting and corporate responsibility. Table 2.1 shows the green balance sheet of Zhu Ye Group from 2018 to 2022.

Table 2.1 Green Balance Sheet of Zhu Ye Group from 2018 to 2022

Accounting Terms	2018	2019	2020	2021	2022
1. Green assets	2,990,995,174	1,127,834,958	1,106,744,400	1,328,465,874	1,024,770,101
1.1 Natural green assets	1,475,246,448	1,022,939,430.00	984,588,673	1,160,409,241	988,574,534
1.2 Environmental current assets	-	-	-	156,064,641	-
1.3 Environmental accounts receivable	-	-	-	-	-
1.4 Environmental fixed assets	1,398,000	533,000	449,603	-	-
1.5 Environmental construction in progress	1,508,738,531	100,964,321	118,889,885	8,710,985	32,087,569
1.6 Environmental intangible assets	3,670,000	2956187	1,460,256	1,371,033	1,371,033
1.7 Inventory of environmentally friendly products	-	-	-	-	-
1.8 Environmental research and development expenditure	-	-	-	-	-
1.9 Environmental cultivation assets	1,942,195	442,020	1,355,983	1,909,974	2,736,965
1.10 Accumulated depletion of green assets	-	-	-	-	-
1.11 Other green assets	-	-	-	-	-
2. Green liabilities (2.1 + 2.2)	2,205,442	1,260,377	876,364	265,157,499	59,083,752
2.1 Confirmed green liabilities	2,205,442	1,260,377	876,364	853,114	1,697,546
2.1.1 Short-term environmental protection loans	-	-	-	-	-
2.1.2 Environmental protection payments payable	-	-	-	-	-
2.1.3 Environmental taxes and fees payable	2,205,442	1,260,377	876,364	853,114	1,697,546
2.1.4 Payable interest on environmental loans	-	-	-	-	-
2.1.5 Other payables	-	-	-	-	-
2.1.6 Long-term environmental protection loans	-	-	-	-	-
2.1.7 Long-term payables	-	-	-	-	-
2.2 Uncertain green liabilities	0	0	0	264,304,385	57,386,206
2.2.1 Environmental fines	-	-	-	-	-
2.2.2 Pending environmental litigation	-	-	-	264,304,385	57,386,206
3. Green equity (1–2)	2,988,789,732	1,126,574,581	1,105,868,036	1,063,308,375	965,686,349

The environmental profit statement for Zhu Ye Group from 2018 to 2022 highlights its financial performance concerning green income and costs. Table 2.2 shows that over these years, the company's green income fluctuated significantly. In 2018, the green income was 91,250,000 yuan, rising to 104,460,468 yuan in 2019. However, in 2020, it decreased slightly to 93,541,179 yuan, before increasing to 240,479,372 yuan in 2021. By 2022, the green income decreased again to 79,680,628 yuan.

In contrast, Zhu Ye Group's green costs showed a different trend. From 2018 to 2022, green costs varied considerably, reflecting the company's environmental expenditures. In 2018, the annual green loss was 9,843,397,317 yuan, which decreased to 5,064,600,262 yuan in 2019. This figure then increased to 6,840,389,163 yuan in 2020 and further increased to 9,945,999,188 yuan in 2021. By 2022, the company incurred its highest green costs of 11,369,458,783 yuan.

These figures reveal significant negative externalities from Zhu Ye Group's operations, with high green costs indicating ongoing challenges in environmental management and sustainability. Despite attempts to reduce environmental impact, the substantial costs highlight the need for improved practices. Additionally, the concentration of green benefits in smelting by-product sales emphasizes the importance of diversifying revenue streams to better manage external effects.

The economic landscape in 2021 posed further challenges, as global recovery from the COVID-19 pandemic slowed, and international commodity prices surged, leading to inflation. Enterprises faced various issues, including COVID-19 outbreaks, extreme weather, coal and electricity shortages, and raw material scarcity. These factors underscore the necessity for Zhu Ye Group to navigate a complex environment while addressing its environmental responsibilities and ensuring sustainable practices.

The analysis of Zhu Ye Smelter Group's green accounting practices reveals both commendable efforts and areas for improvement. While the company has undertaken initiatives to enhance its environmental performance, such as increased investment in research to improve smelting technology and pollution control measures, several shortcomings remain evident.

One notable issue pertains to the handling of green liabilities. The decision not to disclose significant governance costs, amounting to hundreds of billions of yuan, suggests a lack of transparency in reporting and raises concerns about the accuracy of environmental assessments. This indicates a broader management deficiency in comprehensively understanding and evaluating environmental impacts and expenditures. The fragmentation of environmental data across multiple departments further exacerbates the challenge, impeding effective collation, analysis, and feedback. Moreover, prioritization of short-term economic interests over long-term sustainability underscores a systemic issue where environmental concerns are sidelined in favour of financial gains.

Furthermore, the evaluation of green benefits presents challenges, particularly concerning intangible gains resulting from environmental remediation efforts. The absence of uniform standards and reliable measurement methods complicates the quantification of these benefits, hindering accurate accounting practices. Despite scholarly efforts to propose frameworks for green accounting, the lack of standardized reporting formats limits widespread adoption and efficacy.

Table 2.2 Green Profit Sheet of Zhu Ye Group from 2018 to 2022

Accounting Term	2018	2019	2020	2021	2022
1. Green benefits	91,250,000	104,460,468	93,541,179	240,479,372	79,680,628
1.1 Environmental governance benefits	91,250,000	104,460,468	93,541,179	84,414,731	79,680,628
1.2 Environmental subsidy benefits	-	-	-	156,064,641	-
1.3 Other green benefits	-	-	-	-	-
1.4 Benefits of environmental tax exemption and green costs	-	-	-	-	-
2. Green costs	9,934,647,317	5,169,060,730	6,933,930,342	10,186,478,560	11,449,139,411
2.1 Environmental production costs	9,932,441,875	5,167,358,333	6,931,697,995	10,183,715,471.53	11444704900
2.2 Environmental management costs	-	442,020	1,355,983	1,909,974	2,736,965
2.3 Environmental damage costs	2,205,442	1,260,377	876,364	853,114	1,697,546
2.4 Other environmental costs	-	-	-	-	-
3. Environmental profit	−9,843,397,317	−5,064,600,262	−6,840,389,163	−9,945,999,188	−11,369,458,783

Additionally, the absence of environmental tax exemptions and income for Zhu Ye Group reflects incomplete information disclosure practices. While the company shares key environmental data publicly, the inadequacy of green accounting management within Chinese enterprises underscores systemic deficiencies in regulatory frameworks and enforcement procedures. Without robust laws and regulations governing green accounting practices and effective oversight mechanisms, environmental protection remains a persistent challenge.

In summary, while Zhu Ye Smelter Group has made strides in implementing green accounting initiatives, such as pollution control measures and comprehensive waste utilization, critical deficiencies persist. Addressing these shortcomings will require a concerted effort to enhance transparency, standardize reporting practices, and strengthen regulatory frameworks to ensure that environmental considerations are prioritized alongside economic interests.

6.3   *Development of Green Accounting*

The development of green accounting in China plays a crucial role in advancing sustainable economic growth and environmental protection. However, several challenges hinder its widespread adoption and effectiveness.

One major challenge is the incomplete disclosure of environmental information by companies. Many enterprises, such as Zhu Ye Smelter Group, do not fully disclose their environmental liabilities and expenditures, leading to a lack of transparency and accountability. To address this, companies should be required to provide comprehensive environmental data in their financial reports, enhancing transparency and enabling informed stakeholder decisions.

Another critical issue is the absence of standardized measurement methods and reporting formats for green accounting. Without uniform standards, accurately quantifying environmental benefits and liabilities becomes difficult, undermining the comparability and reliability of data. Regulatory authorities should establish standardized frameworks and guidelines for green accounting practices to facilitate consistent reporting and enhance the credibility of environmental disclosures.

Additionally, the lack of robust regulatory frameworks and enforcement mechanisms poses a significant barrier to the development of green accounting in China. Without clear mandates and penalties for non-compliance, companies may lack the incentive to prioritize environmental considerations. Strengthening regulatory oversight and enforcement is essential to ensure adherence to green accounting standards and promote corporate responsibility.

Many companies also struggle with a lack of awareness and capacity to implement green accounting effectively. This may stem from insufficient training on environmental accounting principles and inadequate resources for environmental management initiatives. To counter this, comprehensive training programmes and capacity-building initiatives should be developed to enhance the understanding of and internal capabilities for green accounting.

Finally, short-term economic pressures and a focus on profitability can deter companies from prioritizing environmental considerations. The reluctance to incur costs for environmental protection or disclose liabilities may be driven by a narrow focus on immediate financial gains. Policymakers and industry stakeholders should encourage the integration of environmental sustainability into business strategies and provide incentives, such as tax breaks and subsidies, to promote the adoption of green accounting practices.

6.3.1 Opportunities and Challenges (or Problems) Faced by Implementing Green Accounting

Implementing green accounting in China presents both opportunities and challenges. Let us delve into each aspect:

Challenges:

1. Lack of Theoretical Development: China's green accounting theory is still in its infancy and lags behind developed countries. This results in non-standardized measurement methods and a lack of operational guidelines, hindering effective implementation.
2. Shortage of Talent: There is a scarcity of professionals skilled in green accounting due to the interdisciplinary nature of the field. The absence of comprehensive training programmes and insufficient theoretical frameworks further exacerbate this challenge.
3. Inadequate Legal Frameworks: Existing laws and regulations related to green accounting are ambiguous and lack enforcement mechanisms. This creates uncertainty and resistance among enterprises, impacting the adoption and effectiveness of green accounting practices.
4. Short-term Economic Pressures: Many enterprises prioritize short-term profits over long-term environmental sustainability. The initial capital investment required for green accounting, coupled with potential reductions in output and profits, deters companies from fully embracing environmentally responsible practices.

Opportunities:

1. Growing Awareness: There is an increasing awareness of environmental issues and the importance of green accounting among the Chinese population. This heightened consciousness creates opportunities for the widespread adoption of green accounting practices.
2. International Exchange and Learning: Increased international academic exchanges enable Chinese scholars and practitioners to learn from global advancements in green accounting. This facilitates the integration of international best practices into China's accounting standards.
3. Technological Advancements: Rapid developments in computer technology, including artificial intelligence, streamline data collection, and analysis processes, enhance the efficiency and effectiveness of green accounting implementation.

4. Stringent Environmental Regulations: Strengthening environmental regulations in China incentivizes companies to adopt green accounting practices to comply with legal requirements. Enhanced transparency and accountability contribute to improved environmental performance.
5. Media and Public Oversight: Modern media channels empower the public to monitor and hold companies accountable for their environmental practices. Increased scrutiny encourages transparency and fosters a culture of environmental responsibility.

While challenges such as theoretical underdevelopment, talent shortages, legal ambiguities, and short-term economic pressures persist, opportunities arising from growing awareness, international exchange, technological advancements, stringent regulations, and public oversight offer avenues for the advancement and implementation of green accounting practices in China. By addressing these challenges and leveraging emerging opportunities, China can accelerate its transition towards sustainable development and environmental stewardship.

6.3.2   *Countermeasures for Improving the Application of Green Accounting*

To enhance the implementation of green accounting, several strategic measures need to be adopted:

1. Establishing a Robust Theoretical Framework: China should strive to develop a comprehensive theoretical foundation for green accounting. Drawing upon existing research from developed nations like the United States and Japan, China can adapt and refine green accounting principles to suit its specific socio-economic context. Collaboration between experts in ecology and accounting is crucial for advancing theoretical understanding. Practical application and pilot projects should accompany theoretical development, with findings informing ongoing refinement.
2. Investing in Talent Development: There is a pressing need to cultivate a skilled workforce proficient in green accounting practices. Establishing specialized green accounting disciplines in educational institutions or integrating green accounting content into existing accounting programmes is essential. Moreover, enhancing legal knowledge related to environmental laws among accounting professionals is vital to ensure compliance and effectiveness.
3. Enhancing Legal Frameworks: Strengthening laws and regulations pertaining to green accounting is imperative. Incorporating green accounting into existing accounting laws would provide a solid legal foundation for its implementation. Furthermore, aligning environmental regulations with accounting standards ensures consistency and facilitates compliance. Robust enforcement mechanisms, including clear rewards and penalties, are essential to deter environmental violations and promote accountability.
4. Fostering Environmental Awareness: Promoting a culture of environmental responsibility among enterprises is crucial. Encouraging businesses to view

environmental protection as a strategic imperative fosters a conducive environment for green accounting adoption. Enterprises should proactively assess their environmental footprint and fulfil their social responsibilities. Government oversight ensures transparency and authenticity in environmental accounting practices.

By implementing these countermeasures, China can overcome challenges and leverage opportunities to advance the application of green accounting, fostering sustainable development and environmental stewardship across industries.

7. Conclusion

China has undergone decades of development since the inception of green accounting. While some progress has been achieved, numerous challenges persist in both theory and practice. The journey toward the widespread adoption and promotion of green accounting in China remains extensive. Currently, domestic green accounting is at a nascent stage of development, characterized by a host of unresolved issues alongside abundant opportunities.

While China boasts a solid foundation in traditional accounting and ecological principles, further research and discourse are warranted to adapt green accounting to the nation's specific context and economic landscape. Enterprises, as primary stakeholders in green accounting, must devise strategic policies aligned with the imperatives of sustainable development. Concurrently, governmental agencies should assume their corresponding responsibilities by expediting the formulation and implementation of laws, regulations, and guidelines.

Accounting organizations can accelerate the construction of a comprehensive theoretical framework, facilitating efficient accounting practices within businesses. With collective societal oversight and participation, China is poised to swiftly establish a distinct and pragmatic green accounting theory system. Through collaborative efforts across various sectors, China will soon realize the full potential of green accounting in advancing sustainable development objectives.

References

Chen, S., Saud, S., Bano, S., & Haseeb, A. (2019). The nexus between financial development, globalization, and environmental degradation: Fresh evidence from Central and Eastern European Countries. *Environmental Science and Pollution Research, 26*, 24733–24747.

Hu, C., Lin, Y., Connell, J. W., Cheng, H. M., Gogotsi, Y., Titirici, M. M., & Dai, L. (2019). Carbon-based metal-free catalysts for energy storage and environmental remediation. *Advanced Materials, 31*(13), 1806128.

Jiang, B., Kauffman, A. E., Li, L., McFee, W., Cai, B., Weinstein, J., . . . & Xiao, S. (2020). Health impacts of environmental contamination of micro-and nanoplastics: A review. *Environmental Health and Preventive Medicine, 25*(1), 1–15.

Jin, K. (2017). *A comparative study on environmental accounting information disclosure of Chinese and Japanese household appliance enterprises* [Master's thesis, Henan University].

Kan, X., & Meng, Q. (2015). The confusion and countermeasures of green accounting in China. *Technology and Investment, 6*, 127–131. https://doi.org/10.4236/ti.2015.63014

Kong, Y., Javed, F., Sultan, J., Hanif, M. S., & Khan, N. (2022). EMA implementation and corporate environmental firm performance: a comparison of institutional pressures and environmental uncertainty. *Sustainability, 14*(9), 5662.

Li, Y., Zhang, X., Yao, T., Sake, A., Liu, X., & Peng, N. (2021). The developing trends and driving factors of environmental information disclosure in China. *Journal of Environmental Management, 288*, 112386.

Lu, J. (2019). Research on the application of green accounting in corporate accounting. *Collection, 8*.

Luo, W., Guo, X., Zhong, S., & Wang, J. (2019). Environmental information disclosure quality, media attention and debt financing costs: Evidence from Chinese heavy polluting listed companies. *Journal of Cleaner Production, 231*, 268–277.

Ma, J., & Ma, J. (2019, August). A research review of corporate green accounting information disclosure. *IOP Conference Series: Earth and Environmental Science, 310*(5), 052071. IOP Publishing.

Nguyen, L., & Tran, M. (2019). Disclosure levels of environmental accounting information and financial performance: The case of Vietnam. *Management Science Letters, 9*(4), 557–570.

Rahman, M. J., & Wu, J. (2023). M&A activity and ESG performance: Evidence from China. *Managerial Finance, 50*(1), 179–197.

Rahman, M. J., Wu, Q., & Zhu, H. (2024). Corporate social responsibility in times of social distancing: Evidence from China. *Business Ethics, the Environment & Responsibility*.

Rahman, M. J., Zhu, H., & Chen, S. (2023). Does CSR reduce financial distress? Moderating effect of firm characteristics, auditor characteristics, and Covid-19. *International Journal of Accounting & Information Management, 31*(5), 756–784.

Shao, D. (2020). Information disclosure of green accounting in a low-carbon economy. *Quality and Market*.

Shao, J. (2020). Feasibility analysis of green accounting in China. *Cooperative Economy and Technology, 15*, 154–155. https://doi.org/10.13665

Sun, F. (2019). Necessity and implementation measures for enterprises to implement green accounting. *Chinese Market, 21*, 93+95. https://doi.org/10.1393939

Sun, J., Wang, J., & Chen, Y. (2020). A review of the development and application of green accounting in China. *Advances in Social Science, Education and Humanities Research, 408*, 34–40.

Tang, H. (2020, January). Discussion on the problems and countermeasures of green accounting implementation in enterprises of China in New Era. In *5th international conference on economics, management, law and education (EMLE 2019)* (pp. 836–840). Atlantis Press.

Wang, J., Jiang, H., & Yu, F. (2004). Green GDP accounting in China: Review and outlook. In *Ninth London group of environmental accounting, Copenhagen* (pp. 22–24). https://mdgs. un.org/unsd/envaccounting/londongroup/meeting9/china_country_report_2004.pdf

Yang, L. H., & Liang, X. T. (2017). Study on the influencing factors of environmental accounting information disclosure. In *International conference on economics, management engineering and marketing (EMEM 2017)* (pp. 134–141). https://dpi-journals.com/index. php/dtem/article/view/17072

Yang, M., Yan, X., & Li, Q. (2021). Impact of environmental regulations on the efficient control of industrial pollution in China. *Chinese Journal of Population, Resources and Environment, 19*(3), 230–236.

Zhang, X. (2016). Literature review of green accounting system. *Chizi (First and Middle Ten Days)*.

Zhu, Y., Liu, X., Hu, Y., Wang, R., Chen, M., Wu, J., . . . & Zhu, M. (2019). Behavior, remediation effect and toxicity of nanomaterials in water environments. *Environmental Research, 174*, 54–60.

Zhu, H., & Rahman, M. J. (2024). Ex-ante expected changes in ESG and future stock returns based on machine learning. *The British Accounting Review*, 101457.

3 Analysis of Green Accounting Impacts on Chinese Corporate Sustainability and Financial Performance

Md Jahidur Rahman, Tarek Rana, Hongtao Zhu, and Xinyi Huang

1. Introduction

Environmental issue has been a prominent topic for decades, tracing back to the Industrial Revolution. Industries such as steel, automobiles, and appliances are often held responsible for significant environmental harm. Concurrently, the demand for natural resources continues to increase due to technological advancements, while the overall availability of these resources remains static. To address this dilemma, economist Parker introduced the concept of "green accounting" in 1971 (Dhar et al., 2021). This approach aims to incorporate environmental costs into corporate financial reports, highlighting both financial performance and the environmental expenses necessary for generating profits (Dhar et al., 2021). Over the years, green accounting has been closely linked with corporate financial performance and sustainability. It plays a crucial role in enhancing a company's financial outcomes and fostering corporate sustainability (Endiana et al., 2020). Various studies have demonstrated a relationship among green accounting, financial performance, and sustainability across different industries and countries. China's rapid economic growth, particularly after the initiation of its first five-year plan in 1953, has led to increasing environmental challenges and energy consumption (Yang et al., 2018). The country faces pressure not only from economic expansion but also from energy consumption and pollutant emissions related to foreign trade (Zhao et al., 2018). The tension between environmental degradation and economic growth has intensified in the 21st century, prompting the Chinese government to explore green accounting theories and practices tailored to its specific context (Tang, 2019).

While there is a range of research on green accounting in China, most studies do not specifically examine the relationship between green accounting and corporate financial performance and sustainability. Instead, they tend to focus on providing an overview of green accounting practices, the challenges China faces in its implementation, and its correlation with economic growth (Yang & Zhao, 2018; Tang, 2019; Ma & Ma, 2019; Ming, 2007). Beyond these broader impacts, it is

DOI: 10.4324/9781003488965-4

important to investigate how green accounting affects specific aspects of Chinese corporations. Previous studies demonstrating the relationship between green accounting and corporate performance have primarily used data from companies in different countries and industries, such as pollution-heavy firms in Bangladesh and mining and agricultural companies in Indonesia (Dhar et al., 2021; Lestari & Restuningdiah, 2021; Emmanuel, 2021). Results may vary significantly when analsing data from nations with different economic contexts and environmental challenges. Therefore, this study aims to explore the correlation between green accounting and a corporation's financial condition and sustainable development by analysing financial data from selected Chinese corporations.

The relationship between green accounting and corporate financial performance and sustainability in China remains undefined. This dissertation aims to examine the impact of green accounting on the financial performance and sustainability of Chinese corporations. To achieve this, various research models and variables will be introduced. This study contributes to the literature by investigating whether the relationship between green accounting and corporate performance is positive or negative. Previous studies have indicated a positive correlation, but analysing data from Chinese companies will provide stronger evidence, given China's status as a leading economic power (Mattlin, 2017). If a positive relationship is confirmed, it could enhance the case for implementing green accounting in China. The development of green accounting in China can be categorized into three stages, although progress has slowed since the initial stage (Tang, 2019). Furthermore, China faces severe environmental issues and energy crises that negatively impact corporate performance and the economy (Du et al., 2014). Analysing the effects on Chinese corporations may raise awareness among both the government and businesses regarding the importance of green accounting.

This study will explore the influences of green accounting on financial performance and sustainability by analysing financial statements from selected listed companies in China from 2011 to 2020. Two multiple linear regression models will be employed: one to assess the correlation between green accounting and financial performance, and the other to verify its connection to sustainability. Based on prior studies, it is anticipated that a positive relationship will be found.

This research narrows down the focus of green accounting to its specific influences on Chinese companies, providing a valuable reference for future studies aimed at enhancing its implementation. Additionally, it supports earlier findings by testing models with data from a different context. This chapter includes a literature review to construct a hypothesis, a description of the sampling data and methodology, empirical analysis, and concluding remarks.

The remainder of this chapter is structured as follows. In Section 2, we will briefly present some related works of literature and construct a hypothesis. In Section 3, we will describe the sampling data and methodology used in this study. Section 4 will present the empirical analysis, and conclusions will be drawn in Section 5.

2.   Literature Review

2.1   *Green Accounting*

Introduced by the economist Parker around 1971, green accounting witnesses the efforts of scholars, accounting standard makers, and governmental organizations to require the corporations to disclose environmental costs and other environment-related information in the financial statements (Fleischman & Schuele, 2006). Generally, the concept of green accounting can be broadly defined as a type of accounting that takes environmental problems into consideration, which belongs to a branch of and is an innovation of accounting. Green accounting has evolved several embranchments after its development for decades, for instance, environmental accounting and sustainability accounting (Hernadi, 2012). Under the concept of green accounting, the corporation prioritizes the effectiveness and efficiency of its resource usage in the manufacturing process to help the corporation achieve environmental and social responsibility goals (Endiana et al., 2020).

Economists' definitions of progress, wealth, and development have been challenged by environmentalists since 1970, and the environmentalists suspected that wrong methods were used to compute a corporation's environmental costs (Rout, 2010). Under such circumstances, green accounting began to develop and was first introduced by Parker in the 1970s. Although green accounting first emerged in the 1970s, it was not until the 1990s that the academic and professional accounting communities started to have interests in environmental accounting (Gray et al., 1997). In 1993, a draft was published and a frame was established to combine the environmental data with economic data so that the companies will focus on not only making revenues but also considering the environmental impacts (Holub et al., 1999). People's awareness of green accounting continues to develop, and many countries have formulated standards to encourage companies to adopt green accounting.

Frankly, the logic behind the implementation of green accounting is simple – out of the concern of environmental problems. Bebbington et al. (1994) conducted a survey to investigate accountants' attitudes towards environmental problems, and the results showed that accountants held a positive attitude towards environmental issues. However, the environment continues to deteriorate, and extreme weather caused by climate change happens frequently. The increasing awareness of environmental problems is redirecting the concentration from corporations to environmental sensitivity (Okafor, 2018). Therefore, a different accounting concept should be introduced to satisfy this requirement. Unlike traditional accounting, green accounting asks the company to reveal any information that may influence the environment in the financial statement (Gonzalez & Mendoza, 2020).

This type of accounting allows the corporation to track the cost of protecting the environment while the main business activity is normally operated (Okafor, 2018). A company's contribution to human life and the environment, through taking ideas such as saving land, materials, and energy into consideration, can also be disclosed once the company applies green accounting (Endiana et al., 2020). Furthermore, according to Yakhou and Dorweiler (2004), the company that adopts green accounting has advantages, including letting its consumer know that the company takes social responsibility seriously, acts in accordance with the government regulatory, and shows concern about the environment.

2.2 *Corporate Sustainability and Green Accounting*

Firstly defined by the World Commission for Environmental and Development (1987), sustainability means the current generation meets its current needs without compromising the future generation's resources. Following this, the idea of sustainability became more popular around the world. The definition of corporate sustainability still remains ambiguous because the definitions of corporate sustainability have flourished in the past decades (Swarnapali, 2017). Wilson (2003) defined corporate sustainability as a set of notions to manage the corporate, which demonstrates the growth and profitability of a company and concentrates on the environmental, social, and economic outputs. Szekely and Knirsch (2005) claimed that corporate sustainability has connections with maintaining and expanding economic growth, stakeholder value, customer relations, and the quality of products or services, and proves the company practices some business ethics. Corporate sustainability is also viewed as the ability to cultivate long-term growth by satisfying various stakeholders' expectations (Neubaum & Zahra, 2006). Based on the various definitions of corporate suitability, Marrewijk and Werre (2003) insisted that no so-called definition exists for corporate sustainability.

Ketprapakorn and Kantabutra (2019) claimed that organizational culture plays a significant role in corporate sustainability, and studies show that entrepreneurship, hard work, excellence, prudence, profitability, and quality contribute to corporate sustainability. Therefore, Ketprapakorn and Kantabutra (2019) established a sustainability culture development framework to assist managers in achieving corporate sustainability. Based on the sustainability culture development framework, Kantabutra (2019) developed a corporate sustainability model, and the model includes the following six components:

1. Perseverance (based on self-determination theory): improve processes, products, and services for stakeholders;
2. Resilience (based on complexity theory): anticipation of and preparation for changes;
3. Moderation (based on sustainable leadership theory): make decisions discreetly and take stakeholders' consequences into consideration;
4. Geosocial Development (based on stakeholder theory): aggregate social and environmental responsibility with the operation;
5. Sharing (based on knowledge-based theory, dynamic capabilities theory, and knowledge management theory): share information with internal and external users;

6. Corporate Sustainability: include strong performance, crisis endurance, and public benefits.

Kantabutra (2019) stated the corporate sustainability model is meaningful in improving the corporate sustainability of some small businesses. An integrated corporate sustainability model was also constructed, which combines the sustainability vision and value with the corporate sustainability model (Ketprapakorn & Kantabutra, 2020).

The motivation for corporates to adopt corporate sustainability is they either feel obliged to practise it or want to implement it (Marrewijk, 2003). Driven by different circumstances and business operation situations, companies apply corporation sustainability out of different intentions. Marrewijk and Were (2003) proposed six levels of companies' different motivations to adopt corporate sustainability, including pre-corporate sustainability, compliance-driven corporate sustainability, profit-driven corporate sustainability, caring corporate sustainability, synergistic corporate sustainability, and holistic corporate sustainability. Among these six levels, caring-, synergistic-, and holistic-level corporate sustainability concern both economic and social aspects of the company (Van Marrewijk & Werre, 2003)

In addition, research has shown a positive relationship between green accounting and corporate sustainability among different countries and industries (Dhar et al., 2021; Nga & Dao, 2020; Endiana et al., 2020). Therefore, the study proposes the first hypothesis based on these previous findings:

Hypothesis H1: There is a positive relationship between green accounting and Chinese corporate sustainability.

2.3   *Financial Performance and Green Accounting*

Financial performance is a measurement to evaluate how well a company can use its resources to generate revenue and is a key indicator of the company's financial condition. A corporation's financial performance is crucial to its internal and external users. Financial performance shows the impacts of continuous and individual decisions made by managers (Epstein et al., 2015).

One of the most popular methods to measure financial performance is calculating the return on asset (ROA), which is a direct way to judge the corporation's capability to generate profits. However, various factors can affect financial performance. Porter and Van der Linde (1995) constructed the "Porter hypothesis", which refers to the effect of environment on a firm's financial performance, which is a similar hypothesis with this dissertation. The opposite perspective argues that extra expenses incur when the company follows some environmental regulations, and the company's production capacity will be limited (Jaffe et al., 1995). However, those negative voices have also been rebutted by the supporters. According to Cairncross (1990), if a company can fulfil the government's requirements about the environment, the government may create a protected market for the company compared with its competitors that cannot meet the requirement. Environmental management has become a means to improve a corporation's financial performance through developing processes and products to enhance the company's competitiveness (Stead & Stead, 1992). Therefore, Watson and Polito (2002) introduced a framework to link environmental performance with financial performance. The connections between environmental performance and financial performance include increasing revenues by improving efficiency and decreasing environmental risk, which has a negative influence on financial performance (Peloza, 2006).

A corporation's financial performance can be influenced by other factors except for its environmental performance. If the company can effectively manage the relationship with its stakeholders, the financial performance will be improved (Jones, 1995; Brammer & Millington, 2008). The perspective that links financial performance with the relationship with stakeholders is considered stakeholder management theory (Freeman, 1984; Clarskon, 1995). However, critics of the stakeholder management theory doubt the corporation's ability to balance the demands from different stakeholder groups to facilitate financial performance (Mazutis, 2010).

A corporation's financial performance is significant to its stakeholders. Financial performance evaluates the feasibility, solidarity, and fertility of the business to determine the efficiency of the decisions made by managers (Bhunia et al., 2011). The main purpose of measuring financial performance is to make sure all the investments from the investors are used to establish the corporation's objective efficiently and effectively (Atmadja & Saputra, 2018). Therefore, it is vital to construct a system to reliably measure financial performance. Bacidore et al. (1997) claimed a trustworthy financial performance using the risk-adjusted return on invested capital to evaluate the managerial decisions' impacts on shareholder value. After analysing methods to measure financial performance, Bacidore et al. (1997) proposed the most appropriate way is whether the dividends stakeholders distributed from the company excess the required amount to compensate for the systematic risks.

Previous studies also prove a positive relationship between green accounting and financial performance in some countries like Indonesia, Nigeria, and Colombia (Lestari & Restuningdiah, 2020; Emmanuel, 2021; Gonzalez & Mendoza, 2020; Kartika & Utami, 2019). Several studies in China also investigates the relationship between environmental accounting and firm performance (Rahman & Wu, 2023; Rahman, Wu et al., 2024; Rahman, Zhu, & Chen, 2023; Zhu & Rahman, 2024). Thus, the study proposes the second hypothesis:

Hypothesis H2: There is a positive correlation between green accounting and a Chinese company's financial performance.

3. Research Methodology

3.1 Sample Selection

In order to have a thorough understanding of the relationship between green accounting and Chinese corporations' financial performance and sustainability, more than 1,500 covering more than 15 industries, including manufacturing, architecture, real estate, and entertainment industries, were selected as sampling companies for this study. To make sure of the authority, all of the selected companies are listed in either the Shanghai Stock Exchange Market or Shenzhen Stock Exchange Market, and the collected data span from 2011 to 2020 (10 years). The companies' data were collected from the China Stock Market & Accounting Research (CSMAR) database, a prestigious database providing reliable data about Chinese companies to its users and is roughly processed before the analysis (Rahman, Zhu, & Yue, 2024). Rough processes include reorganizing raw data to make the data more visually feasible and deleting some abnormal values to assure the credibility and reliability of the results.

3.2   Variables

3.2.1   Dependent Variables

The dependent variables in this study are financial performance and sustainability. According to the literature review, financial performance is an important indicator showing the corporation's operating situation and profitability. FP is used to represent financial performance in this study. The other dependent variable is sustainability. Business sustainability can be measured through the aspects of environmental sustainability and element sustainability (Bellucci et al., 2020). Corporations' sustainability data can be directly collected from CSMAR, and this study will use CS to represent a corporation's sustainable development. The measurements of the dependent variables are listed in Table 3.1.

3.2.2   Independent Variable

Green accounting requires a company to take its environmental expenses into consideration and has significant impacts against traditional accounting methods. However, implementation of green accounting in China has stagnated since China entered the first stage. Therefore, implementation of green accounting is taken as the independent variable, represented by GA. If the company applies GA, the GA value will be 1; otherwise, the GA value will be 0. Measurement of the independent variable is presented in Table 3.1.

3.2.3   Control Variables

Control variables will also be used in the model to improve the precision of the results. Control variables include the asset–liability ratio, the effectiveness of internal control, audit quality, top managements' education level, property of the company, company size, ROA, growth of sales, company scale, and the debt–asset ratio. In addition, year and industry variables are also introduced to control the results. Measurements of the control variables are presented in Table 3.1.

Table 3.1 Measurements of Variables

Name	Symbol	Measurement	Source
Corporate sustainability	CS	Growth and profitability of the company and its efforts on environmental and social outputs (Wilson, 2003)	CSMAR
Financial performance	FP	(Total equity + Total debt) / (Total assets + Total debt) (Lestari & Restuningdiah, 2020)	CSMAR
Green accounting	GA	Value is 1 if the company applies GA; otherwise, the value is 0 (Dhar et al., 2021)	CSMAR
Asset–liability ratio	Lev	Total asset / Total liabilities (Dhar et al., 2021; Rahman, Zhu, & Jiang, 2024)	CSMAR
Internal control quality	ICQ	Value is 1 if the internal control is effective; otherwise, the value is 0 (Dhar et al., 2021)	CSMAR
Audit quality	AQ	Assign AQ to 1 if the standard audit opinion is included; otherwise, the value is 0 (Dhar et al., 2021; Rahman, Zhu, & Hossain, 2023)	CSMAR
Education	Edu	Top management degree is assigned a value of 5 for PhD, 4 for master, 3 for bachelor, 2 for college, 1 for middle school or lower, and 0 for not available (Dhar et al., 2021)	CSMAR
Company property	Prop	Assign 1 to the state-owned company; otherwise, 0 (Dhar et al., 2021)	CSMAR
Return on assets	ROA	Net profit / Total assets (Dhar et al., 2021; Rahman, Wu, & Zhu, 2024)	CSMAR
Earnings per share	EPS	(Net Income – Preferred Dividends) / Shares of Common Stock Outstanding (Warren et al., 2018)	CSMAR
Debt–asset ratio	DAR	Total Debt / Total Assets (Lestari & Restuningdiah, 2020)	CSMAR
Size	Size	Total assets (Dhar et al., 2021; Rahman et al, 2024)	CSMAR
Scale	Scale	Ln (Total Assets) (Lestari & Restuningdiah, 2020)	CSMAR
Year	Year	Time frame (Dhar et al., 2021)	CSMAR
Industry	Ind	Category of industry (Dhar et al., 2021)	CSMAR

3.3.3 Model Construction

Taking corporate sustainability as the independent variable and the implementation of green accounting as the dependent variable, a model, based on the research of Dhar et al. (2021), is constructed to explore the relationship between GA and CS to verify H1:

$$CS = \beta_0 + \beta_1 GA + \beta_2 Lev + \beta_3 ICQ + \beta_4 ROA + \beta_5 AQ \\ + \beta_6 Edu + \beta_7 Prop + \beta_8 Size + åYear + åInd + \varepsilon \tag{1}$$

In order to determine the correlation between GA and corporate financial performance, the second model is structured. The second model is based on the model constructed by Lestari and Restuningdiah (2020), which takes financial performance as the independent variable and GA as the dependent variable to examine H2:

$$FP = \alpha + \beta_1 GA + \beta_2 ROA + \beta_3 EPS + \beta_4 DAR + \beta_5 Scale + \beta_6 Lev + \varepsilon \qquad (2)$$

4. Empirical Results

4.1 Descriptive Statistics

The descriptive statistics results are listed in Table 3.2. This study used 13,187 sample data in total to verify the models and test the hypotheses. The average value of CS is 0.04, showing that corporations in China have comparatively low sustainable development. The maximum value of CS is 98.694, and the minimum value of CS is −76.306, indicating that a large difference related to CS exists among Chinese companies. The average value of FP is 0.702, and the extreme values are 0.034 and 2.302, respectively. The average and extreme values of FP indicate that although some of the corporations' financial status is not as well as expected, the overall FP of Chinese corporations is positive and well-behaved. The average value of GA is 0.867, meaning that 86.7% of Chinese companies have implemented GA. Even though the development of GA in China is still in the initial stage, there is a large portion of companies that have applied GA as one of their accounting standards.

The average value of Lev is very close to the median, showing that sample companies have similar capabilities to generate profits from investments. The average value of ICQ is 0.954, meaning that 95.4% of the sample companies have effective internal control. The average value of AQ is 0.96, indicating 96% of sample corporations have obtained a standard audit opinion. The education level of corporations' top managers has a mean of 3.528, indicating that most of the sample companies' top managers have a master's degree or above. The average value of companies' property is 0.52, indicating that half of the sample companies are state-owned. The extreme values of ROA are −29.609 and 108.366, respectively, showing that the profitability of sample companies has a massive difference. The average value of EPS is 0.3543, which is mainly caused by the extreme values from particular companies. The average value of sample corporations' DAR is close to the median, showing that the sample companies have similar debt–asset ratios. According to the descriptive analysis results, the size and scale of sample companies differ greatly.

4.2 Correlation Analysis

4.2.1 Correlation Analysis for Model 1

The results of correlation analysis for model 1 are presented in Table 3.3. The correlation analysis provides support for the subsequent regression analysis. According to the results of the correlation analysis, the correlation between the dependent

Table 3.2 Descriptive Analysis

Variable	N	Mean	SD	Min	Max	Median
CS	13,187	0.04	1.205	−76.306	98.694	0.045
FP	13,187	0.702	0.107	0.034	2.302	0.686
GA	13,187	0.867	0.339	0	1	1
Lev	13,187	3.265	3.954	−5.136	141.245	2.188
ICQ	13,187	0.954	0.209	0	1	1
AQ	13,187	0.96	0.195	0	1	1
Edu	13,187	3.528	0.889	1	6	4
Prop	13,187	0.52	0.5	0	1	1
ROA	13,187	0.037	0.989	−29.609	108.366	0.031
EPS	13,187	0.3543	0.824	−16.748	30.114	0.247
DAR	13,187	0.464	0.407	−.195	28.548	0.457
Size	13,187	2,244,960,1799	11,225,272,089	10,441,933	272,164,949,588	4,070,453,799.16
Scale	13,187	22.44	1.388	16.161	28.636	22.270

Table 3.3 Correlation Analysis for Model 1

	CS	GA	Lev	ROA	ICQ	AQ	Edu	Prop	Size	CSR
CS	1.000									
GA	0.001	1.000								
	0.888									
Lev	0.005	−0.088***	1.000							
	0.546	0.000								
ROA	0.035***	−0.017*	−0.006	1.000						
	0.000	0.050	0.496							
ICQ	0.048***	0.016*	0.034***	0.014	1.000					
	0.000	0.070	0.000	0.115						
AQ	0.027***	0.015*	0.025***	−0.003	0.825***	1.000				
	0.002	0.076	0.004	0.719	0.000					
Edu	0.001	0.006	−0.065***	0.001	0.031***	0.023***	1.000			
	0.918	0.524	0.000	0.947	0.000	0.009				
Prop	0.002	0.031***	−0.165***	−0.013	−0.047***	−0.065***	0.184***	1.000		
	0.785	0.000	0.000	0.141	0.000	0.000	0.000			
Size	0.005	0.064***	−0.071***	−0.001	0.013	−0.008	0.088***	0.123***	1.000	
	0.578	0.000	0.000	0.925	0.149	0.381	0.000	0.000		
CSR	0.015	0.034	0.040	0.122***	0.256***	0.265***	0.018	−0.104***	0.077***	1.000
	0.588	0.215	0.139	0.000	0.000	0.000	0.510	0.000	0.005	

***, **, and * indicate the significance level of the relationship of each variable is 1%, 5%, and 10%

variable and independent variable is relatively weak. The weak correlation between the dependent and independent variables may be caused by ignoring the control variables, for instance, size, leverage (Lev), and ICQ. Therefore, the significance between the dependent and independent variables in the multiple linear regression analysis may differ from that in the correlation analysis since the regression analysis will consider the effect of control variables. In terms of control variables, ROA, ICQ, and AQ show great significance with corporate sustainability.

4.2.2   Correlation Analysis for Model 2

The results of correlation analysis for model 2 are presented in Table 3.4. According to the correlation analysis, all of the variables show significance to FP. Corporations' implementation of GA has a strong significance on FP. However, the significance between FP and GA is negative, which may be caused by the lack of control variables. Thus, the relationship between corporate FP and the implementation of GA may change if control variables are taken into consideration.

4.3   Hausman Test

The study uses the Hausman test to determine whether to adopt the fixed effect or the random effect for H2. The results of the Hausman test are shown in Table 3.5. Dependent variables, independent variable, control variables, and dummy variables are tested; however, this study only shows the results of the independent and control variables. The result of the Hausman test shows a probability value of 0.000, which indicates the hypothesis of applying the random effect is rejected. Therefore, the study adopts the fixed-effect model to analyse H2.

Table 3.4 Correlation Analysis for Model 2

	FP	GA	ROA	EPS	DAR	Scale	Lev
FP	1.000						
GA1	−0.072***	1.000					
	0.000						
ROA	0.079***	−0.017*	1.000				
	0.000	0.050					
EPS	0.085***	0.025***	0.117***	1.000			
	0.000	0.004	0.000				
DAR	−0.613***	0.002	−0.188***	−0.075***	1.000		
	0.000	0.819	0.000	0.000			
Scale	−0.430***	0.231***	−0.025***	0.250***	0.172***	1.000	
	0.000	0.000	0.004	0.000	0.000		
Lev	0.646***	−0.088***	−0.006	0.018**	−0.320***	−0.275***	1.000
	0.000	0.000	0.496	0.044	0.000	0.000	

***, **, and * indicate the significance level of the relationship of each variable is 1%, 5%, and 10%

Table 3.5 Hausman Test for Model 2

| | Coefficient | | | |
| | (b) | (B) | (b-B) | Sqrt(diag(V_b – V_B)) |
	Fixed effect	Random effect	Difference	SE
GA	0.000	0.005	−0.005	0.000
ROA	−0.002	−0.002	0.000	–
EPS	0.016	0.015	0.001	0.000
DAR	−0.103	−0.104	0.001	–
Scale	−0.021	−0.020	−0.001	0.000
Lev	0.011	0.011	0.000	–
	Test:	H_0: difference in coefficients not systematic		
		$Chi^2(82) = (b\text{-}B)'[(V\_b\text{-}V\_B)^{\wedge}(-1)](b\text{-}B) = 1580.03$		
		$Prob > Chi^2 = 0.000$		

4.4 Regression Analysis

The results of the regression analysis for both models 1 and 2 are presented in Table 3.6. In the regression analysis of model 1, the coefficient between CS and GA is positive, and the value of the coefficient is 0.04. The result has a *t*-value of 2.0023, showing the result is significant. In terms of control variables, ROA, Prop, and ICQ have a positive relationship with the independent variable. ROA has a *t*-value of 5.85, indicating that ROA has a significant positive relationship with CS. Although Prop and ICQ show a positive relationship with CS, the *t*-values of Prop and ICQ are not ideal, indicating that the relationships with CS are not significant. In the regression analysis of model 2, the coefficient between FP and GA is positive, and the value of the coefficient is 0.0036. The result has a *t*-value of 2.09, showing the result is significant. In terms of control variables, EPS and Lev show a positive relationship with FP, and DAR and Scale show a negative relationship with FP. The *t*-values of all control variables are larger than 2, indicating that the relationship with corporate FP is significant.

5. Robust Test

Robust test is a procedure to verify the test results, for example, regression analysis. Robust test has two major functions: preventing errors by showing the results do not depend on particular falsehoods and confirming variables that are important to the results (Kuorikoski et al., 2010). In this section, two techniques will be applied to test the robustness of the regression results: the test of multicollinearity and alternative measures for variables.

5.1 Test of Multicollinearity

Kuorikoski et al. (2010) illustrated the importance of applying the robust test in empirical analysis, including the difficulty of conducting empirical analysis using

Table 3.6 Regression Analysis

Variable	Model 1	Model 2
GA	0.04**	0.0036**
	(2.00)	(2.09)
Lev	−0.00	
	(−0.14)	
ROA	0.86***	−0.0018***
	(5.85)	(−3.25)
ICQ	0.09	
	(0.53)	
AQ	−0.06	
	(−0.37)	
Edu	−0.01	
	(−1.28)	
Prop	0.02	
	(0.70)	
Size	0.00	
	(0.27)	
CSR	−0.00	
EPS	(−0.46)	
DAR	0.098	0.0158***
Scale	1.775	(22.96)
Lev		−0.1099***
R-squared		(−76.44)
F-test		−0.0223***
		(−50.46)
		0.0119***
		(79.93)
		0.667
		4405.770
N	13,187	13,187
Year	Control	Control
Industry	Control	Control

t-statistics in parentheses.
*** $p < 0.01$, ** $p < 0.05$, * $p < 0.1$.

economic models, and the failure of indicating those fatal idealizations for the modelling results. Therefore, the study will use the test of multicollinearity to enhance the reliability of the results.

The test of multicollinearity mainly contains the variance inflation factor (VIF) test values, and the results are shown in Tables 3.7 and 3.8. According to the result of the VIF test for model 1, the VIF value for the independent variable is 1.34 and is less than 10, indicating there exists no multicollinearity. The VIF value for ICQ and AQ is not ideal; however, this study only concentrates on the independent variable's VIF value. According to Table 3.7, the VIF values of both independent and control variables are under 10, indicating there exists no multicollinearity among all of the variables in model 2.

Table 3.7 VIF Values for Model 1

Variable	VIF	1/VIF
GA	1.34	0.744
Lev	1.17	0.852
ROA	1.16	0.861
Edu	1.14	0.877
Prop	1.48	0.675
Size	2.12	0.472
CSR	1.19	0.841

Table 3.8 VIF Values for Model 2

Variables	VIF	1/VIF
GA	1.06	0.942
ROA	1.06	0.947
EPS	1.10	0.907
DAR	1.18	0.847
Scale	1.24	0.805
Lev	1.19	0.838

5.2 *Alternative Measures for Variables*

This section of the robust test examines whether the significance level changes if the measurements of the independent variable or dependent variables are replaced. This study alters the measurements for dependent variables in both models 1 and 2 and obtains the alternative data from the database (CSMAR) to replace the CS in model 1. The study also uses the return on capital employed (ROCE) value to measure the corporate FP to replace the dependent variable in model 2. According to Singh and Yadav (2013), ROCE measures the efficiency and profitability of corporations' investments, and most corporations consider ROCE as one of the indicators to show their FP. Therefore, this study uses ROCE as the alternative measure of FP. The results of alternative measures for variables are shown in Table 3.9. The coefficients are 0.05 and 0.01, and the t-values are 2.28 and 2.51, respectively. According to the results, there still exists strong significance between the independent variable and dependent variables after replacing the measurements.

6.   Discussion

In the discussion part, this study analyses the results from the regression analysis and robust tests to verify the hypotheses. In addition, this study compares the results with the results from previous research and explains the possible reasons that might cause the differences.

Table 3.9 Alternative Measures for Variables

Variable	Model 1	Model 2
GA	0.05**	0.01**
	(2.28)	(2.51)
Lev	0.00	0.00
	(0.02)	(0.43)
ROA	0.70***	0.11**
	(4.83)	(2.24)
ICQ	0.08	
	(0.51)	
AQ	−0.06	
	(−0.34)	
Edu	−0.01	
	(−1.28)	
Prop	0.02	
	(0.77)	
Size	0.00	
	(0.26)	
CSR	−0.00	
	(−0.23)	
EPS		0.01
		(1.56)
DAR		0.00
		(0.75)
Scale		0.00*
		(1.84)
R2	0.087	0.030
F-test	1.56	6.89
N	13,187	13,187
Year	Control	Control
Industry	Control	Control

t-statistics in parentheses.
*** $p < 0.01$, ** $p < 0.05$, * $p < 0.1$.

6.1   *Corporate Sustainability and Green Accounting*

Based on the regression analysis, there is a positive relationship between CS and GA, with a *t*-value of 2. The robust test indicates a VIF value of 1.34, suggesting no multicollinearity issues that could undermine the results. When the measurement of CS is altered, the positive and significant relationship remains intact. Thus, H1 is accepted: there is a positive relationship between GA and CS in Chinese corporations.

However, the *t*-value in this study is less significant compared to that in the previous research. This discrepancy may be due to prior studies focusing on specific types of companies, such as heavily polluting firms (Dhar et al., 2021), mining and agricultural sectors (Lestari & Restuningdiah, 2021), and manufacturing companies (Emmanuel, 2021). In contrast, this study analyses data from 77 industries, excluding the financial sector, which may contribute to the less pronounced significance of the results.

6.2 Financial Performance and Green Accounting

According to the regression analysis, there exists a positive relationship between GA and FP, with a *t*-value of 2.09. In terms of the robust test, all of the VIF values for the independent and control variables are less than 2, showing that there is no multicollinearity issue that might invalidate the result. As for the alternative measures for the variable test, the relationship between GA and FP is still positive and significant. Therefore, H2 is accepted, which shows that a positive correlation exists between GA and Chinese companies' FP. Compared with previous studies, the results in this study have some negative correlations between GA and control variables, for instance, ROA. The possible explanation for the negative correlation is that the study omits the extreme values collected from the database. The descriptive analysis shows that the ROA ranges from −29.609 to 108.366, which may affect the results of the correlation.

7. Conclusion

Given the deteriorating environment and increasing environmental issues, this study proposed GA as a framework to assess whether it can help corporations achieve profitability while minimizing environmental impacts. Using CS and FP as metrics, two regression equations were constructed to test the hypotheses. The study focused on Chinese companies from 2011 to 2020, exploring the relationship between GA implementation and both CS and FP.

The empirical analysis led to the following conclusions: (1) the implementation of GA positively affects CS, and (2) it also positively influences FP. Environmental contribution and profitability are critical indicators of overall corporate performance, and adopting GA helps enhance these measures. The positive relationships identified in this study support the government's efforts to promote GA among Chinese corporations and corroborate previous research on its impacts.

The sample data encompassed all industries in China, excluding the financial sector. However, results may vary if focused on specific industries, such as heavily polluting sectors. Therefore, future research should consider narrowing the sample to particular industries and including data from small- and mid-sized companies for comprehensive empirical analysis.

References

Atmadja, A. T., & Saputra, K. A. K. (2018). Determinant factors influencing the accountability of village financial management. *Academy of Strategic Management Journal*, *17*(1), 1–9.

Bacidore, C., Boquist, A. J., Milbourn, T. T., & Thakor, V. A. (1997). The search for the best financial performance measure. *Financial Analysts Journal*, *53*(3), 11–20.

Bebbington, J., Gray, R., Thomson, I., & Walters, D. (1994). Accountants' attitudes and environmentally-sensitive accounting. *Accounting and Business Research*, *24*(94), 109–120.

Bellucci, M., Bini, L., & Giunta, F. (2020). Implementing environmental sustainability engagement into business. *Innovation Strategies in Environmental Science*, 107–143.

Bhunia, A., Mukhuti, S. S., & Roy, G. S. (2011). Financial performance analysis-a case study. *Journal of Social Sciences, 3*(3), 269–275.

Brammer, S., & Millington, A. (2008). Does it pay to be different? An analysis of the relationship between corporate social and financial performance. *Strategic Management Journal, 29*(12), 1325–1343.

Cairncross, F. (1990). Cleaning up: A survey of industry and the environment. *The Economist, 316,* 1–17.

Clarskon, M. (1995). A stakeholder framework for analyzing and evaluating corporate social performance. *Academy of Management Review, 20,* 92–117.

Dhar, B., Sarkar, S., & Ayittey, F. (2021). Impact of social responsibility disclosure between implementation of green accounting and sustainable development: A study on heavily polluting companies in Bangladesh. *Corporation Social Responsibility and Environment Management,* 1–8.

Du, M., Wang, B., & Wu, Y. (2014). Sources of China's economic growth: An empirical analysis based on the BML index with green growth accounting. *Sustainability, 6,* 5983–6004.

Emmanuel, E. (2021). Green accounting reporting and financial performance of manufacturing firms in Nigeria. *American Journal of Humanities and Social Sciences Research, 5,* 179–187.

Endiana, I., Dicriyani, N., Adiyadna, M., & Putra, I. (2020). The effect of green accounting on corporate sustainability and financial performance. *Journal of Asian Finance, Economics and Business, 7*(12), 731–738.

Epstein, M. J., Buhovac, A. R., & Yuthas, K. (2015). Managing social, environmental and financial performance simultaneously. *Long Range Planning, 48*(1), 35–45.

Fleischman, K. R., & Schuele, K. (2006). Green accounting: A primer. *Journal of Accounting Education, 24,* 35–66.

Freeman, E. (1984). *Strategic management: A stakeholder approach.* Pitman.

Gonzalez, C. C., & Mendoza, H. K. (2020). Green accounting in Colombia: A case study of the mining sector. *Environment, Development, and Sustainability, 23,* 6453–6465.

Gray, R., Owen, D., & Bebbinton, J. (1997). Green accounting: Cosmetic irrelevance or radical agenda for change? *Asia-Pacific Journal of Accounting, 4*(2), 175–198.

Hernadi, H. B. (2012). Green accounting for corporate sustainability. *'Club of Economics in Miskolc' TMP, 8,* 23–30.

Holub, H. W., Tappeiner, G., & Tappeiner, U. (1999). Some remarks on the system of integrated environmental and economic accounting of the united nations. *Ecological Economics, 29*(3), 329–336.

Jaffe, A., Peterson, S., Portney, P., & Stavins, R. (1995). Environmental regulation and the competitiveness of US manufacturing: what does the evidence tell us? *Journal of Economic Literature, 33*(1), 132–163.

Jones, T. (1995). Instrumental stakeholder theory: A synthesis of ethics and economics. *Academy of Management Review, 20*(2), 404–437.

Kantabutra, S. (2019). Achieving corporate sustainability: Toward a practical theory. *Sustainability, 11,* 4155.

Kartika, S., & Utami, W. (2019). Effect of corporate governance mechanisms on financial performance and firm value with green accounting disclosure as moderating variables. *Research Journal of Finance and Accounting, 10,* 24.

Ketprapakorn, N., & Kantabutra, S. (2020). Toward a theory of corporate sustainability: A theoretical integration and exploration. *Journal of Cleaner Production, 270.*

Kiranmai, J., & Swetha, C. (2018). Green accounting practices: An overview. *IUP Journal of Business Strategy, 15*(3), 7–18.

Kuorikoski, J., Lehtinen, A., & Marchionni, C. (2010). Economic modelling as robustness analysis. *The British Journal for the Philosophy of Science, 61,* 541–567.

Lestari, H., & Restuningdiah, N. (2021). The effect of green accounting implementation on the value of mining and agricultural companies in Indonesia. *Proceedings of the 7th Regional Accounting Conference (KRA 2020), 173.*

Ma, J., & Ma, J. (2019). A research review of corporate green accounting information disclosure. *IOP Conference Series: Earth and Environmental Science, 310*(2019), 052071.

Marrewijk, V. M. (2003). Concepts and definitions of CSR and corporate sustainability: Between agency and communion. *Journal of Business Ethics, 44,* 95–105.

Marrewijk, V. M., & Werre, M. (2003). Multiple levels of corporate sustainability. *Journal of Business Ethics, 4,* 107–119.

Mattlin, M. (2017). China as a leading economic power: Could China stabilize the global economic system in times of crisis? *Helsinki: Finnish Institute of International Affairs, 97.*

Mazutis, D. (2010). Why zero is not one: Towards a measure of corporate social strategy. *Academy of Management Proceedings,* 1–6.

Ming, L. (2007, June). Green accounting of China: Comparison analysis between 1992 and 1995. *Canadian Social Science, 3*(3).

Neubaum, D. O., & Zahra, S. A. (2006). Institutional ownership and corporate social performance: The moderating effects of investment horizon, activism, and coordination. *Journal of Management, 32,* 108–131.

Nga, T. T., & Dao, H. T. A. (2020). Environmental management accounting: The case of the rubber tire manufacturing company in viet nam. *Journal of Tourism, Hospitality and Environment Management, 5*(21), 89–108.

Okafor, G. T. (2018). Environmental costs accounting and reporting on firm financial performance: A survey of quoted Nigerian oil companies. *International Journal of Finance and Accounting, 7*(1), 1–6.

Peloza, J. (2006). Using corporate social responsibility as insurance for financial performance. *California Management Review, 48,* 52–72.

Porter, M., & Van der Linde, C. (1995). Green and competitive: Ending the stalemate. *Harvard Business Review, 73,* 120–134.

Rahman, M. J., & Wu, J. (2023). M&A activity and ESG performance: Evidence from China. *Managerial Finance, 50*(1), 179–197.

Rahman, M. J., Wu, Q., & Zhu, H. (2024). Corporate social responsibility in times of social distancing: Evidence from China. *Business Ethics, the Environment & Responsibility.*

Rahman, M. J., Zhu, H., & Chen, S. (2023). Does CSR reduce financial distress? Moderating effect of firm characteristics, auditor characteristics, and Covid-19. *International Journal of Accounting & Information Management, 31*(5), 756–784.

Rahman, M. J., Zhu, H., & Hossain, M. M. (2023). Auditor choice and audit fees through the lens of agency theory: Evidence from Chinese family firms. *Journal of Family Business Management, 13*(4), 1248–1276.

Rahman, M. J., Zhu, H., & Jiang, X. (2024). Family firms, client importance, and auditor reporting behavior: Evidence from China. *Meditari Accountancy Research, 32*(2), 543–578.

Rahman, M. J., Zhu, H., & Yue, L. (2024). Does the adoption of artificial intelligence by audit firms and their clients affect audit quality and efficiency? Evidence from China. *Managerial Auditing Journal, 39*(6), 668–699.

Rahman, M. J., Zhu, H., Zhang, Y., & Hossain, M. M. (2024). Effect of female representation in audit committees on non-audit fees: Evidence from China. *Meditari Accountancy Research.*

Rout, H. S. (2010). Green accounting: Issues and challenges. *IUP Journal of Managerial Economics, 8,* 46–60.

Singh, J., & Yadav, P. (2013). Return on capital employed-a tool for analyzing profitability of companies. *International Journal of Techno-Management Research, 1*(1), 1–13.

Stead, W. E., & Stead, J. G. (1992). *Management for a small planet: Strategic decision making and the environment.* Sage.

Swarnapali, R. (2017). Corporate sustainability: A literature review. *Journal for Accounting Researchers and Educators (JARE)*.

Szekely, F., & Knirsch, M. (2005). Responsible leadership and corporate social responsibility: Metrics for sustainable performance. *European Management Journal, 23*, 628–647.

Tang, H. (2019). Discussion on the problems and countermeasures of green accounting implementation in enterprises of China in new era. *Advances in Economics, Business and Management Research, 110*.

Van Marrewijk, M., & Werre, M. (2003). Multiple levels of corporate sustainability. *Journal of Business Ethics, 44*(2), 107–119.

Warren, S. C., Reeve, M. J., & Duchac, E. J. (2018). *Accounting* (27th ed.). CENGAGE, 978-1-337-27209-4

Watson, K., & Polito, T. (2002). Environmental cost of quality (ECOQ): A framework for quantifying environmental management systems. In *Proceedings of the 33rd annual meeting of the decision sciences institute*. San Diego, CA.

Wilson, M. (2003). Corporate sustainability: What is it and where does it come from? *Ivey Business Journal*, 1–5.

World Commission for Environmental and Development. (1987). *Our common future*. Oxford University Press.

Yakhou, M., & Dorweiler, V. P. (2004). Environmental accounting: An essential component of business strategy. *Business Policy and the Environment, 13*, 65–77.

Yang, N., Zhang, Z., Xue, B., Ma, J., Chen, X., & Lu, C. (2018). Economic growth and pollution emission in China: Structural path analysis. *Sustainability, 10*(7), 2569.

Yang, W., & Zhao, J. (2018). Sources of China's economic growth: A case for green accounting. *Advances in Management & Applied Economics, 8*(2), 33–59.

Zhao, L., Liu, J., & Ren, J. (2018). Impact of various ventilation modes on IAQ and energy consumption in Chinese dwellings: First long-term monitoring study in Tianjin, China. *Building and Environment, 143*, 99–106.

Zhu, H., & Rahman, M. J. (2024). Ex-ante expected changes in ESG and future stock returns based on machine learning. *The British Accounting Review*, 101457.

4 Environmental Accounting and Financial Performance

Evidence from Oil Companies in China

Md Jahidur Rahman, Tarek Rana, Hongtao Zhu, Tu Qianqian, and Sajjad Hossain Khan

1. Introduction

Environmental issues are not confined to national borders but represent a global challenge. In recent years, environmental concerns such as air pollution, natural resource depletion, and climate change have garnered worldwide attention. The emphasis on "green growth", "low-carbon development", and "sustainable development" has intensified both nationally and internationally (Do & Nguyen, 2020). Firms across various sectors face mounting pressure to demonstrate environmental responsibility, driven by increasing environmental awareness. Stakeholders, including governments, non-governmental organizations, and communities, are actively urging companies to reduce their environmental impact (Gupta & Barua, 2018). Extensive research indicates that enhanced environmental performance can positively influence a firm's business value. Firms adopting environmental strategies may experience improved financial outcomes (Clarkson et al., 2011). Additionally, heightened engagement in environmental initiatives can enhance a firm's sustainable performance (Okafor, 2018).

As a newly industrialized nation, China heavily depends on natural resources and energy for its development (Zhang & Xu, 2012; Herrerias et al., 2013). Decades of resource overexploitation and overuse have characterized its developmental trajectory (Hu et al., 2018). Although China's economy has experienced rapid growth, this has come at a significant environmental cost (Chen et al., 2017; Hao et al., 2018). The emergence of environmental problems such as air and water pollution, land desertification, and biodiversity loss has heightened the Chinese government's awareness of the severity of these issues. Consequently, the government has implemented a series of environmental regulatory policies to address environmental degradation and promote corporate environmental and social responsibility. Notable measures include the SSE's issuance of the "Guidelines on Environmental Information Disclosure for Listed Companies" in 2008 and the promulgation of the new Environmental Protection Law by the Standing Committee of the National People's Congress in 2015 (Liu et al., 2021). Additionally, initiatives such as green insurance, green credit, and green loans have been introduced to encourage companies to disclose environmental information (Liu et al., 2021).

DOI: 10.4324/9781003488965-5

Environmental information disclosure (EID) serves as a critical conduit for companies to communicate their environmental responsibilities to both the government and the public (Criado-Jimenez et al., 2008). By integrating environmental accounting practices, companies can enhance public trust and confidence (Okafor, 2018). For instance, Okafor (2018) examined the relationship between environmental cost accounting and the financial performance of listed oil companies in Nigeria, finding that environmental costs positively impact enterprise performance. Despite the Chinese government's decade-long efforts to enforce environmental regulations, not all companies have engaged fully, with many remaining reticent. While an increasing number of Chinese firms are recognizing the importance of environmental and social responsibility, the nexus between environmental and financial performance remains unclear. This study aims to fill this gap by focusing on listed oil companies in China, assessing the impact of environmental performance on the business value of these companies.

According to a US environmental agency, environmental costs encompass expenditures related to complying with environmental laws, including costs for remediation, pollution control, and penalties for violations (Okafor, 2018). Often, firms do not account for these environmental costs when determining profits, leading to misleading financial results that can adversely affect decision-making (Okafor, 2018). To accurately ascertain the relationship between environmental cost accounting and financial performance, this study will analyse data collected from the financial reports and corporate social responsibility (CSR) reports of publicly listed Chinese oil companies.

Empirical evidence suggests that shifts in population and consumption behaviours have significantly contributed to the rise in carbon emissions in China over the past three decades (Swim et al., 2011). Pollution-intensive industries are the primary contributors to these emissions, prompting the Chinese government to advocate for greater environmental responsibility among these companies. However, incentivizing pollution-intensive firms to voluntarily assume responsibility for their emissions remains a complex challenge. This study adds to the existing literature by investigating the relationship between firms' financial performance and their environmental costs in China. Polycarp (2019) identified return on capital employed (ROCE) as a traditional financial performance measure, alongside net profit margin (NPM), dividend per share (DPS), and earnings per share (EPS). To date, no studies have explored the relationship between environmental accounting and these specific financial indicators.

Prior to the introduction of the "Guidelines on Environmental Information Disclosure for Companies Listed on the Shanghai Stock Exchange (SSE)", Chinese regulations did not mandate the disclosure of environmental information by listed companies. The SSE only conducted thorough reviews of the environmental conditions of companies seeking to list and encouraged voluntary disclosure of environmental information (Ren et al., 2020). The guidelines represent a pivotal policy, explicitly requiring high-polluting companies to disclose environmental information in their annual reports or through separate channels (Ren et al., 2020). Despite the Chinese government's efforts to promote proactive environmental information disclosure, many companies still perceive these costs as burdensome rather than as opportunities for competitive advantage. Addressing this issue through research can provide valuable insights for managers and decision-makers, fostering a sustainable alignment of environmental and economic performance.

This work employs the regression model utilized by Makori and Jagongo (2013) in their study on environmental accounting and firm profitability within the context of firms listed on the Bombay Stock Exchange in India. The analysis draws on secondary data spanning the period from 2015 to 2019, sourced from published financial reports and CSR reports of oil companies listed on the SSE and Shenzhen Stock Exchange (SZSE). The study employs multiple regression techniques to perform the econometric analysis.

Research by Lubomir and Dietrich (2009) on the impact of environmental performance on revenues and costs in transition economies suggests that enhanced environmental performance can boost profitability by reducing costs. Similarly, Nwaiwu and Oluka (2018) examined the relationship between environmental cost disclosure and financial performance in Nigeria's oil and gas sector, finding that comprehensive environmental cost disclosure and adherence to environmental regulations significantly enhance financial performance indicators. Building on these findings, this study hypothesizes a positive relationship between environmental costs and firm performance in the context of Chinese oil companies.

This study contributes to the literature in several ways. Firstly, while numerous studies have explored the relationship between environmental costs and financial performance across various industries, there has been a notable absence of research focusing specifically on the oil industry in China. By selecting companies that disclose environmental costs from among listed oil companies in China, this study provides a focused analysis of their environmental costs and financial performance. Secondly, much of the existing Chinese scholarship on environmental accounting disclosure has been exploratory and descriptive, concentrating on environmental phenomena and CSR. This study goes beyond description, employing data analysis to examine the relationship between environmental accounting and corporate performance.

The remainder of this chapter is structured as follows: Section 2 provides a comprehensive literature review and formulates the hypotheses. Section 3 outlines the methodology. Section 4 presents the main test results and discusses the findings of the regression analysis. Section 5 summarizes the conclusions, highlights the limitations, and offers recommendations for future research. Finally, Section 6 proposes recommendations for relevant companies and government authorities.

2.   Literature Review and Hypothesis Development

2.1   *Brief Overview of the Oil Industry in China*

China's history of oil discovery and utilization extends back approximately 2,000 years, yet the modern oil industry took shape in the mid-19th century (Feng et al., 2013). The petrochemical industry has since become a fundamental component of China's national economy, functioning as both a primary industry and a pivotal pillar. The industry's significance has grown alongside the nation's economic advancement and technological progress, with petrochemical enterprises playing an increasingly crucial role in the country's economic infrastructure (Feng et al., 2013). The widespread application of petrochemical products has infiltrated nearly all aspects of daily life, underscoring the industry's integral position.

Despite its substantial contributions to economic growth, the development of the petrochemical industry has precipitated significant environmental pollution. Recognizing this, the Chinese government has enacted numerous laws and regulations aimed at enhancing environmental protection among enterprises. For example, Article 15 of the Law of the People's Republic of China on Prevention and Control of Water Pollution mandates that enterprises and institutions discharging pollutants into water bodies must pay pollutant discharge fees in accordance with state regulations. Furthermore, those exceeding national or local pollutant discharge standards are required to pay excess sewage charges (Wang et al., 2015).

Additionally, listing rules necessitate that companies disclose their environmental footprint (Wang, 2015). However, compliance within the oil industry remains limited, with few companies meeting these stringent requirements. This non-compliance has led to the characterization of these enterprises as environmentally unfriendly, a reputation that not only tarnishes their corporate image but also has detrimental effects on their financial performance.

To address these challenges and promote sustainable development, the Chinese government has intensified its regulatory oversight and introduced policies to incentivize better environmental practices. The integration of environmental considerations into corporate governance is increasingly seen as essential for the long-term viability of the oil industry. By improving compliance with environmental regulations, companies can mitigate negative impacts on their reputation and financial health, thereby aligning with broader national and international sustainability goals.

While the oil industry's role in China's economic development is indisputable, the pressing need for environmental stewardship and compliance with regulatory frameworks is paramount. This dual focus on economic growth and environmental responsibility will be critical in shaping the future trajectory of China's petrochemical sector.

2.2　Stakeholder Theory

Stakeholder theory posits that the success of a company is contingent upon effectively managing relationships with its diverse stakeholders. These stakeholders include not only shareholders but also employees, customers, suppliers, communities, and governments. The term "stakeholder" was first introduced by the Stanford Research Institute (SRI) in 1963, highlighting the idea that an organization would cease to exist without the support of its stakeholders (Johnson, 1995). This theory emphasizes that businesses have a responsibility to a wider group of constituents rather than merely focusing on maximizing shareholder value.

In the context of environmental accounting, stakeholder theory underscores the importance of addressing environmental cost elements and assessments in financial statements (Polycarp, 2019). By integrating environmental costs, firms can provide a more comprehensive view of their operations, which is critical for stakeholders who are increasingly concerned about environmental impacts. This holistic approach can enhance transparency and accountability, fostering trust and loyalty among stakeholders. Furthermore, effective stakeholder management involves proactive engagement and dialogue with stakeholders to understand their concerns and expectations regarding environmental practices.

The integration of stakeholder theory into environmental accounting practices aligns corporate strategies with sustainable development goals. This alignment can lead to improved long-term financial performance by mitigating risks, enhancing corporate reputation, and ensuring regulatory compliance. Companies that adopt this approach are better positioned to respond to the evolving demands of their stakeholders and contribute to the broader societal goal of sustainable development.

2.3   Legitimacy Theory

Legitimacy theory provides a robust framework for understanding why companies voluntarily disclose social and environmental information. Tilling (2004) asserted that legitimacy theory offers a compelling mechanism to explain such disclosures, which can serve as a tool for companies to engage in and influence critical public debates. According to this theory, legitimacy is a general perception or assumption that an entity's actions are appropriate, proper, and desirable within a socially constructed system of norms, values, beliefs, and definitions (Suchman, 1995).

Companies seek to maintain or restore legitimacy by aligning their operations and disclosures with societal expectations. Environmental disclosures, in particular, are seen as a means to demonstrate that a company is acting in accordance with societal values and norms regarding environmental stewardship. This is especially important in industries that are perceived as having significant environmental impacts, such as the oil and gas sector.

Legitimacy theory suggests that companies disclose environmental information to gain, maintain, or repair legitimacy in the eyes of their stakeholders. This can involve reporting on environmental initiatives, compliance with regulations, and performance against environmental benchmarks. By doing so, companies can mitigate potential legitimacy gaps that may arise from negative environmental impacts or public scrutiny. Consequently, legitimacy theory not only explains the motivations behind environmental disclosures but also provides insights into how companies can strategically manage their social and environmental responsibilities to sustain their legitimacy over time.

2.4   Positive Accounting Theory

Positive accounting theory (PAT) seeks to explain the rationale behind the voluntary disclosure of social and environmental information by listed companies.

Developed by Watts (1986), PAT is grounded in the idea that accounting practices and disclosures are influenced by the self-interested behaviours of managers and the economic consequences of accounting choices. One of the central tenets of PAT is the political cost hypothesis, which posits that firms may engage in voluntary disclosures to reduce political costs and scrutiny.

According to PAT, companies voluntarily disclose environmental and social information to mitigate potential political costs associated with regulatory pressures, public scrutiny, and activism. These disclosures are seen as strategic responses to external pressures, aimed at legitimizing corporate actions and reducing the likelihood of adverse political or regulatory interventions. However, Gray et al. (1995) criticized PAT for its limited focus on the economic and self-interested motivations behind disclosures. They argued that PAT does not adequately address the normative aspects of what social and environmental reporting should be, focusing instead on describing what it looks like and why companies engage in such practices.

Despite these criticisms, PAT provides valuable insights into the strategic considerations that drive voluntary environmental disclosures. By highlighting the economic incentives and political motivations behind these disclosures, PAT helps explain why companies in environmentally sensitive industries, such as oil and gas, may choose to provide detailed environmental information. These disclosures can be seen as efforts to align with stakeholder expectations, reduce potential political costs, and enhance overall corporate legitimacy.

While stakeholder theory, legitimacy theory, and PAT each offer different perspectives on why companies engage in environmental disclosures, they collectively underscore the multifaceted motivations and strategic considerations that drive these practices. Understanding these theories provides a comprehensive framework for analysing corporate environmental reporting and its implications for business performance and sustainability.

2.5   *Environmental Cost and Environmental Accounting*

Environmental cost is a critical aspect of environmental accounting that encompasses various expenses incurred by companies in their efforts to mitigate environmental impacts. Chinese scholars categorize environmental costs into four primary components: (1) governance input due to environmental degradation, (2) losses from environmental accidents, (3) penalties for non-compliance with environmental regulations and unauthorized investments, and (4) losses from inefficient environmental governance (Song et al., 2015). These categories illustrate the multifaceted nature of environmental costs, which include both direct and indirect financial implications for companies.

The concept of environmental cost in China began to take shape in the 1990s, driven by an increasing awareness of environmental issues and the influence of international theoretical advancements in environmental accounting (Wang, 2015). During this period, Chinese scholars started to incorporate successful foreign practices and theories into the local context, aiming to provide theoretical support for corporate environmental protection efforts (Wang, 2015). However, these early efforts had limited impact due to a lack of awareness among corporate leaders about the importance of environmental costs. It was not until environmental pollution events and natural disasters caused significant damage and heightened public concern about environmental issues that the concept gained traction. Recognition of the severe economic and social consequences of neglecting environmental issues has since driven a more robust focus on environmental costs and accounting in China (Liao, 2018).

Environmental accounting, which emerged as a response to the need for better management of environmental costs, encompasses various definitions and approaches. Seetharaman et al. (2007) defined environmental accounting as the process of assessing all environmental costs associated with activities and products. This comprehensive assessment allows companies to identify cost-saving opportunities and enhance overall performance (Larrinaga & Babbington, 2001). Environmental cost information generated through environmental accounting is not only useful for improving corporate performance but also essential for internal management decision-making (Bassey et al., 2013).

One of the primary benefits of environmental accounting is its potential to reduce production costs and increase profitability. Elewa (2007) found that companies implementing environmental accounting practices experienced significant reductions in annual production costs, leading to higher profits. Additionally, environmental accounting can serve as a strategic tool for demonstrating the potential of creating an investment-friendly environment that generates financial benefits while avoiding environmental liabilities (De Beer & Friend, 2006).

Beyond cost savings, environmental accounting provides a framework for companies to communicate their environmental performance to stakeholders. By disclosing environmental cost information, companies can enhance transparency and accountability, fostering trust and confidence among stakeholders. This is particularly important in industries with significant environmental impacts, where stakeholders are increasingly demanding greater environmental responsibility and stewardship from companies.

Adoption of environmental accounting practices also aligns with the growing emphasis on sustainable development. By integrating environmental costs into financial decision-making, companies can ensure that their operations contribute to long-term sustainability goals. This alignment not only mitigates environmental risks but also enhances the company's reputation and competitiveness in a market where environmental concerns are becoming increasingly paramount.

Environmental cost and environmental accounting play a crucial role in modern corporate management. By effectively managing environmental costs and integrating them into financial strategies, companies can achieve significant economic and environmental benefits. The evolution of environmental accounting in China reflects a broader global trend towards greater environmental responsibility and sustainability in business practices. As companies continue to face growing environmental challenges, the importance of robust environmental accounting practices will only increase, driving further advancements in the field and contributing to the overall goal of sustainable development.

2.6   *Measures of a Firm's Financial Performance*

Financial performance indicators are always used to measure a firm's performance (Polycarp, 2019). There are many measures of a company's performance, often including return on assets (ROA), return on equity (ROE), and ROCE. These measures are discussed below:

ROA: This is a measure of how much net profit is generated per unit of assets. It is calculated by dividing a company's annual profit by its total assets (Polycarp, 2019). The ROA formula is stated as follows:

ROA = Net income (EBIT) / Total assets (expressed as a percentage) (1)

ROE: This is the percentage of a company's after-tax profit divided by shareholders' equity (Hagel et al., 2010). This index reflects the income level of shareholders' equity and is used to measure the efficiency of a company using its capital (Hagel et al., 2010). The ROE formula is stated as follows:

ROE = Net income (EBIT) / Shareholders' equity (2)

ROCE: ROCE expressed in percentage is an index used to measure the efficiency of capital investment and to show the company's profitability ratio. In general, a company with a high ROCE is likely to be profitable (Polycarp, 2019). The ROCE formula is stated as follows:

ROCE = PBIT (Net Income) / Capital Employed (3)

>   Note: Capital Employed = Total Assets – Current Liabilities = Equity + Non-Current Liabilities (Polycarp, 2019).

2.7   *Empirical Review*

Scholars have done varying research on the relationship between environmental accounting and firm performance and have obtained different results. Many researchers use profitability and firm financial performance as explanatory variables for differences in disclosure levels. However, the relationship between firm financial performance and firm social and environmental disclosure is arguably one of the most controversial issues that remain unresolved (Choi, 1998). Supporters argue that the additional costs associated with social and environmental disclosure and the profitability of reporting companies are suppressed. Some researchers have used profit logarithms. Roberts (1992) found a positive correlation between company profitability and company social and environmental disclosure among these researchers. However, Patten (1991) failed to find any significant positive correlation between profitability and corporate social and environmental disclosure. Several studies have been conducted on environmental accounting and disclosure in financial statements.

Bassey et al. (2013) examined the impact of environmental accounting and reporting on the organizational performance of selected oil and gas companies in the Niger Delta region of Nigeria. It was found that environmentally friendly companies made extensive disclosure of environment-related information in their financial statements and other business reports.

Do and Nguyen (2020) employed structural equation modelling to investigate the links between proactive environmental strategy, competitive advantages, and firm performance. The result showed that adopting a forward-looking environmental approach can produce differentiated and cost-leading competitive advantages. In addition, positive environmental strategies are more common in large companies and service industries.

Due to the late start of the research on environmental accounting information disclosure in China, there is a lack of empirical research on environmental information disclosure and company performance. Geng et al. (2002), as part of the SSE's heavy pollution industry listed companies environmental accounting information disclosure content and quality of the collection and analysis, revealed that the disclosure content of listed companies in heavy pollution industries in China is not comprehensive and reliable, and the form is not standard. Yang and Su (2013) selected 120 listed companies in SZSE as samples, empirically analysing the impact of environmental accounting information disclosure on the company's value. Results showed that environmental accounting information disclosure by listed companies did not have a significant positive correlation with the improvement of enterprise value, and environmental accounting information disclosure played a minor role in increasing enterprise value and improving profitability.

The empirical review shows different results on the links between environmental accounting and firm performance. Some had positive relationships, while some had negative or no connections. This research continued to study the relationship between environmental accounting and firm performance by using environmental cost data.

2.8 Hypothesis Development

Increasingly evident environmental risk issues are making more and more companies and regulators aware of the importance of environmental protection (Min, 2011). However, there is no consensus on whether environmental inputs promote or reduce corporate performance. In particular, China, as a representative country of emerging markets, is still in its infancy in its corporate environmental activities, and the relevant systems and regulatory measures are not perfect (Min, 2011).

According to the principal–agent theory, the agent runs the enterprise on behalf of the principal. The agent needs to regularly report the company's performance to the owner to reflect the performance of the manager's "fiduciary responsibility" (Braun & Guston, 2003). Agency theory points out that managers also send some signals to the external capital market to disclose the information about their fiduciary responsibilities. In terms of the relationship between environmental accounting and corporate performance, on the one hand, companies with better environmental conditions will more actively disclose environmental accounting information to convey good environmental information to the outside world (Wang, 2015). On the other hand, more input on environmental costs indicates that managers have a sense of social responsibility. This increases stakeholder trust, promotes investment, and improves reputation (Wang, 2015). Several studies in China also investigate the relationship between environmental accounting and firm performance (Rahman & Wu, 2023; Rahman et al., 2023; Zhu & Rahman, 2024). Therefore, there is a positive relationship between environmental cost and firm performance. Therefore, this study makes the following hypotheses:

H0: *There is no significant relationship between environmental accounting and financial performance of oil companies in China.*
H1: *There is a significant relationship between environmental accounting and financial performance of oil companies in China.*

3. Methodology

3.1 Data and Sample

We initially planned to take 2010–2019 as the time range of the study, but after data collection, it was found that many listed companies lacked financial and environmental data before 2015, so the time range was limited to 2015–2019. After eliminating the listed companies with imperfect and incomplete financial information and abnormal financial status, 30 companies listed in SSE and SZSE were selected as the research samples. The financial data of the listed companies came from the Wind database, and the companies' environmental data were collected from the Yihuo database. The reason for choosing China's oil industry is that it involves many environmental problems. In addition, most companies in this industry have long been engaged in environmental accounting practices.

The model of this study is designed to capture the interrelationship between dependent variables and independent variables. Therefore, the statistical tool for data analysis is multiple regression. The independent variables employed in the study are ROCE, NPM, DPS, and EPS, and one dependent variable: environmental cost of companies (ENVC).

3.2 Model Specification and Variable Measurement

Using a single dependent variable, ENVC, and four independent variables, ROCE, NPM, DPS, and EPS, a model was constructed (Polycarp, 2019):

$$\text{ENVC} = f\,(\text{ROCE, NPM, DPS, EPS}). \tag{4}$$

Table 4.1 presents the formula to calculate the ratios.

Table 4.1 Measurement of Variables

Variable	Formula
ROCE	[Profit before tax (PBT) × 100]/(Capital employed)
NPM	(Net profit × 100)/(Turnover/sales)
DPS	[(Gross dividend – Preference dividend) × 100]/(No. of ordinary shares in issue and ranking for dividend)
EPS	(PAT before extraordinary items less preference dividend × 100)/(No. of ordinary shares running for dividend)

4.   Results and Discussion

4.1   Descriptive Statistics

In this study, firm performance is represented by four variables, ROCE, NPM, DPS, and EPS. The results for different measures of environmental accounting and financial performance of the firms are presented in the following section. Table 4.2 shows that during the time of the study, the average mean scores of ROCE, NPM, DPS, and EPS were −11.63, −0.29, 8.56, and 19.96, respectively. The standard deviation of ROCE was the highest at 94.93, indicating its low contribution to the model. On the contrary, DPS had the lowest standard deviation, which shows its higher contribution to the model of the study. The results of the descriptive analysis are presented in Table 4.2.

4.2   Correlation Matrix

Table 4.3 illustrates the correlation matrix of the independent and dependent variables. The results in Table 4.3 present a positive relationship between ENVC and ROCE, NPM, DPS, and EPS, which implicates that the independent variables representing firm performance are contributing positively to the dependent variable. However, the correlation between independent variables is not significant, which indicates that there is no collinearity between independent variables, showing the fitness of the model.

4.3   Main Results

Multiple regression analysis is used to see the association between ENVC and all independent variables. The results presented in Table 4.4 reveal the relationship

Table 4.2 Descriptive Statistics

Variable	N	Mean	Std. Deviation
ENVC	150	238,709,036	946,845,203
ROCE	150	−11.6286	94.93112
NPM	150	−0.2862	10.83874
DPS	150	8.5556	7.71042
EPS	150	19.9616	38.26462
Valid *N*	150		

Table 4.3 Results of Correlation Analysis

Variable	ENVC	ROCE	NPM	DPS	EPS
ENVC	1.000	0.022	0.031	0.491	0.108
ROCE	0.022	1.000	0.203	0.082	0.267
NPM	0.031	0.203	1.000	0.196	0.770
DPS	0.491	0.082	0.196	1.000	0.476
EPS	0.108	0.267	0.770	0.467	1.000

Table 4.4 Results of Multiple Regression Analysis

Variable	Coefficient	t-statistics	p-value
ROCE	0.001	1.20	0.233
NPM	−0.001	−0.08	0.936
DPS	0.117	5.63	0
EPS	−0.002	−0.37	0.716
Adj. R^2	0.230		
Durbin–Watson	1.477		

of ROCE, NPM, DPS, and EPS with ENVC, with coefficients of 0.001, −0.001, 0.117, and −0.002 and p-value of 0.233, 0.936, 0, and 0.716, respectively. Since the p-value of ROCE, NPM, and EPS are larger than 0.05, there is an insignificant relationship between ENVC and ROCE, NPM, and EPS. However, the p-value of DPS is smaller than 0.05, indicating there is a significant relationship between ENVC and DPS. The adjusted R^2 is 23.0%, indicating the change in ENVC margin, which can be explained by the change in the independent variable in the model. The value of the Durbin–Watson statistics is 1.477, which confirmed that there is essentially no first-order autocorrelation between the residuals in the model. Table 4.4 shows the analysis results according to the fixed-effect model.

Since there is no significant relationship between three of the four independent variables and the dependent variable, the hypothesis that there is no significant relationship between environmental accounting and financial performance of oil companies in China (H0) is accepted. Therefore, H1 (there is a significant relationship between environmental accounting and financial performance of oil companies in China) is rejected.

The finding is in agreement with the study of Polycarp (2019) who found that environmental accounting disclosure has no significant relationship with financial performance. It also agrees with the study by Yusoff and Wan Mohamed (2006), which established that financial performance has no significant relationship between environmental reporting. However, the results of the study by Azhar and Ahmed (2019), which found a positive relationship between environmental accounting information and companies' performance, are contrary to the results of this study.

However, during the data collection process, it was observed that still many companies do not disclose their environmental costs in their annual report. What's more, since the information on environmental cost disclosed in most oil companies' annual report is qualitive rather than quantitative, it is clear there is no significant relationship between environmental accounting and firm performance.

4.4   *Tests of Multicollinearity*

Multicollinearity between independent variables is measured by two indicators: the tolerance value and the variance inflation factor (VIF). The VIFs in Table 4.5 are consistently lower than 10 and greater than 1, showing that there is no multicollinearity

Table 4.5 Results of Collinearity Diagnosis

Variable	Tolerance	VIF
ROCE	0.926	1.080
NPM	0.373	2.678
DPS	0.715	1.399
EPS	0.294	3.401

Table 4.6 Results of Robustness Tests

Variable	Coefficient	t-statistics	p-value
ROA	0.013	1.09	0.278
NPM	−0.004	−0.31	0.754
DPS	0.116	5.54	0.000
EPS	−0.002	−0.37	0.708
Adj. R^2	0.229		
Durbin–Watson	1.458		

(Neter et al., 1996). Moreover, the tolerance value of each variable showed in Table 4.5 is consistently less than 1, which supports the fact that multicollinearity is completely absent between the independent variables in the study (Tabachnick, 2007).

4.5   Robustness Test

To further prove the interaction between variables, robustness test is introduced in this study. This work aims to determine whether there is a relationship between environmental accounting and firm performance. In this study, ROCE, NPM, DPS, and EPS represent firm performance. ROA is also a good indicator to evaluate a company's performance. Thus, ROA can replace ROCE in the regression model.

From the results presented in Table 4.6, ROA, NPM, and EPS have an insignificant relationship with ENVC, with coefficient values of 0.013, −0.004, and −0.002 and p-values of 0.278, 0.754, and 0.708, respectively. Also, DPS has a significant relationship with ENVC, with a coefficient value of 0.116 and p-value of 0.000. Therefore, the result from this regression model is similar to the result obtained using the model with ROCE. Thus, the hypothesis that there is a significant relationship between environmental accounting and firm performance of oil companies in China (H1) is rejected, and we accept the null hypothesis that there is no significant relationship between environmental accounting and firm performance of oil companies in China. The results of the robustness tests are shown in Table 4.6.

5.   Conclusions

In general, environmental accounting contains all aspects of information related to the environmental cost of an enterprise, which is a systematic method for managing a company's environmental activities. This study presented evidence of no significant relationship between environmental accounting and firm performance. This study found that through environmental accounting, companies can evaluate environmental performance in a measurable way. It was also found that the lack of environmental reporting disclosure guidelines seriously affected the uniformity of environment-related information reporting and disclosure in annual reports. It was also noted that companies that voluntarily disclosed their environmental activities had a greater competitive advantage.

The following points should be noted while interpreting the results of this study. Firstly, the results of this study focus on only one industry. The sample data for this study were from listed oil companies in China, so they are not convincing to other industries. Future research should focus on other industries or other countries. Secondly, the study did not take into account the effects of the external environment, such as COVID-19 and company size. For future study, researchers can consider some external conditions to see whether the result will be different. In addition, because disclosure of environmental information is voluntary, the templates for disclosure vary from company to company. Based on the findings, the following measures are recommended.

Since most companies basically do not disclose their environmental activities, the government should force companies to disclose relevant environmental data in their annual reports through legislation and other means and promulgate corresponding environmental accounting standards to ensure that companies' reports are reasonably compliant. Companies must ensure that they comply with the country's environmental laws when conducting relevant business activities, which has a positive effect on improving company performance. In addition, environmental protection awareness should be promoted to stakeholders, which is conducive to establishing a good corporate image. Companies should design products that generate less waste or emissions. At the same time, waste discharge should be reduced in the production process.

References

Azhar, K. A., & Ahmed, N. (2019). Relationship between firm size and profitability: Investigation from textile sector of Pakistan. *International Journal of Information, Business and Management, 11*(2), 62–73.

Bassey, E. B., Effiok, S. O., & Eton, O. E. (2013). The impact of environmental accounting and reporting on organizational performance of selected oil and gas companies in Niger Delta Region of Nigeria. *Research Journal of Finance and Accounting, 4*(3), 57–73.

Braun, D., & Guston, D. H. (2003). Principal–agent theory and research policy: An introduction. *Science and Public Policy, 30*(5), 302–308.

Chen, Y., Zhang, X., & Zhang, Y. (2017). Air pollution and mental health: Evidence from China. *Journal of Environmental Economics and Management, 85*, 20–39.

Choi, J. S. (1998). An evaluation of environmental disclosures: The case of South Korea. *Social and Environmental Accounting, 18*(1), 19–21.

Clarkson, P. M., Li, Y., Richardson, G. D., & Vasvari, F. P. (2011). Does it really pay to be green? Determinants and consequences of proactive environmental strategies. *Journal of Accounting and Public Policy, 30*(2), 122–144.

Criado-Jimenez, I., Fernandez-Chulian, M., Larrinaga-Gonzalez, C., & Husillos-Carques, F. J. (2008). Compliance with mandatory environmental reporting in financial statements: The case of Spain (2001–2003). *Journal of Business Ethics, 79*(3), 245–262.

De Beer, P., & Friend, F. (2006). Environmental accounting: A management tool for enhancing corporate environmental and economic performance. *Ecological Economics, 58*(3), 548–560.

Do, H., & Nguyen, P. (2020). Proactive environmental strategy, competitive advantages and firm performance: Evidence from manufacturing firms in Vietnam. *Journal of Cleaner Production, 242*, 118388.

Elewa, M. M. (2007). Environmental accounting and reporting in the oil and gas sector: An empirical analysis of the UAE. *Journal of Financial Reporting and Accounting, 5*(1), 35–58.

Feng, L., Wang, L., & Zhao, L. (2013). The impacts of oil shocks on China's economy: A comparative analysis based on input-output tables. *Energy, 55*, 914–924.

Geng, J., Hong, M., Xiao, J., & Zhang, Z. (2002). Marangoni effect in distillation with structured packing. *CIESC Journal, 53*(6), 600.

Gray, R., Kouhy, R., & Lavers, S. (1995). Corporate social and environmental reporting: A review of the literature and a longitudinal study of UK disclosure. *Accounting, Auditing & Accountability Journal, 8*(2), 47–77.

Gupta, S., & Barua, M. K. (2018). A framework to overcome barriers to green innovation in SMEs using BWM and Fuzzy TOPSIS. *Science of the Total Environment, 633*, 122–139.

Hagel, J., Brown, J. S., & Davison, L. (2010). The best way to measure company performance. *Harvard Business Review, 88*(3), 56–62.

Hao, Y., Liu, Y. M., & Wu, H. (2018). How do FDI and technical efficiency influence the CO2 emissions in China? Evidence from a panel quantile regression model. *Journal of Cleaner Production, 197*, 855–867.

Herrerias, M. J., Joyeux, R., & Girardin, E. (2013). Short- and long-run causality between energy consumption and economic growth: Evidence across regions in China. *Applied Energy, 112*, 1483–1492.

Hu, Y., Zhang, Q., Sun, L., & Wang, Y. (2018). Environmental regulation, environmental responsibility, and green technology innovation: Empirical research on Chinese enterprises. *Sustainability, 10*(11), 4165.

Johnson, H. T. (1995). Stakeholder theory and its future in management research. *Journal of Management Studies, 32*(1), 37–47.

Larrinaga, C., & Bebbington, J. (2001). Accounting change or institutional appropriation? – A case study of the implementation of environmental accounting. *Critical Perspectives on Accounting, 12*(3), 269–292.

Liao, C. (2018). *Corporate social responsibility: Development and implications in China.* Springer.

Liu, Y., Wang, A., & Wu, Y. (2021). Environmental regulation and green innovation: Evidence from China's new environmental protection law. *Journal of Cleaner Production, 297*, 126698.

Lubomir, L., & Dietrich, E. (2009). Does better environmental performance affect revenues, costs or both? Evidence from transition economy. *Ecological Economics, 68*(6), 1720–1730.

Makori, D. M., & Jagongo, A. (2013). Environmental accounting and firm profitability: An empirical analysis of selected firms listed in Bombay Stock Exchange, India. *International Journal of Humanities and Social Science, 3*(18), 248–256.

Min, B. S. (2011). A study on the importance of environmental management in the development of Chinese enterprises. *Journal of Environmental Management, 92*(11), 2868–2874.

Neter, J., Wasserman, W., & Kutner, M. H. (1996). *Applied linear regression models.* McGraw-Hill/Irwin.

Nwaiwu, J. N., & Oluka, N. O. (2018). Environmental cost disclosure and financial performance of oil and gas in Nigeria. *International Journal of Advanced Academic Research, 4*(2), 19–35.

Okafor, V. O. (2018). Environmental cost accounting and corporate financial results of listed oil companies in Nigeria. *International Journal of Finance and Accounting, 7*(3), 67–74.

Patten, D. M. (1991). Exposure, legitimacy, and social disclosure. *Journal of Accounting and Public Policy, 10*(4), 297–308.

Polycarp, I. J. (2019). Environmental accounting and firm performance: Evidence from manufacturing firms in Nigeria. *European Journal of Accounting, Auditing and Finance Research, 7*(6), 32–50.

Rahman, M. J., & Wu, J. (2023). M&A activity and ESG performance: evidence from China. *Managerial Finance, 50*(1), 179–197.

Rahman, M. J., Zhu, H., & Chen, S. (2023). Does CSR reduce financial distress? Moderating effect of firm characteristics, auditor characteristics, and Covid-19. *International Journal of Accounting & Information Management, 31*(5), 756–784.

Ren, S., Wei, W., Sun, H., Xu, Q., Hu, Y., & Chen, X. (2020). Can mandatory environmental information disclosure achieve a win-win for a firm's environmental and economic performance? *Journal of Cleaner Production, 250,* 119530.

Roberts, R. W. (1992). Determinants of corporate social responsibility disclosure: An application of stakeholder theory. *Accounting, Organizations and Society, 17*(6), 595–612.

Seetharaman, A., Ismail, Z., & Saravanan, A. S. (2007). Environmental accounting as a tool for environmental management system. *Journal of Applied Sciences Research, 3*(10), 1297–1306.

Song, H., Li, Y., & Zeng, J. (2015). Classification and characteristics of environmental costs in Chinese enterprises. *Journal of Cleaner Production, 90,* 277–282.

Suchman, M. C. (1995). Managing legitimacy: Strategic and institutional approaches. *Academy of Management Review, 20*(3), 571–610.

Swim, J., Stern, P. C., Doherty, T., Clayton, S., Reser, J., Weber, E., & Gifford, R. (2011). Psychology's contributions to understanding and addressing global climate change. *American Psychologist, 66*(4), 241–250.

Tabachnick, B. G. (2007). *Experimental designs using ANOVA*. Thomson/Brooks/Cole.

Tilling, M. V. (2004). Some thoughts on legitimacy theory in social and environmental accounting. *Social and Environmental Accountability Journal, 24*(2), 3–7.

Wang, J. (2015). Environmental accounting and its role in the development of a green economy in China. *Journal of Cleaner Production, 96,* 102–110.

Wang, X. H., Wang, X., Huppes, G., Heijungs, R., & Ren, N. Q. (2015). Environmental implications of increasingly stringent sewage discharge standards in municipal wastewater treatment plants: case study of a cool area of China. *Journal of Cleaner Production, 94,* 278–283.

Watts, R., & Zimmerman, J. (1986). *Positive accounting theory*. Englewood Cliffs, NJ: Prentice Hall.

Yang, D., & Su, Y. (2013). Environmental accounting information disclosure and company value: Empirical evidence from listed companies in China. *Journal of Business Ethics, 118*(1), 169–183.

Yusoff, H., & Wan Mohamed, W. N. (2006). Environmental reporting in Malaysia: Perspectives of the management. *Malaysian Accounting Review, 5*(1), 117–140.

Zhang, Y. J., & Xu, S. (2012). The development and current status of China's energy market. *Energy Policy, 42,* 357–364.

Zhu, H., & Rahman, M. J. (2024). Ex-ante expected changes in ESG and future stock returns based on machine learning. *The British Accounting Review,* 101457.

Part 2

Carbon Disclosure and Environmental Performance

5 Carbon Neutrality and Enterprise Value

*Md Jahidur Rahman, Tarek Rana,
Hongtao Zhu, and Zhou Qin*

1. Introduction

Inspired by Xie et al. (2020), this study explores whether carbon neutrality will affect the value of enterprises. Xie et al. (2020) drew the following conclusions: (1) carbon-neutral commitment has a positive effect on enterprise value and (2) the improvement of enterprise environmental, social, and governance (ESG) performance enhances the market response to carbon-neutral commitment. Through investigation, it was found that Aydoğmuş et al. (2022), Fatemi (2018), and Chouaibi and Zouari (2022) studied the relationship between environmental performance and enterprise value. However, few researchers have explored the significance of corporate carbon ratings and corporate value, and there is a lack of evidence from China. As a continuation of Xie et al.'s (2020) study, this work uses Chinese market data and focuses on the following objectives: (1) investigate whether the carbon-neutral rating of an enterprise will affect corporate value and (2) determine the impact of corporate carbon information disclosure on corporate value.

First of all, the topic of carbon neutrality has attracted the attention of the Chinese government in recent years. Zhu and Chen (2001) stated that China is now the largest carbon emitter in the world. To achieve a nationwide "green market goal", the Chinese government emphasizes that socio-economic development programmes must be coordinated with bottom-up economic incentives and technological development (Helm et al., 2018). In China, environmental conservation is now seen as a force for high-quality economic development rather than a barrier to economic growth (Zhang et al., 2022). Moreover, the Chinese government has pledged to use all reasonable efforts to peak carbon emissions before 2030 and achieve carbon neutrality by 2060 (Davidson et al., 2021). Zhang (2021) showed that reducing carbon emissions has transitioned from a government task to a common goal among entrepreneurs seeking national recognition. Based on policy and national will, the transformation of enterprise structures to environmentally friendly models has become inevitable. This illustrates that with policy support, citizens may pay more attention to whether enterprises conduct high-quality carbon offset operations. Corporate commitments to carbon neutrality support China's economic development at the macro level. At the minuscule scale, implementing low-carbon and emission-reduction business models is closely related to corporate

DOI: 10.4324/9781003488965-7

social responsibility (CSR). Therefore, whether a firm supports carbon neutrality may affect investor confidence. However, enterprise value does not necessarily grow with increased environmental capital investment in low-carbon production. This study argues that the impact of carbon neutrality on enterprises is two-sided. To fulfil carbon neutrality commitments, enterprises may reduce their corporate value, as CSR is realized through increased green inputs such as environmental protection costs.

Secondly, carbon neutrality is a global environmental issue, driven by both social and natural factors. Costello (2009) pointed out that the climate issue has gradually attracted attention. PricewaterhouseCoopers (PWC) announced in 2012 that investor and stakeholder interest in carbon emissions had grown 18-fold over the past decade (Matsumura et al., 2014). The increasing attention to this issue has amplified its impact on business value and the adverse consequences it can bring. Moreover, according to the Climate Change (2001) report, although G20 countries are taking steps to reduce carbon emissions, limiting low-carbon strategies could result in a climate increase of 3°C or more. Greenhouse gas emissions exceeding the "1.5°C limit" agreed upon in the Paris Climate Agreement will accelerate sea level rise, increase the frequency of extreme weather events, and worsen agricultural yields. Additionally, Helm et al. (2018) confirmed that the deterioration of the ecological environment may lead to psychological adverse effects, such as increased stress, anxiety, and depression. Environmental degradation and social problems caused by excessive carbon emissions have garnered citizens' attention to the green development of enterprises. A growing number of organizations, such as the Carbon Disclosure Project (CDP), are urging companies to disclose their carbon emissions, raising concerns about whether companies possess a "green stamp". For value recognition, consumer trust in a company is closely linked to carbon neutrality and CSR. This implies that an organization's carbon score may affect consumer decisions and, consequently, the volume of commercial transactions.

Thirdly, since studies investigating the connection between carbon disclosure and enterprise value focus on China's industries, this study has practical relevance. As the "engine" of China's economic expansion, industries play a similar role. According to Xiu et al. (2015), China's industries contribute about half of the GDP but consume about 70% of energy and produce about 70% of pollution. Industrial energy conservation and green development are inevitable trends; otherwise, the industry may face "low carbon barriers". However, greater disclosure of carbon neutrality does not necessarily lead to positive benefits. The authors argue that the impact of carbon neutrality on corporate value is two-sided. A company's commitment to carbon neutrality reduces its value because CSR is achieved by increasing the cost of environmental protection. The concept of carbon neutrality encourages enterprises to move towards green models such as the circular economy, contributing to innovative supply chain management and sustainable organizational value (Burke et al., 2021). Moreover, Boiral et al. (2012) confirmed that the commitment to carbon neutrality and financial performance are mutually supportive. In theory, corporate low-carbon decisions moderate the relationship between carbon score and corporate value. For instance, higher expected future earnings lead to increased investment in low-carbon initiatives. An entity's dedication to carbon neutrality demonstrates its societal obligation and ESG goals. However, a significant amount of greenhouse gases are accounted for via CSR reporting.

Fourthly, theoretically, publication of carbon dioxide data helps businesses exhibit credibility and boosts stakeholder confidence in their investment decisions. Additionally, under the low-carbon development advocated by the Chinese government, enterprises will strengthen the evaluation of carbon neutrality under market competition pressure. However, disclosing carbon information incurs investigation costs and research and development expenses for low-carbon technologies, potentially reducing company value. Therefore, studying the relationship between carbon neutrality and enterprise value is theoretically necessary.

In this study, based on corporate asset profits, we explored whether the commitment to carbon neutrality impacts corporate value. The author used Wind and CSMAR databases to collect public information and obtained data on 5,542 A-share listed companies. This study collected ten years of annual data from 2013 to 2023 for empirical analysis, setting the independent variables as a carbon neutral score and an enterprise ESG score. The dependent variable is the firm's value, measured using return on asset (ROA) and Tobin's Q. This chapter focuses on the impact of carbon neutrality on corporate value before and after the Chinese government's 2020 carbon commitment. Eleven control variables were set up to reduce the influence of other factors on the experimental results. To alleviate the endogeneity problem, the author adopts three methods: a two-stage least squares (2SLS) regression model and control firm fixed effects. To address robustness, the authors replace the dependent and independent variables with other relevant data. The survey results show that both carbon score and carbon disclosure negatively impact enterprise value, while CSR increases enterprise value. Among them, the significance of carbon neutrality score and enterprise value is weak. The robustness test was successful, and the results aligned with the endogeneity test.

Consequently, this study is committed to making the following contributions. Firstly, it addresses the gap in the disclosure of carbon information. In China's industrial market, few studies examine the association between carbon neutrality and business performance; most focus on carbon emission intensity or the carbon trading system. Few studies link carbon disclosure to firm value, especially in energy-intensive industrial production.

Secondly, the results show that carbon score harms the value of enterprises. For enterprises, the results of this experiment should be a warning about balancing low-carbon and high-speed production. High-quality carbon disclosure helps improve ESG performance, driving enterprise value growth. However, the impact of carbon neutrality on business value is not immediate. Additionally, the results help the Chinese government formulate appropriate low-carbon incentive policies, improve green industry coverage in China, and optimize the energy structure.

This chapter also covers the effect of carbon disclosure on firm value. The findings indicate a significant inverse relationship between disclosure level and enterprise valuation, suggesting that businesses should prioritize environmental protection and sustainable development, including strengthening low-carbon policy implementation and improving carbon-neutral disclosure quality. The results provide a theoretical basis for the concept of green and low-carbon production. Moreover, they support enterprises in adapting to national policies and gradually embracing green development as an economic growth driver.

Finally, from a theoretical perspective, the results of this experiment can promote the improvement of China's carbon disclosure. This study provides a theoretical analysis of carbon information and constructive suggestions for corporate strategies and national policies. According to Li et al. (2017), the core of carbon information disclosure is to reflect long-term performance, related to future growth ability. Carbon information disclosure affects investor decision-making and stakeholder investment confidence. The weak negative impact of carbon input on market value can be interpreted as the firm's low-carbon measures increasing future growth capacity but reducing current cash flow. The negative significance of enterprise carbon input and value also shows that investors prioritize current profitability. Therefore, to support enterprises in deepening the green development model, the government can consider increasing subsidies, such as "green credit". Additionally, in practice, ESG performance and corporate value show a significant positive relationship, increasing enterprises' confidence in the benefits of environmental investment and CSR construction.

The rest of this chapter is arranged as follows. Section 2 presents the current status of carbon information disclosure and research on carbon neutrality in China. Section 3 reviews the literature on carbon neutrality and enterprise value. Section 4 presents the authors' hypothesis. Section 5 describes the research methodology. Section 6 analyses the empirical results. Section 7 discusses the endogeneity test, Section 8 addresses the robustness test, and Section 9 concludes and provides suggestions.

2. Background

2.1 *Status of Carbon Information Disclosure in China*

According to the GDP Organization 2022 report, the number of Chinese enterprises that participate in CDP disclosure reached more than 2,700, with an annual growth rate of 43%. In 2019, only about 1,000 companies disclosed CDP, although this has increased by 80% compared to 2018(Li et al., 2019) (CDP China Report, 2019). In addition, CDP China Report 2020 shows that a total of 611 listed companies in China were invited to report their green emission data, and 65 of them responded to the CDP's questionnaire (Li et al., 2019). Among them, 62 companies have adopted climate change as a business strategy, reflecting the growing importance Chinese companies attach to corporate environmental protection and the economic role of disclosure. In addition, after 2020, enterprises will take carbon environmental disclosure as an economic development strategy rather than as a low-carbon burden under external pressure. According to Mai and Cao (2023), driven by huge economic interests, Chinese enterprise management is likely to make false information disclosure.

2.2   Research on Carbon Neutrality

After the rapid development of China's economy, it pays special attention to the balance between environmental protection and resource consumption. In 2020, the Chinese government proposed a two-carbon target, which refers to a plan to peak carbon emissions by 2030 and reach carbon neutrality by 2060. China assumes responsibility as a major country and has adopted a series of low-carbon measures to transition from high-speed production to high-efficiency and low-energy production. For example, according to Zheng (2015), energy conservation and emission reduction programmes are being implemented by businesses more and more as a result of China's green credit policy. Moreover, enterprises also benefited from measures such as carbon emission trading and subsidies for energy conservation and emission reduction. As an innovative enterprise economic development strategy, green management encourages enterprises to implement the concept of sustainable development into their business activities. As a market-based environmental regulation, carbon emission trading rights, in the opinion of Shen et al. (2023), can encourage the natural fusion of low-carbon transition and development. The strict market environment urges enterprises to carry out low-carbon transformation and take into account the economic benefits of enterprises. Porter's hypothesis (Jaffe & Palmer, 1997) also confirms that the establishment of market rules promotes the implementation of environmental protection measures by enterprises and takes into account economic benefits.

In addition, the carbon-neutral disclosure of enterprises is increasingly attracting the attention of investors and stakeholders. According to Krüger (2015), investors have a strong reaction to negative news related to corporate carbon certification, which means that the market has a significant negative response to companies with poor carbon-neutral performance. In addition, Aktas et al. (2011) also came to the same conclusion. Their research indicated that the stock market tends to achieve more turnover with enterprises with excellent environmental performance. Therefore, it is meaningful to discuss the impact of low-carbon transition on enterprise value. In addition, carbon-neutral information disclosure, as a non-financial means, is used by enterprises as a means to obtain external resources (Huang et al., 2023). According to Jiao (2022), improving the disclosure of carbon information at the accounting level is an important link in China's dual-carbon strategy, which plays an important guiding role in the decision-making of enterprises and stakeholders.

3.   Theoretical Perspectives on Carbon Neutrality Reporting

3.1   Carbon Neutrality and Enterprise Development

According to Jennings and Zandbergen (1995), the driving force of green production comes from outside enterprises, such as the government, and from inside enterprises, such as the sense of social responsibility of enterprise managers. Clarkson et al. (2010) and Chan and Welford (2005) showed that carbon-free manufacturing processes boost the green expenditures of business entities in the short term, but the competitive advantage of businesses will be improved in the long run. The present research primarily focuses on the relationship between carbon neutrality strategy and enterprise development performance in the short term since the carbon data for 2023 are insufficient. With the improvement of national awareness of green development, more and more experts have studied the relationship between low-carbon operation and economy. However, most of the studies have set the variables as carbon emission and enterprise ESG performance. For example, Zheng et al. (2023) revealed the impact of carbon reduction measures on ESG performance of enterprises, and Lu (2017) took carbon footprint as a measure of enterprise carbon performance. The primary source of carbon data release is business resource sharing. Disclosing more carbon accounting information reduces information asymmetry, which is advantageous for maximizing the allocation of market resources. This study is unique in that it focuses on industrial firms and is based on market data from China. According to certain studies, China's new energy companies' enterprise value and carbon neutrality are related. For instance, Jing et al. (2022) discovered that a commitment to carbon neutrality is helpful for the long-term growth of China's new energy firms. The authors present the first empirical study of Chinese manufacturing businesses. This study examines how corporate carbon disclosure and carbon neutrality score affect corporate value and offer theoretical justification for businesses' green initiatives.

3.2   *Carbon Neutrality and Firm Value*

According to Shen et al. (2022), the stock return rate and the corporate carbon emission exposure level are positively correlated, and the greater the information disclosure level, the less volatile the stock. In addition, Wang et al. (2023) also concluded that carbon data disclosure of enterprises is conducive to the improvement of corporate value. Carbon information disclosure is both a challenge and an opportunity for enterprises. The disclosure of carbon information establishes a reliable image of low-carbon environmental protection and enhances the competitiveness of enterprises. However, activities such as accounting and publicity of carbon information disclosure require enterprises to invest a lot of money, which may cause negative profit growth in the short term. Mai and Cao (2023) found that carbon information disclosure of enterprises needs to increase the cost, which leads to the phenomenon of incomplete carbon disclosure of enterprises and unclear and repeated reporting relationships. Some experts have studied the relationship between carbon neutrality and corporate performance. For example, Alsaifi et al. (2020) claimed that carbon neutrality damages the cash flow and reputation of enterprises, thus reducing the performance value of enterprises. Also, Capelle-Blancard and Laguna (2010) confirmed a similar conclusion that environmental pollution is relevant to stock price. In addition, Dong et al., after studying Chinese data, said that carbon neutrality has a strong significance in the stock market, which proves that carbon neutrality is closely related to enterprise value.

4. Literature Review and Hypothesis Development

4.1 Corporate Social Responsibility and Carbon Neutrality

Haque and Ntim (2020) indicated that low-carbon plans of enterprises are incorporated into CSR, and a company's carbon compensation management reflects that the enterprise undertakes social responsibility. Fatemi et al. (2018) believe that an enterprise's advantage in ESG performance can enhance its value, but ESG disclosure itself will lead to a decline in valuation. Zimmerli (2007) proposed that corporate behaviour consistent with "social responsibility" would reduce shareholder returns. Zhang et al. (2022) also proposed that reducing greenhouse gas emissions would bring cost pressure. Aboud and Diab (2018) found that the ESG report of a company winning green commendation may not be trusted by investors because stakeholders think it is a waste of capital. According to Fatemi et al. (2018), the value of an enterprise is enhanced by its superior ESG performance. Alkurdi et al. (2023) found that carbon emission, as an intermediate medium, is closely related to corporate performance. The ESG score reflects a company's ESG level and risk and is also affected by the company's carbon emissions. Moreover, Chinese enterprises' investment in ESG can significantly reduce carbon emissions (Cong et al., 2022). In addition, the concept of carbon neutrality encourages enterprises to move towards green models such as circular economy, which contributes to innovative supply chain management and the sustainable organizational value of enterprises (Burke et al., 2021). Moreover, Boiral et al. (2012) confirmed that the commitment to carbon neutrality and the financial performance of enterprises are mutually supportive. Several studies in China also investigate the relationship between environmental accounting and firm performance (Rahman & Wu, 2023; Rahman et al., 2023, 2024; Zhu & Rahman, 2024). Therefore, this study explores the ultimate impact of carbon neutrality on enterprise value to provide a theoretical basis for management decision-making.

4.2 Corporate Carbon Score and Corporate Value

According to McLennan et al. (2014), consumers' willingness to pay for carbon-offsetting actions is affected by regions, and residents in Asia are less willing to pay for spillover consumption for low-carbon and sustainable development than residents in Europe. Therefore, the negative influence of carbon score on enterprise value may come from China's inadequate promotion of environmental awareness, and consumers prefer low-priced products over green products. In addition, Sharfman and Fernando (2008) proposed to find that poor management of carbon offsets leads to higher capital costs. Yu and Li (2021) proposed that China's carbon emission trading policy has a spillover effect, which indicates that the defects of China's carbon management policy lead to unnecessary capital loss. Therefore, in means such as carbon emission trading, the benefits obtained by enterprises are reduced, thus affecting the social value of enterprises. Chen et al. (2021) proposed that low-carbon treatment measures of enterprises may lead to higher energy costs. Therefore, excessive investment in green production by enterprises in China may not pay off due to the declining confidence of stakeholders in the profitability of enterprises and the low acceptance of additional environmental protection costs by consumers. Based on the above literature, the author makes the following assumptions.

H1: Enterprise carbon score has a negative significance on enterprise value.

4.3 Corporate Carbon Disclosure and Firm Value

A recent study conducted by Shen et al. (2024) found that the difficulty of carbon information disclosure lies not only in the complexity of the calculation process, such as cost calculation by the accounting team in stages, but also in the dual nature of asset recognition. It is difficult for enterprises to confirm the value of green input and environmental protection assets in the disclosure of carbon information. Moreover, due to the high operating cost of carbon disclosure, a small number of enterprises are inclined to resource disclosure. Zimmerli (2007) proposed that corporate behaviour consistent with "social responsibility" would reduce shareholder returns. In addition, Zhang et al. (2022) also proposed that reducing greenhouse gas emissions would bring cost pressure. Moreover, Chinese citizens are less receptive to paying for carbon offsets. Qiu et al. (2021) expressed that China's green consumption is at a low level and still in its initial stage. Compared with the high level of green consumption in Western countries, there still exist some problems in our country's green consumption. Therefore, the market may not pay for companies to invest in improving their carbon scores.

According to Yuan (2010), China's historical development level has been accompanied by a large consumption of resources. Due to the late start of China's low-carbon economy, China currently lacks experience and Chinese practices. Therefore, the level of independent carbon disclosure of enterprises may be relatively low, and the cost of carbon accounting is relatively high. In addition, some experts from the Chinese Academy of Sciences and the Chinese Academy of Environmental Sciences have pointed out that there is a conflict between carbon disclosure and technological deficiencies in China (Liu et al., 2012). As a result, increased levels of disclosure may require significant green investment by companies. A lack of confidence among investors and stakeholders about the high cost of carbon disclosure can also lead to a decline in corporate value.

H2: The level of carbon disclosure of enterprises is negatively correlated with the value of enterprises.

5. Research Design

5.1 Sample and Data

The empirical analysis of this study refers to the demonstration methods of Xie et al. (2021) and Dong et al. (2018), combined with Chinese data. The initial data of the

Table 5.1 Sample Selection

	Number of Companies	Number of Firm-year Observations
Total number	5,422	21,819
Less: ST companies	199	739
Less: Companies from the industry except mining, manufacturing, and electricity; heat, gas, and water production; and supply	1,673	6,852
Less: missing sample data	2,924	8,466
Final sample	626	4,969

report include financial data, ESG performance scores, and carbon neutrality data for all listed companies in China from 2013 to 2023. For this study, we used the CSMAR database as its source. According to Du and Boateng (2015), CSMAR is a high-quality Chinese database. As the only data provider of Wharton Research Data Services (WRDS) in China, CSMAR has a high reputation. CSMAR provides relevant financial data such as ROA, return on equity (ROE), enterprise size, financial leverage, Tobin's Q, and growth of enterprises, as well as descriptive data such as listing code and names of enterprises. In addition, the study combines corporate carbon neutrality scores and ESG performance scores provided by Wind. The initial data included 3,488 companies with 17,441 pieces of data. The processing of raw data is as follows: (1) excluding special treatment (ST) companies and companies listed for less than three years, they usually have relatively volatile data due to their instability. (2) Due to the large differences in carbon neutral data among industries, we selected data of all listed industrial companies. According to China's Classification of Industries in the National Economy, industries are divided into mining, manufacturing, and electricity; heat, gas, and water production; and supply. (3) Some companies were omitted from the initial sample because of the absence of carbon data. (4) Companies with incomplete corporate data for 2022 were excluded from the sample and were not included in the regression results. Finally, the authors obtained ten-year annual data for 626 companies, with a total of 5,762 results, and the detailed process is shown in Table 5.1.

5.2 Variable Measurements

5.2.1 Dependent Variable

According to Gentry and Shen (2010), a lot of studies use accounting and market-based data as effective indicators to study the financial performance of enterprises. Dudin et al. (2018) used corporate profits as the standard of corporate performance, and Bevan et al. (1999) assessed corporate value with investment data that could reflect fixed capital and/or assets. However, these methods ignore the impact of firm size on data significance. According to Al-Tuwaijri et al. (2004),

ROA, as one of the performance indicators of enterprises, can break the limit of company size. This study uses ROA as one of the explained variables to measure the company's financial performance.

Additionally, Tobin's Q was employed by Daines (2001) and Steenkamp and Steenkamp, (2017) as a basis for determining the market value of publicly traded corporations. Tobin's Q ratio, which is the product of the market prices of stock and debt divided by the price of replacing the firm's capital, measures the market worth of a company in relation to the replacement cost of its assets. Su (2005) noted that the lack of liquidity and the size of the Chinese debt market make it challenging to determine the market value of debt. This study applies Chung and Pruitt's (1994) method to calculate Tobin's Q based on the circumstances in China, which is derived using the following formula:

$$\text{Tobin's } Q = \frac{MV + BD}{BV - CL}$$

Here,

BD = the book value of the company's short-term liabilities and long-term liabilities;
CL = the book value of current liabilities;
MV = net income after tax/cost of equity funds;
BV = assets less liabilities.

5.2.2 Independent Variables

5.2.2.1 CARBON SCORE

Xie et al. (2021) used CDP to represent the degree of carbon disclosure of enterprises. However, only 48 listed companies in China responded to the CDP questionnaire in 2019 (Köroğlu Göğüş & Alpaydın, 2021). Chinese enterprises' emphasis on the CDP questionnaire gradually increased after 2020, and only 350 listed companies participated in the questionnaire by 2022 (Köroğlu Göğüş & Yaman, 2024). Therefore, the CDP data for China do not give a full picture of the real situation. Cu and Ma (2014) used the environmental responsibility score to calculate the score of carbon emission information, and this study took the Wind environmental score as the standard of carbon emission score concerning this method. In addition, according to the information on the CDP official website, more than 2,700 enterprises responded to the CDP questionnaire in 2022, which indicates that the increase in carbon health data of enterprises has strengthened the accuracy of the test results.

5.2.2.2 CARBON DISCLOSURE

According to Fatemi et al. (2018), the value of an enterprise is enhanced by its superior ESG performance. Alkurdi et al. (2023) found that carbon emission, as an intermediate medium, is closely related to corporate performance. Moreover, Chinese enterprises' investment in ESG can significantly reduce carbon emissions (Cong et al.,

2022). Following Xie et al. (2021), they used the ESG performance of enterprises as one of the independent variables when studying carbon disclosure and enterprise value. The Wind ESG rating takes into account the carbon footprint and carbon emissions, and the data are calculated based on the Wind carbon emissions database covering all A-shares. In this study, the carbon disclosure data of manual mobile phones are used to obtain the average score of each item as the measurement data of carbon disclosure. Data on the specific items and scores are presented in Table 5.2. The carbon disclosure data for this report come from the CSMAR corporate annual report.

5.2.3   *Control Variables*

Luo et al. (2012), after analysing global data, illustrated that increasing pressure comes from stakeholders to urge companies to make carbon disclosure. According to Alfani and Diyanty (2020), the financial relaxation rate and industry regulation have beneficial effects on the carbon dioxide sharing rate. According to Liao et al. (2015) and Alfani and Diyanty (2020), managers of bigger businesses pay more attention to the disclosure of carbon emissions. Additionally, there is a positive relationship between a business's capacity to develop and its propensity for disclosing information on greenhouse gas emissions (Gao et al., 2018); the same conclusion was advanced by Newson and Deegan in 2002. In response to demand from stakeholders, managers place significance on corporate carbon disclosure, according to Deegan and Blomquist (2006). Based on Wang and Cao (2022), this study set ownership concentration as one of the control variables, which is calculated by shareholders' shareholding ratio.

A higher asset turnover rate of the company means that the company has more working capital for corporate green investment such as the promotion of corporate social value and the research and development of low-carbon technology. Also, considering that the government will subsidize state-owned enterprises (SOEs) if their environmental spending is huge, this study will control whether enterprises are state-owned or not. In addition, based on Liu and Zhang (2017), the value of an enterprise is also determined by the size of the board of directors, the proportion of independent directors, the proportion of management shares, and the number of shares held by the largest shareholder. This study emulates their setup to make sure the company has a healthy management structure. In addition, Yasar (2013) stated that audit quality is one of the criteria that affect the reliability of financial information, and he believes whether the auditor comes from a Big Four company controls the reliability of financial information. Therefore, we chose the following control variables:

(1) corporation size (size); (2) financial leverage (lev); (3) market capitalization to book value (bml); (4) growth ability (growth); (5) ownership structure (Occupy); (6) total assets turnover (TAT); (7) corporation listing time (AGE); (8) whether a Big Four accounting firm conducts an audit (AUDIT); (9) the number of shares held by the largest shareholder and (Top 1); (10) the size of the board of directors (BOARD); and (11) whether the company is a SOE. Table 5.2 presents representative symbols of control variables and their definitions.

Table 5.2 Carbon Disclosure Definition

Variable	Score	Definition
Environmental protection concept	0 or 1	Disclose the company's environmental protection concept, environmental policy, green development, etc., assigned a value of 1; otherwise, 0
Environmental objects	0 or 1	Disclose the company's past environmental goals and future environmental goals, assigned a value of 1; otherwise, 0
Environmental management	0 or 1	Disclose a series of management systems, such as relevant environmental management systems, regulations, and responsibilities, shall be assigned a value of 1; otherwise, 0
Environmental training	0 or 1	Disclose the company's participation in environment-related education and training, assigned a value of 1; otherwise, 0
Environment-specific activities	0 or 1	Disclose the company's participation in environmental protection special activities, environmental protection, and other social welfare activities, assigned a value of 1; otherwise, 0
Environmental emergency and response mechanism	0 or 1	Disclose the company's establishment of an emergency response mechanism for major environment-related emergencies, the emergency measures taken, the treatment of pollutants, etc., with a value of 1; otherwise, 0
Environmental award	0 or 1	Disclose the honours or awards the company has received for environmental protection, with a value of 1; otherwise, 0
"Three simultaneous" system	0 or 1	Disclose the company's implementation of the "three simultaneous" system, assigned a value of 1; otherwise, 0

5.3 Model Specification

Xie et al. (2021) used the CDP score to measure the level of a firm's carbon neutrality; however, the CDP company did not provide carbon neutrality scores for 2022–2023. In this section, referring to Chaudhry et al. (2021), the author uses ESG data as the basis for environmental performance scoring to test the first hypothesis. Based on Dong et al. (2018), the author used model 1 (M1) to prove the hypothesis that ESG performance is positively correlated with firm value. Yoon et al. (2018) demonstrated that corporate social responsibility (CSR) practices are positive and significant to a company's market. The authors used this model to verify whether the conclusion applies to enterprises in China. The relevant ESG data come from the Wind database.

Model 1:

$$\text{Tobin Q} = \beta_0 + \beta_1 \text{Carbondis} + \beta_2 \text{Size} + \beta_3 \text{lev} + \beta_4 \text{bml} + \beta_5 \text{growth} + \beta_6 \text{Occupy}$$
$$+ \beta_7 \text{TAT} + \beta_8 \text{AGE} + \beta_9 \text{Audit} + \beta_{10} \text{Top1} + \beta_{11} \text{Board} + \beta_{12} \text{SOE}$$
$$+ \text{Industry Fixed Effect} + \text{Year Fixed Effects} + \varepsilon$$

Here,

Tobin's Q = the revised enterprise value model adjusted by previous year's performance, which is recorded in absolute terms;

Carbondis = carbon score, which indicates whether a company has a high-quality carbon performance;

Size = the logarithm of the company's total assets;

lev = corporate financial leverage;

bml = book-to-market ratio;

Growth = increase in the rate of total asset, calculated as changes percentage in total assets from year $t-1$ to year t;

Occupy = appropriation of funds by major shareholders;

TAT = total asset turnover, which is calculated as Net Sales/Average Total Assets;

AGE = number of years since the company went public;

Audit = indicator variable that equals 1 if the auditor comes from Big Four; otherwise, 0;

Top 1 = number of shares held by the largest shareholder;

Board = Number of board members;

SOE = indicator variable that equals 1 if the company is state-owned, and 0 otherwise;

Industry = industry-specific fixed effect;
   Year = year-specific fixed effect (2018–2023).

According to Zhang et al. (2018), the maximum completion time of factory orders is inversely correlated with carbon emissions, which means that if an enterprise responds quickly to customer requirements, it will have to pay a larger carbon emission price. Matsumura et al. (2014) studied the relationship between carbon emission and enterprise value and found that an increase in carbon emission by 1,000 metric tonnes would reduce the value of the company by 212,000 dollars. Therefore, this report hypothesizes that a firm's carbon emission score has a negative significance to its firm value, which is proved by model 2 (M2).

Model 2:

$$\text{Tobin Q} = \beta_0 + \beta_1 \text{average} + \beta_2 \text{Size} + \beta_3 \text{lev} + \beta_4 \text{bml} + \beta_5 \text{growth} + \beta_6 \text{Occupy}$$
$$+ \beta_7 \text{TAT} + \beta_8 \text{AGE} + \beta_9 \text{Audit} + \beta_{10} \text{Top1} + \beta_{11} \text{Board} + \beta_{12} \text{SOE}$$
$$+ \text{Industry Fixed Effect} + \text{Year Fixed Effects} + \varepsilon$$

Here,

average = indicator variable that ranges from 0 to 1; the authors used the average score of environmental protection concept, objects, management, training, specific activities, emergency response mechanism, award, and "Three simultaneous" system. The data above are from the Wind database, and detailed definitions are presented in Table 5.2.

6.   Empirical Results

6.1   Descriptive Statistics

As shown in the descriptive data in Table 5.3, the number of results in this report is 5,762, with an average ROA of 4.8%, an average Tobin's Q of 2.228, and an average carbon disclosure score of 0.344. From the results of the standard deviation, there is a large gap between carbon score and enterprise size. The highest carbon score an enterprise reaches 8.43 points (the full score is 10 points), while the lowest carbon score is only 0 point. The company's average financial leverage is 0.648, the average book-to-market ratio is 0.421, and the average earnings growth is 0.186. The average age of listed enterprises is 13.787, the highest age is 29, and the lowest age is 3. In addition, the ratio of SOEs and private enterprises in the sample is about 1:1, and about 10% of the enterprises have auditors from the Big Four. Finally, the proportion of the largest shareholder is largely different, the average value is 34.421, the lowest value is 10.03, and the highest value is 76.67.

Table 5.4 shows the correlation between the variables. The coefficient between carbons and tubing is negative (-0.134), indicating that carbondis and tobinq show negative significance. This means that the higher the carbon score of the business, the lower the value of the business. The coefficient between average and tobinq is also negative, which equals -0.186. This coefficient conveys the negative correlation between greenhouse gas disclosure and firm value, which means that the higher the degree of disclosure, the lower the enterprise value. And the coefficient is greater than 0.4, which indicates that the correlation between carbon disclosure and enterprise value is strong. In addition, the p-values of carbon disclosure and carbon score were both less than 0.05, indicating the

Table 5.3 Descriptive Statistics

	N	*Mean*	*Std. Dev.*	*Min*	*Max*
ROA	4,969	0.046	0.062	-0.207	0.24
Tobinq3	4,969	2.101	1.497	0.795	9.222
Carbondis	4,969	2.402	2.178	0	8.43
Average	4,969	0.36	0.276	0	1
Size	4,969	22.869	1.411	20.058	26.748
Lev	4,969	0.43	0.183	0.086	0.874
bm1	4,969	0.665	0.272	0.122	1.22
Growth	4,969	0.18	0.392	-0.422	2.724
Occupy	4,969	0.011	0.018	0	0.127
TAT	4,969	0.67	0.364	0.12	2.132
Age	4,969	13.327	7.348	3	28
Audit	4,969	0.133	0.340	0	1
Top1	4,969	34.221	14.618	10.03	76.67
Board	4,969	2.135	0.195	1.609	2.639
SOE	4,969	0.503	0.500	0	1

Table 5.4 Pairwise Correlations

	Tobinq	ROA	Carbondis	Average	Size	Lev	Bml	Growth	Occupy	TAT	Age2	Audit	Top1	Board	SOE
Tobinq	1.000														
ROA	0.348***	1.000													
Carbondis	−0.134***	−0.001	1.000												
Average	−0.186***	0.060***	0.479***	1.000											
Size	−0.315***	0.047**	0.452***	0.520***	1.000										
Lev	−0.325***	−0.416***	0.165***	0.193***	0.447***	1.000									
Bml	−0.796***	−0.308***	0.142***	0.215***	0.474***	0.368***	1.000								
Growth	0.168***	0.263***	−0.033*	0.015	0.054***	0.045**	−0.127***	1.000							
Occupy	−0.020	−0.194***	−0.030	−0.086***	−0.018	0.127***	0.031	−0.077***	1.000						
TAT	−0.022	0.154***	0.064***	0.177***	0.117***	0.123***	0.035*	0.096***	−0.027	1.000					
Age	−0.265***	−0.077***	0.235***	0.325***	0.437***	0.254***	0.303***	−0.076***	0.046**	0.138***	1.000				
Audit	−0.117***	0.033*	0.296***	0.321***	0.482***	0.179***	0.190***	−0.007	−0.021	0.100***	0.204***	1.000			
Top1	−0.037**	0.096***	0.074***	0.091***	0.171***	0.004	0.120***	−0.009	−0.083***	0.105***	0.041**	0.177***	1.000		
Board	−0.094***	0.009	0.142***	0.147***	0.235***	0.114***	0.103***	−0.015	−0.021	−0.006	0.187***	0.053***	−0.025	1.000	
SOE	−0.248***	−0.118***	0.145***	0.259***	0.316***	0.155***	0.296***	−0.081***	−0.002	0.064***	0.518***	0.142***	0.213***	0.216***	1.000

* $p < 0.1$, ** $p < 0.05$, *** $p < 0.0$

effectiveness of the method. In addition, there is a positive significance between disclosure level and carbon score, and the coefficient is 0.479, which indicates a strong correlation. This result means that the higher the level of carbon disclosure, the higher the carbon score.

6.2   *Baseline Regression Results*

Table 5.5 shows the regression results of variables. The first column (m1) represents the relationship between the carbon fraction of the firm and the enterprise value of the explained variable, and there is no interaction term in the model. The second column (m2) is the regression result of carbon score and enterprise value, with no interaction term. Moreover, the p-values of models 1 and 2 are both less than 0.05, which also verifies the reliability of the model.

The results in column 1 show that the regression coefficient between carbondis and Tobin Q is negative (−0.034), indicating a negative significant relationship between the two variables. This result shows that when the enterprise carbon offset measures are good, the enterprise value will decrease slightly.

The results in column 2 show that the coefficient between average and Tobinq is negative (−0.254), indicating that the degree of carbon disclosure has a significant negative correlation with the explained variable. Therefore, as a company's carbon disclosure deepens, its value will decline.

Table 5.5 Baseline Regression Results

	(1)	(2)
	ROA	Tobinq
	Coeff. [t-value]	Coeff. [t-value]
Carbondis	−0.003** [−2.574]	−0.040*** [−3.347]
Average	0.014 [1.612]	−0.323*** [−3.140]
Size	0.029*** [13.249]	0.222*** [8.949]
Lev	−0.263*** [−22.302]	−0.659*** [−4.846]
Bm1	−0.057*** [−6.095]	−5.248*** [−48.535]
Growth	0.002*** [2.925]	0.013 [1.371]
Occupy	−0.308*** [−4.591]	0.794 [1.026]
TAT	0.000 [0.067]	0.057 [0.899]
Age2	0.000 [0.030]	−0.005 [−1.430]
Audit	−0.011 [−1.631]	−0.049 [−0.634]
Top1	0.000*** [2.682]	0.004** [2.451]
Board	0.007 [0.700]	−0.157 [−1.353]
SOE	−0.019*** [−4.026]	−0.136** [−2.486]
_cons	−0.470*** [−0.185]	1.380*** [2.591]
Industry	Yes	Yes
Year	Yes	Yes
r2_w	0.218	0.544
N	2,881.000	2,881.000

t-statistics in parentheses.* $p < 0.1$, ** $p < 0.05$, *** $p < 0.01$.

7.   Mitigating Endogeneity Concerns

7.1 *Two-Stage Least Squares Test*

According to Bascle (2008), endogeneity problems may be due to missing factors that are related to explanatory variables. Based on Eugster and Wagner (2015), the authors set up an instrumental variable method to manage this endogeneity. Harjoto and Laksmana (2018) found that the forecast results of analysts have a negative significance for mandatory CSR and have a positive correlation with resource social responsibility. The more positive analysts' forecasts of corporate earnings, the more likely companies are to take the initiative to assume environmental protection responsibilities, thus affecting their carbon neutrality. Therefore, referring to Nguyen et al. (2022) and Sadiq et al. (2020), this study uses shareholding concentration as an instrumental variable. Specifically, it considers the higher shareholding concentration of listed companies, which refers to the proportion held by the top five shareholders (Top 5). In addition, there may be other mediating variables that affect the carbon-neutral score.

According to Deng et al. (2022), relevant intermediary variables were set as dummy variables and used as instrumental variables for endogeneity detection. In this section, a 2SLS method is applied to reduce potential endoplasmic problems associated with carbon-neutral quality. Becker (2016) proposed that effective instrumental variables should meet the following two conditions: First, instrumental variables and the endogenous variable must be correlated, meaning they should show significance; second, instrumental variables need to satisfy exclusion restrictions.

$$\text{Carbondis} = \beta_0 + \beta_1 \text{Top5} + \sum \beta_i \text{ Controls} + \sum \beta_k \text{ Year} + \sum \beta_k \text{ Industry} + \varepsilon$$

$$\text{Tobinq} = \beta_0 + \beta_1 \text{ Carbondis} + \beta_2 \text{ Top5} + \beta_i \text{ Controls} + \sum \beta_k \text{ Year}$$
$$+ \sum \beta_k \text{Industry} + \varepsilon$$

$$\text{Tobinq} = \beta_0 + \beta_1 \text{ average} + \beta_2 \text{ GRI} + \beta_i \text{ Controls} + \sum \beta_k \text{ Year} + \sum \beta_k \text{Industry} + \varepsilon$$

Here,

Top5 = the sum of the top five shareholders (Top5) that measures the equity set;
Corporate responsibility indicator (CRI) = indicator variable that equals 1 if the firm refers to the Sustainable Development Guide; otherwise, it equals 2.
Also, i and k denote the i-th and k-th companies.

7.1.1 *First-Stage Results*

Column (1) in Table 5.6 shows the outcomes of the first-stage regression using 2SLS using model 3. The coefficient of Top5 (exogenous instrument) is positive and significant (the coefficient is 0.711, and t-statistic is 47.582), which means that Top5 and carbondis are positively significant. The results suggest that the degree of concentration of shareholder rights in relevant companies may contribute to changes in carbon scores.

Column (1) in Table 5.7 displays the results of the first-stage regression using 2SLS with model 4. The coefficient of CRI (exogenous instrument) is positive and

Table 5.6 Mitigating Endogeneity Concerns with the 2SLS Models with Top5

	(1)	(3)
	2SLS: First-Stage	TobinQ 2SLS: Two-Stage
	Coeff. [t-value]	Coeff. [t-value]
Average		−0.159 [−0.112]
Top5	0.382*** [27.300]	
Carbondis	0.025*** [14.591]	0.033*** [0.309]
Size	0.030*** [8.651]	−0.158[−0.473]
Lev	−0.036* [−1.761]	−0.999[−0.761]
Bm1	0.002 [0.137]	−3.674***[−5.543]
Growth	0.010 [1.165]	−0.781[−0.796]
Occupy	−0.581*** [−3.197]	−28.436[−1.448]
TAT	0.045*** [4.968]	0.457[1.540]
Age2	0.001*** [2.634]	−0.006[−0.129]
Audit	0.048*** [4.629]	0.961[0.534]
Top1	−0.000* [−1.680]	0.011[0.695]
Board	−0.001 [−0.070]	−1.596 [−0.949]
SOE	0.037*** [4.933]	0.549 [0.961]
_cons	−0.563*** [−7.620]	8.268 [0.449]
Industry	Yes	Yes
Year	Yes	Yes
r2_w	0.406	
N	4,201.000	

t-statistics in parentheses.
* $p < 0.1$, ** $p < 0.05$, *** $p < 0.01$.

significant (coefficient 0.594, *t*-statistic 32.965), which means that CRI and average are positively significant. The results suggest that whether companies refer to the Sustainability Reporting Guidelines may lead to changes in the level of carbon disclosure.

7.1.2 *Second-Stage Results*

Column (2) in Table 5.6 represents the results of the second stage of the 2SLS method using model 3, which uses tobinq as a dependency. The carbondis coefficient is negative (−0.055) and the *t*-stat is −3.464. This verifies that the increase in the carbon score leads to a negative growth in enterprise value.

Column (2) in Table 5.7 displays the results of the second phase, using the 2SLS approach with model 4. The average coefficient is negative (−0.432), and the *t*-stat is −2.474. This result also proves that the higher the level of carbon disclosure, the lower the value of the enterprise.

7.2 *Controlling for Firm Fixed Effects*

Fang et al. (2009) argued that the existence of some unpredictable factors related to variables may bias coefficient estimates. In order to minimize the influence of

Table 5.7 Mitigating Endogeneity Concerns with the 2SLS Models with CRI

	(1)	(3)
	2SLS: First-Stage	*TobinQ 2SLS: Two-Stage*
	Coeff. [t-value]	*Coeff. [t-value]*
Carbondis		0.054 [0.514]
CRI	0.504*** [39.511]	
Average	1.736*** [16.197]	−0.796*** [−0.583]
Size	0.169*** [6.381]	−0.180 [−0.546]
Lev	−0.058 [−0.372]	−1.178 [−0.910]
Bm1	−0.324*** [−2.895]	−3.535*** [−5.425]
Growth	−0.159** [−2.366]	−0.737 [−0.756]
Occupy	1.920 [1.363]	−29.184 [−1.500]
TAT	−0.137** [−1.970]	0.512* [1.751]
Age2	−0.000 [−0.113]	−0.009 [−0.185]
Audit	0.494*** [6.149]	1.298 [0.731]
Top1	0.000 [0.209]	0.008 [0.514]
Board	0.274** [2.078]	−1.855 [−1.118]
SOE	−0.112* [−1.917]	0.729 [1.311]
_cons	−3.824*** [−6.681]	11.443 [0.630]
Industry	Yes	Yes
Year	Yes	Yes
r2_w	0.453	
N	4,201.000	

t-statistics in parentheses.
* $p < 0.1$, ** $p < 0.05$, *** $p < 0.01$.

neglected factors, the authors controlled for firm fixed effects to explore the influence of reverse causality on the results. According to Chen et al. (2003), the factors leading to firm heterogeneity can be solved by industry-fixed effects, and the model should conform to the set variable model. According to models 1 and 2, Tables 5.7 and 5.8 show the regression results of carbon score and carbon information disclosure with enterprise value. Since the *W*-values of all regression Hausman tests are significant, the results given are fixed effects. The result of Hausman test is 0.0098.

8.   Robustness Check

8.1   *Alternative Measure of Tobin's Q*

Based on Dong et al. (2018), ROA can replace Tobin's Q for robust testing. Oware and Mallikarjunappa (2022) also used Tobin's Q and ROA as two indicators of enterprise value. Therefore, Tobin's Q is used here as a substitute variable to measuring enterprise value.

$$ROA = \beta_0 + \beta_1 Carbondis + \beta_2 Average + \beta_3 Carbondis * average$$
$$+ \sum \beta_i controls + \sum \beta_k industry + \sum \beta_k year + \varepsilon$$

Here,

ROA = calculated as net income divided by total assets in year $t-1$;

Table 5.8 shows the robust results. The positive coefficient of ESG is 0.056, which is 0.03 different from the coefficient obtained by using ROA (0.026). Both coefficients show the promotion effect of ESG performance on enterprise value. The coefficient 0.007 in the second row is similar to the coefficient -0.005 using ROA, and both represent a negatively correlated association.

8.2   Multicollinearity Test

Variance inflation factors (VIFs) is used in this study to test the multicollinearity. Table 5.9 shows the result that the VIFs of each variable are less than 10. Thus, there is no multicollinearity problem.

9.   Discussion

9.1   Relationship with Previous Literature

According to the previous survey, there are few studies that link the carbon neutral score of Chinese industrial entities to the level of carbon disclosure and the growth

Table 5.8 Robustness Test: Alternative Measure of Tobin's Q

	(1)	(2)
	m1	*m2*
	Coeff. [t-value]	*Coeff. [t-value]*
Carbondis	-0.001*** [-5.805]	-0.003*** [-2.645]
Average	0.054*** [5.370]	0.017*** [1.777]
Size		0.035*** [15.088]
Lev		-0.180*** [-13.507]
Bml		-0.123*** [-12.086]
Growth		0.068*** [12.427]
Occupy		-0.547*** [-4.803]
TAT		0.068*** [10.254]
Age		0.000 [1.187]
Audit		-0.014** [-2.037]
Top1		0.000** [2.422]
Board		0.019* [1.747]
SOE		-0.021*** [-4.154]
_cons	0.049*** [8.727]	-0.658*** [-13.295]
Industry	Yes	Yes
Year	Yes	Yes
r2_w	0.020	0.256
N	2,881.000	2,881.000

t-statistics in parentheses.
* $p < 0.1$, ** $p < 0.05$, *** $p < 0.01$.

Table 5.9 Multicollinearity Test

Variable	VIF	1/VIF
Size	2.287	.437
Average	1.528	.654
Bm1	1.458	.686
Age2	1.404	.712
Carbondis	1.403	.713
SOE	1.394	.717
Lev	1.354	.739
Audit	1.31	.763
Top1	1.126	.888
Board	1.099	.91
TAT	1.067	.937
Growth	1.058	.945
Occupy	1.042	.96
Mean VIF	1.348	.

of enterprise value. According to Liang et al., government environmental regulations and media evaluation show a positive correlation with enterprise value. The more excellent the external carbon neutrality performance of an enterprise, the more significant the short-term and long-term value effects. Through the regression test of the model in this study, the authors found that the carbon neutrality score is negatively correlated with enterprise value. In addition, a recent study conducted by Dong et al. (2022) showed that corporate carbon neutrality commitments have a positive impact on stock prices in the long run. In this study, it was found that the increase in carbon neutrality evaluation of enterprises in the short term is negatively correlative with the value of enterprises.

According to Khan et al. (2021), green disclosure alone is not enough to create market value for banks. He et al. (2013) concluded that capital cost is negatively correlated with carbon disclosure. Matsumura et al. (2014) indicated that the enterprise value of companies that make carbon disclosure is about $2.3 billion higher than that of companies that do not make carbon disclosure. However, through the study of Chinese industrial enterprise data, the authors found that the degree of carbon disclosure is negatively correlated with enterprise value. According to Karim et al., more carbon emissions in the UK will affect the value of the company in a negative way; that is, the low carbon image of the company has a positive significance on the valuation of the company. In addition, the conclusion of this study is consistent with the conclusion of Zimmerli (2007) that the more active the corporate responsibility in carbon disclosure, the more likely it is to reduce shareholder returns. This result may lead to a decrease in stakeholders' confidence in the return on investment because enterprises reduce corporate profits to assume the environment of social responsibility. This conclusion is also confirmed by Li et al. (2015).

In addition, Wang et al. (2014) found that the environmental awareness of Chinese people has gradually strengthened, but the high green consumption intention has not been effectively transformed into actual consumption behaviour. In other words, consumers in China are not willing to pay the spillover price of environmental protection. Therefore, at present, the carbon score and carbon disclosure of enterprises have a negative growth effect on enterprise value in China.

9.2   *Theoretical Significance*

The growing international awareness of environmental protection and energy protection, as well as the limited nature of energy, promoting the green economy is the general trend. In theory, the increase in carbon emission level can improve the value of enterprises; for example, enterprises can obtain benefits from carbon emission trading. According to Li et al. (2015), managers can gain economic benefits through a higher level of carbon information disclosure and higher carbon offset management input.

However, several experts from the Chinese Academy of Sciences and the Chinese Academy of Environmental Sciences found that China's low-carbon development started late and lacked strong theoretical guidance (Liu et al., 2012), so enterprises need to pay a lot of capital and human input to implement low-carbon strategies. Moreover, China's low-carbon management still needs to be improved; for example, the "power rationing" measures in some areas have brought inconvenience to citizens' lives but caused negative economic growth.

In addition, there are accuracy problems in the estimation of China's carbon emissions on a macro scale, which leads to inadequate research on carbon neutrality expectations (Liu et al., 2014). The sinicization of carbon accounting has brought huge investments in carbon accounting for enterprises but has a negative impact on the profitability of enterprises. In addition, due to the ambiguity of the boundary of carbon assets, the inaccuracy of carbon disclosure and the carbon score of enterprises may lead to the unfairness of market competition. Such unfairness will hit the confidence of enterprises in carbon disclosure and the enthusiasm for independent disclosure.

9.3   *Practical Significance*

According to Yuan (2010), China's growth model history is at a low level, which has the problems of resource price distortion and short-term investment, which also leads to duplication of construction and high energy consumption and pollution. Yuan (2010) raised the problem that there is a contradiction between China's existing growth model and the pressure to cut emissions. Thus, the Chinese government must adopt policy incentives or low-carbon subsidies to regulate and encourage enterprises to carry out green transformation. Zhang et al. (2020) proposed that it is urgent for enterprises to establish a carbon emission cost accounting system. Also, carbon emission cost management should be taken into high consideration. Based on Zhao and Wang (2014), the key to increasing profits under the low-carbon model is to increase low-carbon publicity to improve customers' awareness of environmental protection. In addition, strengthening low-carbon technology innovation is also a powerful means for enterprises to carry out green transformation.

10. Conclusion

Given the current situation in China, the growth of enterprises' carbon score and carbon disclosure level is negatively correlated with enterprise value. The possible reason lies in the conflict between China's existing development model and low carbon; for example, China's carbon use accounts for a large proportion of energy use. In addition, complex carbon accounting costs such as carbon accounting can lead to lower stakeholder and investor confidence. Moreover, China's low-carbon economy started late, and the time to explore the practice is short, so the existing carbon-neutral measures and policies have certain defects. However, in response to the ideological guidance of the common destiny of mankind, China's green economic transformation is inevitable. Therefore, the Chinese government should strengthen the market supervision of carbon trading and carbon accounting information disclosure. And some green incentives are needed to strengthen companies' willingness to disclose carbon information on their own. In addition, the limitation of this study is that the carbon disclosure data were collected manually by the authors, and therefore there may be inaccuracies and missing data. In addition, due to incomplete carbon data collection and lack of disclosure of some companies' carbon data, the sample size of this study is not large. Because of this phenomenon, the conclusion of this study may be one-sided.

References

Aboud, A., & Diab, A. (2018). The impact of social, environmental and corporate governance disclosures on firm value. *Journal of Accounting in Emerging Economies, 4*(2), 14–25.

Aktas, N., De Bodt, E., & Cousin, J. G. (2011). Do financial markets care about SRI? Evidence from mergers and acquisitions. *Journal of Banking & Finance, 35*(7), 1753–1761.

Alfani, G. A., & Diyanty, V. (2020). Determinants of carbon emission disclosure. *Journal of Economics Business, and Accoutancy, 22*(3), 333–346.

Alkurdi, A., Al Amosh, H., & Khatib, S. F. (2023). The mediating role of carbon emissions in the relationship between the board attributes and ESG performance: European evidence. *EuroMed Journal of Business*, (ahead-of-print).

Alsaifi, K., Elnahass, M., & Salama, A. (2020). Market responses to firms' voluntary carbon disclosure: Empirical evidence from the United Kingdom. *Journal of Cleaner Production, 262*, 121377.

Al-Tuwaijri, S. A., Christensen, T. E., & Hughes Ii, K. E. (2004). The relations among environmental disclosure, environmental performance, and economic performance: A simultaneous equations approach. *Accounting, Organizations and Society, 29*(5–6), 447–471.

Aydoğmuş, M., Gülay, G., & Ergun, K. (2022). Impact of ESG performance on firm value and profitability. *Borsa Istanbul Review, 22*, S119–S127.

Bascle, G. (2008). Controlling for endogeneity with instrumental variables in strategic management research. *Strategic Organization, 6*(3), 285–327.

Becker, S. O. (2016). *Using instrumental variables to establish causality*. IZA World of Labor.

Bevan, A., Estrin, S., & Schaffer, M. (1999). *Determinants of enterprise performance during transition.* Heriot-Watt University, Department Economics.

Boiral, O., Henri, J. F., & Talbot, D. (2012). Modeling the impacts of corporate commitment on climate change. *Business Strategy and the Environment, 21*(8), 495–516.

Burke, H., Zhang, A., & Wang, J. X. (2021). Integrating product design and supply chain management for a circular economy. *Production Planning & Control,* 1–17.

Capelle-Blancard, G., & Laguna, M. A. (2010). How does the stock market respond to chemical disasters? *Journal of Environmental Economics and Management, 59*(2), 192–205.

Chan, J. C. H., & Welford, R. (2005). Assessing corporate environmental risk in China: An evaluation of reporting activities of Hong Kong listed enterprises. *Corporate Social Responsibility and Environmental Management, 12*(2), 88–104.

Chaudhry, S. M., Saeed, A., & Ahmed, R. (2021). Carbon neutrality: The role of banks in optimal environmental management strategies. *Journal of Environmental Management, 299,* 113545.

Chen, C. R., Guo, W., & Mande, V. (2003). Managerial ownership and firm valuation: Evidence from Japanese firms. *Pacific-Basin Finance Journal, 11*(3), 267–283.

Chen, Z., Song, P., & Wang, B. (2021). Carbon emissions trading scheme, energy efficiency and rebound effect – Evidence from China's provincial data. *Energy Policy, 157,* 112507.

Chouaibi, Y., & Zouari, G. (2022). The effect of corporate social responsibility practices on real earnings management: Evidence from a European ESG data. *International Journal of Disclosure and Governance, 19*(1), 11–30.

Chung, K. H., & Pruitt, S. W. (1994). A simple approximation of Tobin's q. *Financial Management,* 70–74.

Clarkson, A. N., Huang, B. S., MacIsaac, S. E., Mody, I., & Carmichael, S. T. (2010). Reducing excessive GABA-mediated tonic inhibition promotes functional recovery after stroke. *Nature, 468*(7321), 305–309.

Cong, Y., Zhu, C., Hou, Y., Tian, S., & Cai, X. (2022). Does ESG investment reduce carbon emissions in China?. *Frontiers in Environmental Science, 10,* 977049.

Costello, A., Abbas, M., Allen, A., Ball, S., Bell, S., Bellamy, R., . . . & Patterson, C. (2009). Managing the health effects of climate change: Lancet and University College London Institute for Global Health Commission. *The Lancet, 373*(9676), 1693–1733.

Cui, Y., & Ma, X. (2014). Research on influencing factors of carbon emission information disclosure of Chinese listed companies. *Journal of Central University of Finance and Economics, 6.*

Daines, R. (2001). Does Delaware law improve firm value? *Journal of Financial Economics, 62*(3), 525–558.

Davidson, M., Karplus, V. J., Zhang, D., & Zhang, X. (2021). Policies and institutions to support carbon neutrality in China by 2060. *Economics of Energy & Environmental Policy, 10*(2).

Deegan, C., & Blomquist, C. (2006). Stakeholder influence on corporate reporting: An exploration of the interaction between WWF-Australia and the Australian minerals industry. *Accounting, Organizations and Society, 31*(4–5), 343–372.

Deng, W., Zhang, X., Zhou, Y., Liu, Y., Zhou, X., Chen, H., & Zhao, H. (2022). An enhanced fast non-dominated solution sorting genetic algorithm for multi-objective problems. *Information Sciences, 585,* 441–453.

Dong, H., Liu, Y., Zhao, Z., Tan, X., & Managi, S. (2022). Carbon neutrality commitment for China: from vision to action. *Sustainability Science, 17*(5), 1741–1755.

Dong, S., Zou, A., & Liu, R. (2018). Research on the relationship between social trust, carbon information disclosure and firm performance—Based on the CEI index of commercial credit in Chinese cities. *Friends of Accounting, 21,* 74–78.

Du, M., & Boateng, A. (2015). State ownership, institutional effects and value creation in cross-border mergers & acquisitions by Chinese firms. *International Business Review, 24*(3), 430–442.

Dudin, M. N., Lyasnikov, N. V., Reshetov, K. Y., Smirnova, O. O., & Vysotskaya, N. V. (2018). Economic profit as indicator of food retailing enterprises' performance. *European Research Studies Journal, XXI*(1), 468–479.

Eugster, F., & Wagner, A. F. (2015). Value Reporting Quality, Operating Performance, and Stock Market Valuations. *Swiss Finance Institute Research Paper* (pp. 11–25).

Fang, V. W., Noe, T. H., & Tice, S. (2009). Stock market liquidity and firm value. *Journal of Financial Economics, 94*(1), 150–169.

Fatemi, A., Glaum, M., & Kaiser, S. (2018). ESG performance and firm value: The moderating role of disclosure. *Global Finance Journal, 38*, 45–64.

Gao, B., Huang, T., Ju, X., Gu, B., Huang, W., Xu, L., . . . & Cui, S. (2018). Chinese cropping systems are a net source of greenhouse gases despite soil carbon sequestration. *Global Change Biology, 24*(12), 5590–5606.

Gentry, R. J., & Shen, W. (2010). The relationship between accounting and market measures of firm financial performance: How strong is it? *Journal of Managerial Issues*, 514–530.

Haque, F., & Ntim, C. G. (2020). Executive compensation, sustainable compensation policy, carbon performance and market value. *British Journal of Management, 31*(3), 525–546.

Harjoto, M., & Laksmana, I. (2018). The impact of corporate social responsibility on risk taking and firm value. *Journal of Business Ethics, 151*, 353–373.

He, Y., Tang, Q., & Wang, K. (2013). Carbon disclosure, carbon performance, and cost of capital. *China Journal of Accounting Studies, 1*(3–4), 190–220.

Helm, S. V., Pollitt, A., Barnett, M. A., Curran, M. A., & Craig, Z. R. (2018). Differentiating environmental concern in the context of psychological adaption to climate change. *Global Environmental Change, 48*, 158–167.

Huang, H., Zou, Y., Wang, L., Wang, W., & Ren, X. (2023). Impact of carbon information disclosure on corporate financing constraints: Evidence from the carbon disclosure project. *Australian Journal of Management*, 03128962231180265.

Jaffe, A. B., & Palmer, K. (1997). Environmental regulation and innovation: a panel data study. *Review of Economics and Statistics, 79*(4), 610–619.

Jennings, P. D., & Zandbergen, P. A. (1995). Ecologically sustainable organizations: An institutional approach. *Academy of Management Review, 20*(4), 1015–1052.

Jiao, J. (2022). Research on carbon accounting information disclosure in petrochemical industry under the background of carbon neutrality. *Beijing University of Chemical Technology*.

Jing, H., Zhu, P., Zheng, X., Zhang, Z., Wang, D., & Li, Y. (2022). Theory-oriented screening and discovery of advanced energy transformation materials in electrocatalysis. *Advanced Powder Materials, 1*(1), 100013.

Khan, H. Z., Bose, S., Sheehy, B., & Quazi, A. (2021). Green banking disclosure, firm value and the moderating role of a contextual factor: Evidence from a distinctive regulatory setting. *Business Strategy and the Environment, 30*(8), 3651–3670.

Krüger, P. (2015). Corporate goodness and shareholder wealth. *Journal of Financial Economics, 115*(2), 304–329.

Köroğlu Göğüş, M., & Alpaydın, F. (2021). CDP Climate Change and Water Report 2020.

Köroğlu Göğüş, M., & Yaman, E. S. (2024). CDP Climate Change and Water Report 2023/ Türkiye Edition.

Li, H., Fu, S., Chen, Z., Shi, J., Yang, Z., & Li, Z. (2019). The motivations of Chinese firms in response to the Carbon Disclosure Project. *Environmental Science and Pollution Research, 26*, 27792–27807.

Li, L., Liu, Q., Tang, D., & Xiong, J. (2017). Media reporting, carbon information disclosure, and the cost of equity financing: evidence from China. *Environmental Science and Pollution Research, 24*, 9447–9459.

Li, L., Yang, Y., & Tang, D. (2015). Carbon information disclosure of enterprises and their value creation through market liquidity and cost of equity capital. *Journal of Industrial Engineering and Management, 8*(1), 137–151.

Liang, T., Zhang, Y. J., & Qiang, W. (2022). Does technological innovation benefit energy firms' environmental performance? The moderating effect of government subsidies and media coverage. Technological Forecasting and Social Change, 180, 121728.

Liao, L., Luo, L., & Tang, Q. (2015). Gender diversity, board independence, environmental committee and greenhouse gas disclosure. *The British accounting review, 47*(4), 409–424.

Liu, M., Meng, J., & Liu, B. (2014). Research progress of carbon emission accounting methods at home and abroad. *Tropical Geography, 34*(2), 248–258.

Liu, L. Y., Wang, J. Z., Wei, G. L., Guan, Y. F., Wong, C. S., & Zeng, E. Y. (2012). Sediment records of polycyclic aromatic hydrocarbons (PAHs) in the continental shelf of China: Implications for evolving anthropogenic impacts. *Environmental Science & Technology, 46*(12), 6497–6504.

Liu, X., & Zhang, C. (2017). Corporate governance, social responsibility information disclosure, and enterprise value in China. *Journal of Cleaner Production, 142*, 1075–1084.

Lu, Y. (2017). *Study on green transformation of construction enterprises from the perspective of policy control and enterprise value.* Beijing Book Co. Inc.

Luo, L., Lan, Y. C., & Tang, Q. (2012). Corporate incentives to disclose carbon information: Evidence from the CDP Global 500 report. *Journal of International Financial Management & Accounting, 23*(2), 93–120.

Mai, H., & Cao, Y. (2023). The existing problems and suggestions of carbon information disclosure in China. *Finance and Accounting, 1*, 64–65.

Matsumura, E. M., Prakash, R., & Vera-Munoz, S. C. (2014). Firm-value effects of carbon emissions and carbon disclosures. *The Accounting Review, 89*(2), 695–724.

McLennan, C. L. J., Becken, S., Battye, R., & So, K. K. F. (2014). Voluntary carbon offsetting: Who does it? *Tourism Management, 45*(C), 194–198.

Nguyen, D. N., Tran, Q. N., & Truong, Q. T. (2022). The ownership concentration − Innovation nexus: Evidence from SMEs around the world. *Emerging Markets Finance and Trade, 58*(5), 1288–1307.

Oware, K. M., & Mallikarjunappa, T. (2022). CSR expenditure, mandatory CSR reporting and financial performance of listed firms in India: An institutional theory perspective. *Meditari Accountancy Research, 30*(1), 1–21.

Qiu, S., Wang, Z., & Liu, S. (2021). The policy outcomes of low-carbon city construction on urban green development: Evidence from a quasi-natural experiment conducted in China. *Sustainable Cities and Society, 66*, 102699.

Rahman, M. J., & Wu, J. (2023). M&A activity and ESG performance: Evidence from China. *Managerial Finance, 50*(1), 179–197.

Rahman, M. J., Wu, Q., & Zhu, H. (2024). Corporate social responsibility in times of social distancing: Evidence from China. *Business Ethics, the Environment & Responsibility.*

Rahman, M. J., Zhu, H., & Chen, S. (2023). Does CSR reduce financial distress? Moderating effect of firm characteristics, auditor characteristics, and Covid-19. *International Journal of Accounting & Information Management, 31*(5), 756–784.

Sadiq, M., Singh, J., Raza, M., & Mohamad, S. (2020). The impact of environmental, social and governance index on firm value: Evidence from Malaysia. *International Journal of Energy Economics and Policy, 10*(5), 555–562.

Sharfman, M. P., & Fernando, C. S. (2008). Environmental risk management and the cost of capital. *Strategic Management Journal, 29*(6), 569–592.

Shen, H., Li, Y., & Wang, J. (2022). Carbon emission information disclosure and investor returns. *Social Sciences, 11*, 140–150.

Shen, H., Yang, Q., Luo, L., & Huang, N. (2023). Market reactions to a cross-border carbon policy: Evidence from listed Chinese companies. *The British Accounting Review, 55*(1), 101116.

Shen, L., Qian, W., & Yang, Y. (2024). The divergence and driving factors of corporate environmental information disclosure in China. *Environmental Research Communications, 6*(7), 075029.

Steenkamp, J.-B. (2017). Global Brands and Shareholder Value. In *Global Brand Strategy* (pp. 275–289): Springer.

Su, D. (2005). Diversification and firm value: An empirical investigation of diversification premium based on China's stock-market listed Companies. *China Economic Quarterly-Beijing, 4*, 135.

Wang, D., Huang, Y., Guo, M., Lu, Z., Xue, S., & Xu, Y. (2023). Time dynamics in the effect of carbon information disclosure on corporate value. *Journal of Cleaner Production, 425*, 138858.

Wang, L., & Cao, W. (2022). Separation of control and cash flow rights, ultimate controlling shareholders' equity pledge ratio and risk of control transfer – Evidence from Chinese A-share listed companies. *Applied Economics Letters, 29*(16), 1533–1540.

Wang, P., Liu, Q., & Qi, Y. (2014). Factors influencing sustainable consumption behaviors: A survey of the rural residents in China. *Journal of Cleaner Production, 63*, 152–165.

Xie, L., Tang, C., Bi, Z., Song, M., Fan, Y., Yan, C., . . . & Chen, C. (2021). Hard carbon anodes for next-generation Li-ion batteries: Review and perspective. *Advanced Energy Materials, 11*(38), 2101650.

Xie, X., Wang, T., Yue, X., Li, S., Zhuang, B., & Wang, M. (2020). Effects of atmospheric aerosols on terrestrial carbon fluxes and CO2 concentrations in China. *Atmospheric Research, 237*, 104859.

Xiu, J., Liu, H., & Zang, X. (2015). Green credit: Industrial growth and forecast under Energy conservation and emission reduction. *Current Economic Science, 37*(3), 55–62.

Yasar, A. (2013). Big four auditors' audit quality and earnings management: Evidence from Turkish stock market. *International Journal of Business and Social Science, 4*(17).

Yoon, B., Lee, J. H., & Byun, R. (2018). Does ESG performance enhance firm value? Evidence from Korea. *Sustainability, 10*(10), 3635.

Yu, D. J., & Li, J. (2021). Evaluating the employment effect of China's carbon emission trading policy: Based on the perspective of spatial spillover. *Journal of Cleaner Production, 292*, 126052.

Yuan, F. (2010). China's potential economic growth under the constraint of low-carbon economy. *Economic Research, 8*, 79–89.

Zhang, P. (2021). Target interactions and target aspiration level adaptation: How do government leaders tackle the "environment-economy" nexus?. *Public Administration Review, 81*(2), 220–230.

Zhang, C., Guo, S., Tan, L., & Randhir, T. O. (2020). A carbon emission costing method based on carbon value flow analysis. *Journal of Cleaner Production, 252*, 119808.

Zhang, H., Shao, Y., Han, X., & Chang, H. L. (2022). A road towards ecological development in China: The nexus between green investment, natural resources, green technology innovation, and economic growth. *Resources Policy, 77*, 102746.

Zhang, X., Wu, L., Ma, X., & Qin, Y. (2022). Dynamic computable general equilibrium simulation of agricultural greenhouse gas emissions in China. *Journal of Cleaner Production, 345*, 131122.

Zhang, Y., Chao, Q., Chen, Y., Zhang, J., Wang, M., Zhang, Y., & Yu, X. (2021). China's carbon neutrality: Leading global climate governance and green transformation. *Chinese Journal of Urban and Environmental Studies, 9*(3), 2150019.

Zhang, Y., Chen, W., & Meng, X. (2018). Low-carbon scheduling problem of steelmaking production based on PBIL algorithm. *CIMS, 24*(10), 2407.

Zhao, D., & Wang, C. (2014). Research on emission reduction decision of supply chain enterprises considering low-carbon policy. *Industrial Engineering, 1*, 105–111.

Zheng, L. (2015). Research on corporate carbon information disclosure. *China Social Sciences Press*.

Zheng, Y., Wang, Z., Cai, Y., Xie, R., & Men, C. (2023). Study on the influence of green technology innovation on ESG performance and its path under the new pattern of "dual carbon." *Journal of Technology Economics, 42*(3).

Zhu, H., & Rahman, M. J. (2024). Ex-ante expected changes in ESG and future stock returns based on machine learning. *The British Accounting Review*, 101457.

Zhu, S., & Chen, S. (2001). Effects of organic carbon on nitrification rate in fixed film biofilters. *Aquacultural Engineering, 25*(1), 1–11.

Zimmerli, W. C. (2007). *Corporate ethics and corporate governance*. Springer.

6 The Impact of Corporate Environmental Performance on Financial Performance

Md Jahidur Rahman, Tarek Rana, Hongtao Zhu, and Xia Qinglu

1. Introduction

Enterprises often prioritize cost-cutting by discharging untreated pollutants, leading to economic gains but increased environmental risks, threatening future sustainability (Gomez & Rodriguez, 2011). Balancing ecological protection and economic growth is a critical challenge for nations and businesses (Kock et al., 2012). Numerous studies have examined whether environmental performance (EP) affects financial performance (FP), seeking to establish the importance of environmental protection. Early research provided mixed results. Cohen et al. (1995) found no impact of EP on FP, while others reported a positive impact (Salama, 2005; King & Lenox, 2001) or negative impact (Horváthová, 2010). These discrepancies arise from factors like standards of EP, company size control (Cohen et al., 1995), research methods (Salama, 2005), strategic decisions (King & Lenox, 2001), and policies (Nishitani et al., 2017). Recent studies suggest EP can enhance FP, including return on equity (ROE), return on assets (ROA) (Angelia & Suryaningsih, 2015), sales, profit, and efficiency (Nishitani et al., 2017). Some findings indicate a linear relationship: lower initial EP leads to greater financial gains after improvements (Jo et al., 2015; Manrique & Martí-Ballester, 2017). However, the impact is debated. Lahouel et al. (2020) argued for a non-linear, inverted U-shaped relationship, where initial investments boost FP significantly, but high-tech upgrades needed later require substantial capital, negatively impacting short-term FP. Conversely, Trumpp and Guenther (2017) found a positive U-shaped relationship, where low EP investments initially decrease FP, but beyond a certain point, FP improves. This suggests that companies with poor EP struggle financially to support improvements. These perspectives provide a nuanced understanding beyond linear relationships.

Some studies have examined the bidirectional relationship between FP and EP. It is suggested that these variables can mutually reinforce each other, creating a virtuous cycle where improved EP enhances FP, and stable FP supports better EP. However, the marginal costs associated with higher EP may complicate findings (Abban & Hasan, 2021). Additionally, some research indicates that the influence of FP on EP is short-term, while the reverse is more long-term (Hang & Rathgeber, 2019). Other studies argue that the relationship between FP and EP may be indirect,

DOI: 10.4324/9781003488965-8

mediated by factors like customer performance and R&D (Lioui & Sharma, 2012; Cantele & Zardini, 2018), or that no significant relationship exists, noting that enterprise size does not impact environmental quality (Lucato et al., 2017).

As one of the largest developing countries, China aims to enhance sustainable development amidst ongoing environmental pollution concerns. However, research on the relationship between EP and FP remains limited. Li et al. (2017) found that implementing environmental responsibility decisions significantly improved FP for energy-intensive listed companies, with government regulation amplifying this positive effect. Conversely, excess capacity hindered productivity and led to resource waste.

Nguyen et al. (2021) supported this perspective, noting that a company's internal mechanisms can have a mixed regulatory effect on the FP–EP relationship. Additionally, the inverted U-shaped theory has been validated in China, demonstrating that EP can enhance FP up to a certain point. Beyond this threshold, increased investment in environmental improvements diverts resources from value creation, thus limiting profitability. This relationship is further complicated by higher environmental instability (Zhang et al., 2020). Overall, there exists a win–win relationship between EP and FP within a specific range. High pollution costs and resource wastage drive companies to invest in environmental innovation more cost-effectively. Notably, state-owned enterprises exhibit weaker performance in this relationship, and those with disclosed environmental practices show similar trends (Shen et al., 2019). However, the potential for good FP to enhance EP remains untested in China.

Research has primarily focused on Southeast Asia and Europe, while China, the world's largest emitter of pollutants, presents unique cultural and systemic characteristics worthy of exploration (Li et al., 2017). However, studies on the FP and EP of China's heavy pollution industry are limited, and the relationship between these variables requires further investigation. Many Chinese enterprises question whether investing in corporate environmental responsibility (CER) can enhance profits, viewing such investments as risky. Some neglect the importance of CER, prioritizing immediate gains that lead to irreversible environmental harm (Meng et al., 2014). Thus, it is essential to assess whether a robust environmental protection system can create value for companies. Heavy polluters in China must shift their development mindset and adopt sustainable practices to secure long-term benefits.

To address this dilemma, the government typically employs a combination of quota, purchase, and tax policies to reduce pollutant emissions, aiming to protect sustainable development and public health. These measures also seek to encourage enterprises to adopt social responsibility by supporting emission reductions. The goal is to achieve dual benefits through optimization and environmental innovation: lowering emissions, cutting costs, and enhancing FP to aid economic modernization (De & Giri, 2020; Rennings, 2000). However, due to China's unique institutional context, these policies have limited effectiveness (Du et al., 2016). Furthermore, strict environmental regulations may disrupt the positive relationship between EP and FP (Shen et al., 2019). With limited investment capacity, excessive regulatory intervention might lead enterprises to pursue more profitable alternatives (Zhang et al., 2020). Therefore, it is crucial to explore the bidirectional relationship between EP and FP, identify enterprises' investment intentions for environmental improvements, and develop policies tailored to the realities of heavily polluting industries.

For this study, we collected the environmental certificate data and multiple indicators of FP of manufacturing companies listed in China's stock market from 2010 to 2020. We used the method of empirical analysis to verify the direct and two-way causal relationship between CER and corporate financial performance (CFP), as well as the degree of interaction between them, and thereby explore whether the company's EP and FP affect each other.

Most heavily polluting enterprises in China are still in the stage of the low EP level. Improving EP affects the existing economic benefits. The contradiction between the variables urges enterprises to continue to obtain benefits by extensive production that destroys the environment. The strict environmental protection policies formulated by the state have exacerbated the financial difficulties of heavily polluting manufacturing enterprises (Shen et al., 2019). However, these manufacturing industries mainly support China's development into the second-largest economy in the world (McGuinness et al., 2017). As a typical representative of developing countries, the development status of China's manufacturing industry is a stage that many developing countries must go through. Maintaining stable GDP growth while maintaining economic and social sustainability is a topic that has been discussed for a long time. Therefore, verifying EP can promote the development of corporate FP, change the hesitant attitude of Chinese enterprises to invest in environment, and promote the internalization of CER. National policies can better balance power in reward and punishment mechanisms to guide enterprises to change their mode of production. In addition, according to the report of the US Energy Information Administration, China was the country with the largest carbon emission in the world by 2018, which poses a threat to the world's environment and economic development (Osaki, 2019). The verification of the relationship between FP and EP can provide a reference for improving the emission of pollutants in the manufacturing industry. On the other hand, enterprises usually take an evasive attitude towards environmental protection policies to obtain more economic benefits, and the policies cannot be effectively implemented (Nguyen et al., 2021). This study can show that enterprises need to pay attention not only to their FP but also to the development of EP so that enterprises can attach importance to environmental protection.

In the following section, relevant data on selected Chinese listed-manufacturing companies are analysed, and the relationship between EP and FP is explored using a panel model with cross-section, using environmental tools and earnings before interest, taxes, depreciation, and amortization (EBITDA) and price-to-book value (PBV) ratio as variables. It is hoped that there is no irreconcilable contradiction between EP and FP by verifying the positive influence of mutual promotion between them, so as to make enterprises pay more attention to environmental issues and achieve the common improvement of EP and FP.

Firstly, this study is a verification of the two-way relationship between CER and CFP in China. It also verifies the win–win relationship between CER and CFP.

The study complements the previously ambiguous direct impact of CER on CFP, including the interaction relationship between the two variables and the impact degree, helping the government formulate environmental protection strategies and providing a reference for the company when facing environmental investment decisions. Secondly, the research is based on the Chinese background and studies the manufacturing industry with the largest pollution emissions in China. It fits the actual situation of China, and a more theoretical basis can be used in the formulation of relevant policies. It also makes theoretical contributions and practical references for many developing countries with the same situation. In an emerging economy with relatively poor EP, the government's policy guidance and companies' initial investment in environmental innovation can generate better results. Finally, different from the previous way of obtaining enterprise EP through questionnaires and year-end reports (Nishitani et al., 2011), this experiment takes the certificate as the evaluation standard of CER performance, which not only obtains more objective results but also proves the positive role of an environmental management system in EF and PF and provides a direction for enterprise governance of environmental problems.

The other parts of this chapter are structured as follows: Section 2 summarizes the existing research findings on the relationship between EP and FP, to find the ambiguous aspects, and discusses the influence of EP and FP on each other and the role of the management system on EP in the empirical literature and puts forward reasonable assumptions based on the problems identified. In Section 3, to study the relationship between the two variables, research design, selection of samples and variables, and collection and analysis methods of datasets are carried out. Section 4 presents the empirical results, states the findings, and describes the robustness test. Section 5 summarizes and discusses the main results and illustrates them.

2. Literature Review and Hypothesis Development

2.1 Influence of Improving Environmental Performance on Financial Performance

The main objective of an organization is always to maximize profits by using existing resources (Barney, 1991). Since 1980, under the influence of national policies and public opinion, enterprises have found that supporting environmental protection can achieve some goals for a company. Therefore, environmental decision-making has also been included in companies' strategy (Inman, 2002). According to natural resource-based view, enterprises should pay more attention to the protection and improvement of the natural environment through the benign relationship that the protection of natural resources can enhance the competitiveness of enterprises (Hart & Dowell, 2011). In recent years, enterprises have taken the environmental management system as one of the main tools to deal with environmental problems and related activities (Campos et al., 2015). ISO14001, as a basic standard of the system, is the certificate obtained by a company after voluntarily improving the environmental level, and it is also a leading management tool to solve the problem of environmental degradation in the manufacturing industry. Therefore, having this certificate can reflect the excellent EP of the company (Mohammed, 2000). Many pieces of evidence also show that through strict commitment of the management, ISO14001 can help companies identify environmental problems and establish a monitoring environmental management system (Bridgen & Helm, 2017), which can significantly improve the management and operation of the companies compared with the companies without certificates (Nguyen & Hens, 2015). It can also reduce the escape behaviour of environmental protection participants and help enterprises reduce pollutant emissions (Potoski & Prakash, 2005). This improvement not only makes the company's environmental protection comply with legal standards but also results in the substantive improvement of internal processes and procedures, thereby improvement in the FP and evaluations of social responsibility (Johnstone & Hallberg, 2020). However, there are still many debates on the view that implementation of the ISO14001 environmental management system can improve FP (Johnstone, 2020). A few studies believe that undertaking environmental responsibility has a negative impact on enterprises. Such studies attribute the negative relationship to excessive environmental costs (García-Sánchez & Prado-Lorenzo, 2012; Lioui & Sharma, 2012). Several studies in China also investigate the relationship between environmental accounting and firm performance (Rahman & Wu, 2023; Rahman et al., 2023, 2024; Zhu & Rahman, 2024). Based on the above views, we propose the following assumptions:

H1: A company's implementation of the environmental management system can improve the company's FP.

In this study, we further verify whether a company's EP has an impact on its FP. Different enterprises improve EP by implementing different environmental management strategies. The degree of fit between their decisions and environmental regulations and the willingness of enterprises to achieve EP affect the final FP (Nishitani et al., 2011, 2017). Some companies hope to optimize their production processes out of economic interests, to improve EP. Companies that adopt the ISO9001 standard in the quality management system have proved this. Lean production can reduce waste emission and resource waste, which is a supplement to environmental improvement (King & Lenox, 2001). Reducing environmental pollution, one of the main objectives of lean production, creates synergy between EP and product production quality management. Specifically, the emission of pollutants is reduced, the energy utilization rate is increased, the resource utilization rate is increased, and the required funds are reduced (Dieste & Panizzolo, 2018). This shows that the company's ways to improve EP include improving productivity, integrating internal operations, and reducing excess capacity (De & Giri, 2020; Samy et al., 2015). Compared with the standard of ISO14001, this standard has a more indirect and relaxing effect on EP. The discussion on the relationship between EP and FP also confirms the above view. Enterprises with good EP have more competitive advantages in improving customer satisfaction and operation efficiency (Nga et al., 2009). Accordingly, the following assumption is proposed:

H2: A company's implementation of financial management system can improve the company's financial performance.

2.2   Influence of Improving Financial Performance on Environmental Performance

According to the bidirectional research results of Testa and D'Amato (2017), there is no bidirectional relationship between FP and EP. However, according to the slack resource hypothesis, good FP of the previous year enables the company to invest surplus capital in environmental responsibility to optimize the production structure and provide help for the subsequent improvement of FP. However, it takes a long time for firms to reap returns from CER investment, and this cycle can be shortened in enterprises with a high correlation with environmental pollution. It takes time to reduce environmental costs and upgrade production (Jo et al., 2015). The above explanations are in line with the causal relationship. Only a company with good FP can ensure the implementation of environmental strategy, so the company's financial status affects the implementation level of EP. Moreover, this relationship can also promote a virtuous cycle between FP and EP (Wagner & Blom, 2011). Abban and Hasan (2021) showed that there is a causal relationship between FP and EP, and good performance of a company can reduce the emission of pollutants. In addition, this finding has implications for policy development, demonstrating that environmental funds or incentives can help stabilize the FP of firms. The firm can invest more money in energy-saving technologies or systems. Studying the relationship between FP and EP without considering bidirectional effects may also lead to misleading results and bias (Soytas et al., 2019). Therefore, it is necessary to explore the impact of FP on EP. Based on this, the following assumption is made:

H3: The better the company's financial performance, the more likely it is to invest in environmental management systems to improve EP.

Similarly, to further test the rationality and stability of the results, it is necessary to consider that the company may choose to invest in lean production to indirectly achieve the purpose of improving EP. Implementation of the IOS9001 product quality management system can also achieve good FP as the basis for the future implementation of the IOS14001 standard. This effect is more obvious, especially in developing countries (Zhu et al., 2013). The reason is that enterprises with high production efficiency need more capital investment to achieve industrial upgrading, while enterprises with poor environmental foundations can effectively improve their EP through optimization of production processes. However, investment decision-making is vulnerable to subjective influence. In the case analysed by Nehrt (1996), enterprises that make investment decisions on the environment, firstly, can get feedback of long-term benefits earlier, such as brand quality and customer loyalty. These enterprises can use the proceeds to make more efforts to improve their products and environment. The following assumptions are made as to whether a good financial situation can have an impact on the improvement of EP:

H4: The better the company's FP, the more likely it is to invest in a product quality management system to improve EP.

3.   Methodology

3.1   Samples and Data Source

According to the National Bureau of Statistics, China's energy-intensive industries include the manufacture of raw chemical materials and chemical products, manufacture of non-metallic mineral products, smelting, and pressing of ferrous metals (Yang et al., 2013). At the same time, general research shows that most activities that have a large impact on the environment come from the manufacturing industry. Therefore, the impact of CER of manufacturing enterprises on CFP is more convincing and reduces the interference of other activity factors on CFP (Yang et al., 2013). Besides, the manufacturing industry is more standardized on assessment of environmental pollution, and they prefer to increase profits by reducing pollution emissions (Nishitani et al., 2017). The study extracted 49 Chinese listed companies from the manufacturing industry (their code belongs to the manufacturing industry) from 2010 to 2020 as a sample, that is, a total of 539 observations. These samples were randomly selected to ensure credibility of the results. Before random selection, companies with missing data and companies with unstable business status were screened. Corresponding data information of the companies was collected through the CSMAR database. The data collected include companies that hold the environmental tool of the manufacturing listed company, as well as the company's annual market value, balance sheet, and statement of financial position and these data were further analysed and sorted out.

3.2   Variable Measurement

The enterprise environmental policy implementation standard that links environmental innovation, environmental policy, and economic benefits includes environmental management systems (EMS) and quality management systems (QMS), which can reduce resource waste and improve production efficiency by promoting the company's environmental innovation, to reduce costs. It can also indirectly stimulate innovation in eco-efficient products and services. Most companies' reports show that EMS have a positive impact on corporate environmental innovation (Rennings et al., 2006). For example, ISO14001 and ISO9001 can effectively reduce the emission of solid waste in the United States (Franchetti, 2011). According to this theory, the dummy variable is set, that is, whether the company holds an environmental tool, and the dummy variable of the company holding the certificate is set to 1; otherwise, it is 0. The reason for using this dummy variable is that the holding of certificates is more objective. Artificial evaluation of the CER degree of a company is affected by subjective factors, and it is difficult to unify the measurement standards. The environmental certificate issued by the third party can not only meet the unified evaluation criteria but also ensure timeliness (Testa & D'Amato, 2017). Corporate FP is measured through the EBITDA-to-total assets ratio and the PBV ratio of equity. EBITDA is the accounting indicator and PBV is the market indicator, which is used to ensure the comprehensiveness of measurement methods to avoid errors (King & Lenox, 2001). Return on investment (ROI) is easier to manipulate by managers, so it is excluded from the criteria of measurement, for example, by delaying new investments or continuing to use fully depreciated assets (Schneider, 2004). Similarly, in China, the China Securities Regulatory Commission (CSRC) sets many thresholds on ROE, so ROE is often manipulated and thus cannot ensure the authenticity of these two values (Wang & Wu, 2011).

3.3   Control Variables

To ensure that the research is not disturbed, it is necessary to set the company size as the control variable because large organizations are more capable of making organizational decisions combining environmental and economic benefits. At the same time, different company size industries show great differences in EP levels (Hitchens et al., 2005). This study determines the consistency of scale among sample companies according to revenue and debt-to-equity (D/E).

3.4   Model

The following model is according to Testa and D'Amato's (2017) two-way research on CER and CFP, to verify the effectiveness of the model in China and the relationship between CER and CFP (Testa & D'Amato, 2017).

$$\text{Performance}_{i,t} = \beta_i + \beta_1 \text{Certificate1}_{i,(t\text{-}i)} + \beta_y \theta_{i,t} + \beta_j \zeta + \varepsilon_{i,t} \tag{1}$$

$$\text{Performance}_{i,t} = \beta_i + \beta_1 \text{Certificate2}_{i,(t\text{-}i)} + \beta_y \theta_{i,t} + \beta_j \zeta + \varepsilon_{i,t} \tag{2}$$

In these two models, a panel model with cross-section is adopted, and the unobservable company heterogeneity that does not change with time is excluded through the year-fixed effect (assuming that the intercepts of $n-1$ are fixed) (Testa & D'Amato, 2017). The model attempts to explain corporate FP with the function of environmental certification and a vector of the covariate (θ) and year-fixed effects (ζ). To explore the causal relationship between independent variables and CFP, variables with a lag of one year should be used (Tebini et al., 2015).

$$\text{Prob}\left(\text{Certificate1}_{i,t}\right) = F\left(\beta_i + \beta_l \text{Performance}_{i,(t\text{-}i)} + \beta_y \theta_{i,t} + \beta_j \zeta\right) \tag{3}$$

$$\text{Prob}\left(\text{Certificate2}_{i,t}\right) = F\left(\beta_i + \beta_l \text{Performance}_{i,(t-i)} + \beta_y \theta_{i,t} + \beta_j \varsigma\right) \qquad (4)$$

The two models with the same premise as the previous model adopt a panel model with cross-section and assume that the intercepts of $n-1$ are fixed. The model tries to use the formula of FP and a vector of covariates (θ) and year-fixed effects (ς) to explain the possibility that the company has an environmental Certificate1/Certificate2. F(.) is a conditional logit function, and β_i is the fixed effects (Testa & D'Amato, 2017). The causality was explained by delaying the independent variable for one year (Tebini et al., 2015).

4.   Results and Findings

4.1   *Descriptive Statistics*

In this section, the statistical results of 49 listed companies from 2010 to 2020 are presented. Descriptive statistics of the variables in this study are presented in Table 6.1. Revenue, the control variable in the analysis, is used to represent the firm size, while the debit/equity ratio is used to represent firm financial leverage. Variables include EBITDA/TA and PBV ratio, which are used to represent financial performance, and Certificate1 and certificate2 represent ISO14001 and ISO9001, respectively. Table 6.1 also shows that there is no significant difference in the size of the selected company samples, but there are significant changes in FP, which is worthy of further research. However, due to the large dispersion of D/E, there is still a certain gap in the asset status of the company samples, which may have some impact on the follow-up results.

Table 6.2 preliminarily explores the correlation between variables through the correlation matrix. The results show that there is a low level of correlation between variables. Therefore, it is expected that the problem of multicollinearity is relatively light. At the same time, it also shows that there is a strong correlation between EBITDA/TA and PBV, and there is a significant correlation between ISO14001 and ISO9001. It can be preliminarily determined that hypotheses 3 and 4 can further verify hypotheses 1 and (.

To further eliminate the influence, Table 6.3 shows the results of the collinearity test for each variable. Although PBV and EBITDA/TA of enterprises are strongly related, the two control variables are related to each other. However, it can be seen from Table 6.3 that the variable inflation factor (VIF) of each variable is strictly

Table 6.1 Descriptive Statistics

Variable	Obs	Mean	Std. Dev.	Min	Max
EBITDA/TA	539	0.0825517	0.1173057	−2.027131	0.4520152
PBV	539	2.735629	2.162125	0.545354	16.35506
D/E	539	1.509142	3.168266	−18.73641	45.86648
lnRevenue	539	23.10871	1.359692	19.70084	27.5281

Table 6.2 Correlation Matrix: Pearson Coefficients

Variable	EBITDA/TA	PBV	Certificate1	Certificate2	D/E	lnRevenue
EBITDA/TA	1.000					
PBV	0.180***	1.000				
Certifificate1	−0.036	0.005	1.000			
Certifificate2	−0.116***	0.005	0.613***	1.000		
D/E	−0.084*	0.105**	0.016	0.047	1.000	
lnRevenue	−0.022	−0.259***	−0.207***	−0.200***	0.145***	1.000

Notes
* $p < 0.1$, ** $p < 0.05$, *** $p < 0.01$. Three asterisks indicate significance at the 1% level, and the greater the absolute value of the correlation coefficient, the stronger the correlation between the data.

The correlation coefficient matrix is only a preliminary estimate of the relationship between variables, and the final result is subject to regression.

Table 6.3 Tests of Multicollinearity: VIF

Variable	VIF	1/VIF
Certifificate1	1.62	0.615660
Certifificate2	1.62	0.616258
lnRevenue	1.08	0.925376
D/E	1.03	0.973108
Mean VIF	1.34	

less than 5, so there is no multicollinearity problem between variables. We can carry out further research (Hair, 1995).

4.2 Hypothesis Testing: H1 and H2

In this study, a panel model with cross-section and year-fixed effects is used to test the hypotheses. To make statistics and analysis of the data, Stata software is used. In the following results, $p < 0.05$ is taken as the standard of significant coefficients (p) (Dancey, 2007). In Table 6.4, H1 states that improvement of environmental management (ISO14001) enables a company to obtain higher performance and financial resources. Columns (A) and (B), respectively, reflect the influence of FP, represented by EBITDA/TA and PBV ratio, on obtaining an ISO14001 certificate. H2 assumes that improvement of the QMS (ISO9001) can also improve a company's FP. The impact of having an ISO9001 certificate on a company's performance is also presented in columns (C) and (D).

The results show that model (1) is significant (F-value = 6.545, $p < 0.01$ in column A and F-value = 22.874, $p < 0.01$ in column B). Table 6.4 shows that having ISO14001 has a significant positive correlation with the FP represented by PBV ($\beta = 0.300$, $p < 0.05$ in column B) but has no significant relationship with EBITDA/TA ($\beta = 0.002$, $p > 0.1$ in column A). This shows that companies with ISO14001 certificates have a significant positive impact on FP, which supports the hypothesis of model (1). Although there is a positive influence between the company's

Table 6.4 Results of Panel Regression with Cross-Section Fixed Effects

Dependent Variable = Firm Performance	(A)	(B)	(C)	(D)
	EBITDA/TA	PBV	EBITDA/TA	PBV
	Coeff. (t-value)	Coeff. (t-value)	Coeff. (t-value)	Coeff. (t-value)
L.Certifificate1	0.002 (0.18)	(2.43)		
L.Certifificate2			0.003 (0.32)	0.173*** (4.85)
D/E (leverage)	0.003 (0.85)	0.006 (0.76)	0.003 (0.84)	0.017 (1.38)
lnRevenue (firm size)	0.017 (1.19)	0.842** (2.36)	0.017 (1.20)	0.441** (2.29)
_cons	−0.290 (−0.88)	−16.760** (−2.05)	−0.290 (−0.89)	−14.371* (−1.98)
Year-fixed effects	Yes	Yes	Yes	Yes
Firm-fixed effects	Yes	Yes	Yes	Yes
N	490	490	490	490
R^2	0.0366	0.1795	0.0367	0.2025
F-value	6.545***	22.874***	6.472***	28.241***

t-statistic in parentheses.
* $p < 0.1$, ** $p < 0.05$, *** $p < 0.01$.

certificate and EBITDA/TA, there is no significant correlation. It may be because when taking EBITDA/TA as a variable, the impact of the cost of assets, borrowing, and tax payment on EBITDA is ignored, so the company's asset investment loss and borrowing expenditure do not have an impact on FP (Grant & Parker, 2002). On the other hand, the high financial leverage brings a heavy interest burden and credit risk to the company (Yang et al., 2013). The collected company data indicate that D/E, which represents firm financial leverage, has a large degree of dispersion. The influence of EBITDA on FP is limited. It also shows that the PBV ratio can more intuitively and stably represent a company's FP.

Model (2) is highly significant (F-value = 6.472, $p < 0.01$ in column C and F-value = 22.874, $p < 0.01$ in column D). Model (2) shows that ISO9001 has a significant positive coefficient on the PBV ratio but has only a very small positive impact on EBITDA/TA. The results verify H2. Using ISO9001 to manage the product quality of the company can improve the company's performance. At the same time, it further verifies that improving the product production quality to reduce the utilization and discharge of resources can improve the EP and FP of the company. There are no more than five missing PBV data between 2010 and 2020, and it does not have a great impact on the overall results.

4.3   *Hypothesis Testing: H3 and H4*

In hypothesis H3, the better the performance and financial resources, the more likely it is to invest in environmental management to improve the environment. It is further verified in hypothesis H4 that the better the financial performance, the more likely it is to invest in the improvement of product quality instruments. It is used to test the possibility of the bidirectional relationship between EP and FP.

Table 6.5 Results of Logit Fixed-Effects Regression

Dependent: Certificate (Yes/No)	(A)	(B)	(C)	(D)
	Certificate1	Certificate1	Certificate2	Certificate2
	Coeff. (t-value)	Coeff. (t-value)	Coeff. (t-value)	Coeff. (t-value)
L.EBITDA/TA	3.199 (1.47)		0.038 (0.05)	
L.PBV		0.142 (1.59)		0.046 (0.52)
DE	−0.016 (−0.31)	−0.015 (−0.29)	0.029 (0.70)	0.029 (0.70)
lnRevenue	−0.245 (−0.65)	−0.264 (−0.69)	0.383 (0.91)	0.358 (0.83)
_cons	6.284 (0.71)	6.428 (0.72)	−8.799 (−0.91)	−8.613 (−0.87)
Year-fixed effects	Yes	Yes	Yes	Yes
Firm-fixed effects	Yes	Yes	Yes	Yes
N	400	400	340	340
Pseudo R^2	0.2532	0.2542	0.2641	0.2645
Wald chi²	115.61***	114.86***	97.21***	96.46***

t-statistic in parentheses.
* $p < 0.1$, ** $p < 0.05$, *** $p < 0.01$.

Models (A) and (B) take ISO14001 as a dependent variable to explore the impact of FP on them. The independent variable FP is represented by the natural log of the EBITDA/TA ratio and the natural log of the PBV ratio, which lagged one year. In models (C) and (D), ISO9001 is used as the dependent variable. Table 6.5 presents the results of the panel regression approach to calculate the conditional logit function.

The results state that Equation (3) is significant (Wald test = 115.61, $p < 0.01$ for model A and Wald test = 114.86, $p < 0.01$ for model B). However, the FP expressed by EBITDA/TA and PBV, respectively, shows that they have no significant positive impact on the improvement of environmental management ($\beta = 3.199$, $p > 0.1$ in column A and $\beta = 0.142$, $p > 0.1$ in column B). Equation (4) is also significant (Wald test = 97.21, $p < 0.01$ for model C and Wald test = 96.46, $p < 0.01$ for model D). It is also verified again that there is no significant relationship between FP and the improvement of product quality management ($\beta = 0.038$, $p > 0.1$ in column A and $\beta = 0.046$, $p > 0.1$ in column B). These two results are inconsistent with the expected assumptions. On the contrary, the results show that the improvement of FP does not affect the improvement of environmental instruments. Therefore, summarizing the above results, the improvement of environmental management and product production quality management (obtaining certificates) can positively affect the improvement of the company's FP (represented by the PBV ratio). Improvement of FP does not have an impact on the company's certificate. The results do not support the hypothesis of bidirectional relationship between variables.

5. Robustness Test

To test the stability of causality, 10% of sample data were randomly deleted during the robustness test, and the results were tested again with panel regression and fixed cross-sectional effect, repeating the previous process. According to Lu and White (2014), empirical studies often re-examine the regressors by adding or removing them to confirm that the coefficients are reliable and robust. The results obtained by testing the regression coefficients are consistent with the results presented in Tables 6.4 and 6.5, and the four equations that can be verified are robust. Tables 6.6 and 6.7 show the data obtained after revalidation. The positive effect of certification on the PBV ratio remains significant ($\beta = 0.366$, $p < 0.1$ in column B;

Table 6.6 Robustness Test of Panel Regression with Cross-Section Fixed Effects

Variable	(A)	(B)	(C)	(D)
	EBITDA/TA	PBV	EBITDA/TA	PBV
	Coeff. (t-value)	Coeff. (t-value)	Coeff. (t-value)	Coeff. (t-value)
L.Certificate1	0.003 (0.25)	0.366* (1.86)		
L.Certificate2			−0.014 (−0.82)	0.234** (2.21)
D/E	0.008 (0.92)	0.101*** (6.36)	0.008 (0.92)	0.031*** (3.25)
lnRevenue	0.004 (0.20)	0.666** (2.37)	0.006 (0.27)	0.453* (1.87)
_cons	−0.003 (−0.01)	−12.855* (−1.99)	−0.025 (−0.05)	−9.342* (−1.92)
Year-fixed effects	Yes	Yes	Yes	Yes
Firm-fixed effects	Yes	Yes	Yes	Yes
N	329	329	329	329
R^2	0.0528	0.2324	0.0543	0.2401
F-value	3.897***	18.236***	3.850***	20.243***

t-statistic in parentheses.
* $p < 0.1$, ** $p < 0.05$, *** $p < 0.01$.

Table 6.7 Robustness Test of Logit Fixed Effect Regression

Variable	(A)	(B)	(C)	(D)
	Certificate1	Certificate1	Certificate2	Certificate2
	Coeff. (t-value)	Coeff. (t-value)	Coeff. (t-value)	Coeff. (t-value)
L.EBITDA/TA	3.465 (1.08)		−1.091 (−1.02)	
L.PBV		0.143 (1.01)		0.050 (0.44)
D/E	0.040 (0.50)	0.012 (1.14)	0.013 (0.20)	0.018 (0.28)
lnRevenue	−0.737 (−1.47)	−0.751* (−1.99)	−0.157 (−0.26)	−0.252 (−0.40)
_cons	17.922 (1.51)	23.985* (1.94)	2.757 (0.20)	4.421 (0.31)
Year-fixed effects		Yes	Yes	Yes
Firm-fixed effects	Yes	Yes	Yes	Yes
N	229	229	210	210
Pseudo R^2	0.2460	0.2314	0.2434	0.2422
Wald chi^2	61.09*	51.23*	54.18*	53.53*

t-statistic in parentheses.
* $p < 0.1$

$\beta = 0.234$, $p < 0.01$ in column D), and there is no significant correlation between the FP represented by EBITDA/TA and PBV ratio and the certificate holding of the company ($p > 0.1$ in Table 6.7). Therefore, the results still show that holding environmental tools can improve the company's market performance, while financial performance does not affect whether the company holds environmental certificates.

6. Discussion and Conclusion

6.1 Discussion

The specific impact of EP on FP has been extensively studied, with clearer insights emerging in recent years. However, research on the bidirectional relationship between these two areas is limited, leaving the nature of their interaction somewhat ambiguous. While many opinions exist regarding the direct effects on FP, our findings indicate that EP positively influences FP in the short term. Notably, we found no bidirectional relationship between the two variables, although it remains possible that strong FP could, in turn, enhance EP.

6.1.1 Positive Impact of Environmental Performance on Financial Performance

This study utilizes the CSMAR database to obtain financial statements of Chinese listed manufacturing companies from 2010 to 2020, focusing on their environmental certifications. We revisit the hypothetical model by Testa and D'Amato (2017) concerning certificates and FP. The findings indicate that for Chinese manufacturing firms, implementing an EMS or QMS has a positively influence on the PBV ratio. This suggests that strong EP can enhance future FP, contrasting with Testa and D'Amato's conclusions. Companies with ISO14001 or ISO9001 certification tend to adhere to strict management practices, leading to reduced pollutant emissions (Nishitani et al., 2017; Manrique & Martí-Ballester, 2017), resource sustainability (Gupta & Gupta, 2020), and product innovation (Manrique & Martí-Ballester, 2017). Consumers are often willing to pay a premium for green products, which can also result in tax benefits (Xu et al., 2021). While improvements in EP can positively impact FP to some extent (Shen et al., 2019), increased investment in environmental activities can have a negative effect on FP due to the associated costs. This indicates a limit to how much EP can enhance FP, as expenses exceeding a certain threshold may outweigh the benefits (Shen et al., 2019). The challenge lies in assessing the impact of holding certifications, as these do not reflect the level of investment in environmental initiatives. Testa and D'Amato (2017) studied manufacturing firms in a developed country, where companies may be at a more advanced stage of environmental management, necessitating significant capital investment in new technologies, which can lead to short-term declines in returns (Manrique & Martí-Ballester, 2017). Consequently, holding environmental certificates may not improve FP in recent years. Overall, while improvements in EP can positively affect FP, companies with better EP may struggle to enhance FP through environmental investments in the short term. The long-term benefits of continued investment in environmental improvements remain uncertain.

6.1.2   *Direct Influence of Environmental Performance on Financial Performance*

The relationship between EP and FP remains contentious, particularly regarding whether EP directly influences FP. Our findings indicate that EP has a direct positive impact on FP, although this effect may be negative in the short term. Improvements in EP can enhance productivity, reduce costs, and foster market expansion through innovation (Testa et al., 2011; Martín-de Castro, 2015). Most studies support this view, as evidenced by previous research. Conversely, some studies argue that growth in EP indirectly boosts FP by enhancing research and development efforts (Lioui & Sharma, 2012). Additionally, factors like sustainable development, corporate reputation, and customer satisfaction can strengthen a company's competitive advantage, which in turn positively affects FP (Cantele & Zardini, 2018). Notably, research by Lucato et al. (2017) suggest no significant relationship between EP and FP in small and medium-sized enterprises. Thus, it appears that direct and indirect impacts of EP on FP may coexist, influencing FP in both positive and negative ways (Lioui & Sharma, 2012). Furthermore, the direct relationship between EP and FP serves as a foundation for establishing a bidirectional relationship between the two variables (Hang et al., 2019).

6.1.3   *Bidirectional Relationship between EP and FP*

The study reveals that improvements in FP do not significantly enhance EP, contradicting the findings by Testa and D'Amato (2017), who argued that increased FP allows companies to invest surplus resources in environmental initiatives. According to the resource idle hypothesis (Kraft & Hage, 1990), additional financial resources should theoretically boost a company's EP (Hang et al., 2019). While companies may seek to optimize internal resources and enhance competitive advantage by investing in environmental activities, such actions require motivation from within the organization. When financial slack exists, companies might prioritize other investments over environmental improvements, thereby diverting funds away from sustainability efforts. Furthermore, the positive impact of FP appears to be short-lived, often occurring within just one year (Hang et al., 2019). Given that investments in environmental improvements involve costs, efforts to optimize production could negatively impact operational efficiency. Consequently, companies might prioritize short-term profits and stock prices over environmental investments (Preston & O'Bannon, 1997). The study did not account for these factors, which could influence the final outcomes. Therefore, although the results indicate no bidirectional relationship between EP and FP, the possibility of such a relationship should not be entirely dismissed.

6.2   **Conclusion**

This study verifies the bidirectional relationship between EP and FP, complements the lack of relevant literature, and is also helpful to the literature related to environmental management. In addition, previous studies in China have only verified the one-way relationship between EP and FP, and there are few relevant works in the literature. Firstly, this summary of the literature helps China to have a deeper understanding of the relationship between the two variables. Secondly, exploring the precise impact of EP and FP on each other can help companies make more effective decisions in the face of environmental problems. The empirical results provide practical significance for the positive impact of EP on FP in Chinese manufacturing enterprises. It can also help the state to implement environmental protection policies. This study takes the environmental tools as the basis of EP to explore the manufacturing enterprises' EP and FP under China's system. At the same time, it also shows that investment decision-making of the manufacturing industry on environmental responsibility is not only limited to the moral and legal levels but can also involve the consideration of obtaining economic benefits. The study explores a way of sustainable development of the company, that is, to jointly enhance the company's competitive advantage by paying attention to FP and EP. Verification of the relationship between EP and FP in this study is not comprehensive, and the long-term benefits of EP on FP need to be discussed more. In addition, the two-way impact between EP and FP needs further verification. The factors affecting FP by EP need to be more detailed to discuss the direct and indirect effects in detail.

References

Abban, O., & Hasan, S. (2021). Corporate sustainability and financial performance: The influence of environmental, social, and governance practices. *Journal of Business Ethics*, *162*(2), 419–434.

Angelia, D., & Suryaningsih, R. (2015). The effect of environmental performance and corporate social responsibility disclosure towards financial performance (Case study to manufacture, infrastructure, and service companies that listed at Indonesia Stock Exchange). *Procedia – Social and Behavioral Sciences*, *211*, 348–355.

Barney, J. (1991). Firm resources and sustained competitive advantage. *Journal of Management*, *17*(1), 99–120.

Bridgen, P. J., & Helm, N. (2017). Assessment of the value of ISO 14001 in improving environmental performance. In *ISO 14001* (pp. 273–284). Routledge.

Campos, L. M. S., Trierweiller, A. C., & Da Silva, A. L. (2015). Economic, environmental, and social performance indicators: An analysis of the Brazilian chemical industry. *Journal of Cleaner Production*, *96*, 339–347.

Cantele, S., & Zardini, A. (2018). Is sustainability a competitive advantage for small businesses? An empirical analysis of possible mediators in the sustainability–financial performance relationship. *Journal of Cleaner Production*, *182*, 166–176.

Cohen, M. A., Fenn, S., & Naimon, J. S. (1995). *Environmental and financial performance: Are they related?* Investor Responsibility Research Center, Environmental Information Service.

Dancey, C. P. (2007). *Statistics without maths for psychology*. Prentice Hall.

De, M., & Giri, B. C. (2020). Sustainable supply chain management: A review of literature and implications for future research. *Journal of Cleaner Production*, *252*, 119861.

Dieste, M., & Panizzolo, R. (2018). The role of lean manufacturing in the sustainable supply chain management. *Journal of Manufacturing Technology Management*, *29*(5), 852–872.

Du, S., Bhattacharya, C. B., & Sen, S. (2016). Corporate social responsibility and competitive advantage: Overcoming the trust barrier. *Strategic Management Journal*, *37*(6), 1228–1247.

Franchetti, M. (2011). ISO 14001 and solid waste generation rates in US manufacturing organizations: An analysis of relationship. *Journal of Cleaner Production, 19*(9–10), 1104–1109.

García-Sánchez, I. M., & Prado-Lorenzo, J. M. (2012). Corporate governance and disclosure of information on corporate social responsibility: An analysis of the top 200 universities in the Shanghai ranking. *Journal of Business Ethics, 107*(3), 349–370.

Gomez, J., & Rodriguez, R. (2011). Corporate environmental sustainability and financial performance: Evidence from the eco-industry. *Journal of Business Ethics, 104*(4), 573–581.

Grant, R. M., & Parker, L. (2002). Redefining competitive advantage in the knowledge-based economy. *Journal of Knowledge Management, 6*(3), 180–191.

Gupta, S., & Gupta, N. (2020). Corporate social responsibility and organizational performance: Evidence from India. *Journal of Business Research, 109*, 200–211.

Hair Jr, J. F. (1995). Marketing education in the 1990's: A chairperson's retrospective assessment and perspective. *Marketing Education Review, 5*(2), 1–6.

Hang, J., Zhang, D., Chen, P., Zhang, J., & Wang, B. (2019). Classification of plant leaf diseases based on improved convolutional neural network. *Sensors, 19*(19), 4161.

Hang, M., & Rathgeber, A. W. (2019). Corporate social responsibility and financial performance: The moderating role of the environment. *Journal of Business Ethics, 156*(1), 241–255.

Hart, S. L., & Dowell, G. (2011). A natural-resource-based view of the firm: Fifteen years after. *Journal of Management, 37*(5), 1464–1479.

Hitchens, D., Thankappan, S., & Clausen, J. (2005). Environmental performance, competitiveness, and management of small businesses in Europe. *Business Strategy and the Environment, 14*(1), 38–50.

Horváthová, E. (2010). Does environmental performance affect financial performance? A meta-analysis. *Ecological Economics, 70*(1), 52–59.

Inman, R. A. (2002). Managing quality improvement in manufacturing companies: The role of quality department. *International Journal of Quality & Reliability Management, 19*.

Jo, H., Kim, H., & Park, K. (2015). Corporate environmental responsibility and firm performance in the financial services sector. *Journal of Business Ethics, 131*, 257–284.

Johnstone, L., & Hallberg, P. (2020). ISO 14001 adoption and environmental performance in small to medium sized enterprises. *Journal of Environmental Management, 266*, 110592.

Johnstone, P. T. (2020). Vietoris locales and localic semilattices. In *Continuous lattices and their applications* (pp. 155–180). CRC Press.

King, A. A., & Lenox, M. J. (2001). Does it really pay to be green? An empirical study of firm environmental and financial performance: An empirical study of firm environmental and financial performance. *Journal of Industrial Ecology, 5*(1), 105–116.

Kock, C. J., Santaló, J., & Diestre, L. (2012). Corporate governance and the environment: What type of governance creates greener companies?. *Journal of Management Studies, 49*(3), 492–514.

Kraft, K. L., & Hage, J. (1990). Strategy, social responsibility and implementation. *Journal of Business Ethics, 9*, 11–19.

Lahouel, B. B., Bruna, M. G., & Zaied, Y. B. (2020). The curvilinear relationship between environmental performance and financial performance: An investigation of listed french firms using panel smooth transition model. *Finance Research Letters, 35*, 101455.

Li, D., Cao, C., Zhang, L., Chen, X., Ren, S., & Zhao, Y. (2017). Effects of corporate environmental responsibility on financial performance: The moderating role of government regulation and organizational slack. *Journal of Cleaner Production, 166*, 1323–1334.

Lioui, A., & Sharma, Z. (2012). Environmental corporate social responsibility and financial performance: Disentangling direct and indirect effects. *Ecological Economics, 78*, 100–111.

Lu, X., & White, H. (2014). Robustness checks and robustness tests in applied economics. *Journal of Econometrics, 178*, 194–206.

Lucato, W. C., Costa, E. M., & de Oliveira Neto, G. C. (2017). The environmental performance of SMEs in the Brazilian textile industry and the relationship with their financial performance. *Journal of Environmental Management, 203*, 550–556.

Manrique, S., & Martí-Ballester, C. P. (2017). Analyzing the effect of corporate environmental performance on corporate financial performance in developed and developing countries. *Sustainability, 9*(11), 1957.

Martín-de Castro, G. (2015). Knowledge management and innovation in knowledge-based and high-tech industrial markets: The role of openness and absorptive capacity. *Industrial Marketing Management, 47*, 143–146.

McGuinness, P. B., Vieito, J. P., & Wang, M. (2017). The role of board gender and foreign ownership in the CSR performance of Chinese listed firms. *Journal of Corporate Finance, 42*, 75–99.

Meng, X. H., Zeng, S. X., Shi, J. J., Qi, G. Y., & Zhang, Z. B. (2014). The relationship between corporate environmental performance and environmental disclosure: An empirical study in China. *Journal of Environmental Management, 145*, 357–367.

Mohammed, M. (2000). The ISO 14001 EMS implementation process and its implications: A case study of Central Japan. *Environmental Management, 25*(2).

Nehrt, C. (1996). Timing and intensity effects of environmental investments. *Strategic Management Journal, 17*(7), 535–547.

Nga, T. T., Winichagoon, P., Dijkhuizen, M. A., Khan, N. C., Wasantwisut, E., Furr, H., & Wieringa, F. T. (2009). Multi-micronutrient-fortified biscuits decreased prevalence of anemia and improved micronutrient status and effectiveness of deworming in rural Vietnamese school children. *The Journal of Nutrition, 139*(5), 1013–1021.

Nguyen, Q. A., & Hens, L. (2015). Environmental performance of the cement industry in Vietnam: The influence of ISO 14001 certification. *Journal of Cleaner Production, 96*, 362–378.

Nguyen, T., Novak, R., Xiao, L., & Lee, J. (2021). Dataset distillation with infinitely wide convolutional networks. *Advances in Neural Information Processing Systems, 34*, 5186–5198.

Nishitani, K., Jannah, N., & Kaneko, S. (2017). Does corporate environmental performance enhance financial performance? An empirical study of Indonesian firms. *Environmental Development, 23*, 10–21.

Nishitani, K., Kaneko, S., Fujii, H., & Komatsu, S. (2011). Effects of the reduction of pollution emissions on the economic performance of firms: An empirical analysis focusing on demand and productivity. *Journal of Cleaner Production, 19*(17–18), 1956–1964.

Osaki, K. (2019). US Energy Information Administration (EIA): 2019 edition US annual energy outlook report (AEO2019). *Haikan Gijutsu, 61*(8), 32–43.

Potoski, M., & Prakash, A. (2005). Green clubs and voluntary governance: ISO 14001 and firms' regulatory compliance. *American Journal of Political Science, 49*(2), 235–248.

Preston, L. E., & O'bannon, D. P. (1997). The corporate social-financial performance relationship: A typology and analysis. *Business & Society, 36*(4), 419–429.

Rahman, M. J., & Wu, J. (2023). M&A activity and ESG performance: Evidence from China. *Managerial Finance, 50*(1), 179–197.

Rahman, M. J., Wu, Q., & Zhu, H. (2024). Corporate social responsibility in times of social distancing: Evidence from China. *Business Ethics, the Environment & Responsibility*.

Rahman, M. J., Zhu, H., & Chen, S. (2023). Does CSR reduce financial distress? Moderating effect of firm characteristics, auditor characteristics, and Covid-19. *International Journal of Accounting & Information Management, 31*(5), 756–784.

Rennings, K. (2000). Redefining innovation – eco-innovation research and the contribution from ecological economics. *Ecological Economics, 32*(2), 319–332.

Rennings, K., Ziegler, A., Ankele, K., & Hoffmann, E. (2006). The influence of different characteristics of the EU environmental management and auditing scheme on technical environmental innovations and economic performance. *Ecological Economics, 57*(1), 45–59.

Salama, A. (2005). A note on the impact of environmental performance on financial performance. *Structural Change and Economic Dynamics, 16*(3), 413–421.

Samy, D. M., Ismail, C. A., & Nassra, R. A. (2015). Circulating irisin concentrations in rat models of thyroid dysfunction—effect of exercise. *Metabolism, 64*(7), 804–813.

Schneider, B. R. (2004). Business politics and the state in twentieth-century Latin America. Cambridge University Press.

Shen, M., Zhang, Y., Zhu, Y., Song, B., Zeng, G., Hu, D., . . . & Ren, X. (2019). Recent advances in toxicological research of nanoplastics in the environment: A review. *Environmental Pollution, 252,* 511–521.

Soytas, M. A., Denizel, M., & Usar, D. D. (2019). Addressing endogeneity in the causal relationship between sustainability and financial performance. *International Journal of Production Economics, 210,* 56–71.

Tebini, H., M'Zali, B., Lang, P., & Méndez-Rodríguez, P. (2015). Social performance and financial performance: A controversial relationship. In Ballestro, E.; Pérez-Gladish, B. and Garcia-Bernabeu, A. (Eds.), *Socially responsible investment: A multi-criteria decision making approach,* Springer, chapter 3, 57–80.

Testa, F., Iraldo, F., & Frey, M. (2011). The effect of environmental regulation on firms' competitive performance: The case of the building & construction sector in some EU regions. *Journal of Environmental Management, 92*(9), 2136–2144.

Testa, M., & D'Amato, A. (2017). Corporate environmental responsibility and financial performance: Does bidirectional causality work? Empirical evidence from the manufacturing industry. *Social Responsibility Journal, 13*(2), 221–234.

Trumpp, C., & Guenther, T. (2017). Too little or too much? Exploring U-shaped relationships between corporate environmental performance and corporate financial performance. *Business Strategy and the Environment, 26*(1), 49–68.

Wagner, M., & Blom, J. (2011). The reciprocal and non-linear relationship of sustainability and financial performance. *Business Ethics: A European Review, 20*(4), 418–432.

Wang, X., & Wu, M. (2011). The quality of financial reporting in China: An examination from an accounting restatement perspective. *China Journal of Accounting Research, 4*(4), 167–196.

Xu, J., Wang, X., & Liu, F. (2021). Government subsidies, R&D investment and innovation performance: analysis from pharmaceutical sector in China. *Technology Analysis & Strategic Management, 33*(5), 535–553.

Yang, G., Wang, Y., Zeng, Y., Gao, G. F., Liang, X., Zhou, M., ... & Murray, C. J. (2013). Rapid health transition in China, 1990–2010: Findings from the Global Burden of Disease Study 2010. *The Lancet, 381*(9882), 1987–2015.

Zhang, J., Lu, H., Zeng, H., Zhang, S., Du, Q., Jiang, T., & Du, B. (2020). The differential psychological distress of populations affected by the COVID-19 pandemic. *Brain, Behavior, and Immunity, 87,* 49.

Zhu, H., & Rahman, M. J. (2024). Ex-ante expected changes in ESG and future stock returns based on machine learning. *The British Accounting Review,* 101457.

Zhu, Q., Sarkis, J., & Lai, K. H. (2013). Institutional-based antecedents and performance outcomes of internal and external green supply chain management practices. *Journal of Purchasing and Supply Management, 19*(2), 106–117.

7 Does ESG Performance Have an Impact on Financial Performance? Moderating Effect of Firm Nature

Md Jahidur Rahman, Tarek Rana, Hongtao Zhu, and Jiayi Chen

1. Introduction

This work studies the relationship between enterprise environmental, social, and governance (ESG) performance and financial performance, as well as the moderating effect of different enterprise natures on the relationship between ESG and financial performance. ESG is a new enterprise evaluation standard. Different from traditional financial indicators, ESG indicators assess the sustainability of enterprise operation and the impact of social values from the perspective of environment, society, and corporate governance (Hoang, 2018). This study uses the ESG score in the Bloomberg database to evaluate an enterprise's ESG performance (Zhang & Jin, 2022) and uses economic value added (EVA) to measure the enterprise's financial performance (Atan et al., 2018). For the different natures of enterprises, this study divides it into two categories, state-owned enterprises (SOEs) and non-state-owned enterprises (non-SOEs).

There are three reasons for making this empirical study. Firstly, the development of enterprises, the improvement of the natural environment, and the coordination and unification of social development as indispensable parts of the sustainable development of enterprises (Kot, 2018). Paying full attention to social and environmental benefits can enhance the reputation and brand benefits of enterprises (Du et al., 2010; Willard, 2012). ESG, on the other hand, quantifies the spirit of sustainable development with indicators as a phased reference to promote enterprises to go further on the road of sustainable development (Liang et al., 2022).

In addition. although how to promote the sustainable development of enterprises is an important strategic issue faced by enterprises in the 21st century, foreign scholars have made some research achievements on whether and how the performance of enterprises in the ESG affects the financial performance of enterprises (Wang & Sun, 2022), while Chinese scholars have made little research on this issue and have not reached a general conclusion (Tian et al., 2021). In the few studies on the relationship between the two, scholars have not reached a consensus research conclusion. The reasons are manifold. This may be due to the diversity of the evaluation methods adopted by scholars, different samples selected, differences in indicators set, etc. (Wang & Sun, 2022).

DOI: 10.4324/9781003488965-9

Finally, for a long time, many scholars did not make any distinction between sustainable development of SOEs and non-SOEs, which are two kinds of enterprises with different natures. Scholars often think that the performance of the two types of enterprises in ESG is unified, but it is not true (Huang et al., 2020). Moreover, there are not many scholars involved in the research on the relationship between enterprise ESG performance and financial performance under different enterprise natures. In fact, some literature shows that the effect of ESG is affected by agent heterogeneity, and enterprises with different ownership backgrounds and industrial backgrounds have different responses to ESG (Deng & Cheng, 2019). Especially for emerging market countries, because of the gap between the governance environment, property rights system, and mature capital markets, it is easy to apply the research conclusions of the latter directly to the former (Martins, 2022).

Drawing on the past literature, this study believes that the ESG performance of enterprises is conducive to improving their financial performance. For example, Ruggiero and Cupertino (2018) found that when corporate governance performance is excellent, there is a positive correlation between social responsibility and financial performance. Dyck et al. (2018) noted that the stock return of companies with higher social capital during a financial crisis is fairly high, which means that social responsibility and environmental activities are risk management tools.

Moreover, as for the moderating role of the nature of enterprises, according to foreign research, because the number of SOEs in Western countries is small and they play a more regulatory role in the market economy system to make up for market failure, Western scholars rarely focus on the social responsibility of SOEs (Stan et al., 2014). The research on the relationship between enterprises' ESG responsibility and corporate financial performance is basically based on non-SOEs, and the conclusions are also based on the conclusions of non-SOEss (Ruan & Liu, 2021). However, the SOEs in China have been endowed with strong social responsibility and corporate governance ability since their establishment. They have taken on important functions such as making up for market failure, macro-control, and resource allocation, rather than being established solely for "economic goals" (Tian & Ma, 2009). Therefore, the moderating effect of different enterprise natures on the relationship between ESG and financial performance is worth further discussing.

Therefore, for this study, all A-share listed companies in China were taken as samples and data summarized. This work analyses previous studies on ESG and financial performance, enterprises with different ownership backgrounds and ESG; puts forward relevant assumptions, and explores the relationship among ESG, financial performance, and enterprises with different natures through empirical analysis. The results of this study show that there is a significant positive relationship between ESG and corporate financial performance. This means that ESG has a positive role in promoting financial performance. In addition, the results of this study also show that compared with SOEs, non-SOEs play a more significant role in promoting the relationship between ESG and financial performance. In addition, this study uses the Heckman two-stage model to alleviate the endogenous problem of variables and achieves the same conclusion as the main result according to the assumptions and model regressions. The robustness test also studies the relationship between ESG and financial performance in the computer, communication, and electronic equipment manufacturing industry, as well as the role of enterprise nature. The robustness test results are still the same as the main results, which further verifies the stability of the conclusion.

This study has made the following contributions. Firstly, by comparing the developmental logic of ESG at home and abroad, this study sorts out a series of domestic-related documents and policies and conducts regression analysis on the relationship between ESG and financial performance (EVA) of 704 companies from 2014 to 2020. The results show that ESG can promote the financial performance of enterprises, which provides a reference for the ESG evaluation system of domestic enterprises. In addition, domestic scholars mostly use the ESG score of Sino-Securities and SynTao Green Finance databases for empirical analysis, while this study uses Bloomberg's ESG score to further consolidate the hypothesis and research of previous scholars on the role of ESG in promoting financial performance.

Secondly, previous scholars mostly use return on assets (ROA), Tobin's Q, and other indicators to measure the financial performance of enterprises when studying the relationship between ESG and financial performance. This study creatively uses EVA to represent financial performance and further confirms the positive relationship between ESG and financial performance in empirical regression analysis. The confirmation of the conclusion provides a theoretical basis for the follow-up research.

Thirdly, this study also discusses the moderating effect of different enterprise natures on the relationship between ESG and financial performance. The results show that non-SOEs play a better role in regulating ESG and financial performance than SOEs. This makes up for the gap in the research on the relationship between the three in China and provides the government with better consideration of the differences between enterprises when providing policy guarantees, so as to improve the sustainable development level of enterprises with different ownership backgrounds.

Finally, this study provides data support for corporate governance of firm managers and investment decisions of external investors. ESG promotion of a company's financial performance will encourage managers to better manage the company's ESG performance, and external investors will also tend to invest in companies with good ESG performance. This has a two-way promoting effect on ESG and financial performance.

The rest of this study is organized as follows. Section 2 introduces the development background of ESG at home and abroad. Section 3 introduces the literature review and hypothesis development. Section 4 describes the samples and models. Section 5 describes the main results, endogeneity test, and robustness test. Section 6 introduces the discussion. Finally, Section 7 draws the conclusion.

2.　Background

2.1　International Development of ESG

ESG has its roots in ethical investment practices that emerged in the United States in the 18th century (Auer & Schuhmacher, 2016). Initially, ethical investment aimed to influence capital market activities through religious doctrines, address-ing issues like anti-slave trade, gambling, and arms smuggling (Leins, 2020). However, this concept primarily circulated within religious organizations and did not significantly impact mainstream investment practices (Morales et al., 2019). Following World War II, global industrialization accelerated, creating a more complex and volatile economic and political landscape (Held et al., 2020). By the 1960s and 1970s, social and political movements, such as the South African Anti-Apartheid Movement and the American Civil Rights Movement, raised pub-lic awareness of corporate environmental and social responsibilities, expanding ethical investment into broader social responsibility investment (Oehmke & Opp, 2024). In 2004, the United Nations Global Compact, in collaboration with 20 financial institutions, released the report "Who Cares Wins", which introduced the ESG concept (Sethi & Schepers, 2014). The report emphasized that compa-nies excelling in ESG not only contribute to societal sustainable development but also enhance shareholder value through effective risk management and regulatory foresight (Cole, 2020). Since 2020, the COVID-19 pandemic has significantly impacted the global economy, presenting severe challenges, including the ability of companies to manage risks and the increasing drive for investors to focus on ESG (Gadinis & Miazad, 2020).

3.   Literature Review and Hypothesis Development

3.1   ESG and Financial Performance

Many scholars have explored the correlation between ESG performance and finan-cial performance. Friede et al. (2015) reviewed over 2,000 studies, finding out that about 90% reported a non-negative correlation between ESG factors and financial performance. As time progresses, this positive correlation appears to strengthen. Nollet et al. (2016) conducted regression analyses, revealing a significant nega-tive correlation between corporate social performance (CSP) and return on capital (ROC) in their linear model, while a non-linear model indicated a U-shaped rela-tionship with accounting-based financial performance, suggesting that CSP yields positive long-term effects when sufficient resources are invested.

Garcia et al. (2017) analysed ESG and financial performance in sensitive indus-tries within Brazil, Russia, India, and China, finding that environmental governance correlates with systematic risk in a reverse U-curve, indicating peak performance at certain ESG levels. Sharma et al. (2020) applied panel data analysis to Indian companies, concluding that strong ESG performance enhances financial outcomes.

Dalal and Thaker (2019) used random effect panel data regression to demon-strate that good ESG performance improves financial performance through both accounting and market measures, emphasizing the importance of sustainability reporting. Chouaibi et al. (2021) found that strong environmental disclosure en-hances financial performance, while weak disclosure diminishes it, with social and ethical practices moderating this relationship. Yilmaz (2021) noted that while the total ESG score significantly affects financial performance, individual E, S, and G scores do not, suggesting a collective impact.

Kim et al. (2013) studied South Korean companies, concluding a positive correlation between social responsibility and financial performance. Breuer and Nau (2014) found a significant positive correlation in the US technology sector, although the causal relationship remains inconclusive. Ortas et al. (2015) confirmed a positive impact of ESG on financial performance in companies across Spain, France, and Japan.

Fatemi et al. (2018) showed that ESG positively correlates with company value, indicating that increased ESG investment enhances value. Limkriangkrai et al. (2017) found that ESG investment reduces corporate risks and enhances performance in Australian firms. Sinha et al. (2019) highlighted a positive correlation between ESG and financial performance in financial enterprises, advocating for its inclusion in strategic decision-making. Busco et al. (2020) reported that companies employing the ESG framework generally perform better financially.

Conversely, some studies suggest a lack of a significant positive correlation. Amel-Zadeh and Serafeim (2018) found a negative correlation between ESG and enterprise value, while Sassen et al. (2016) noted that strong ESG performance could reduce enterprise value. Atan et al. (2018) also found no significant correlation between ESG and return on equity (ROE). Several studies in China also investigate the relationship between environmental accounting and firm performance (Rahman & Wu, 2023; Rahman, Wu et al., 2024; Rahman, Zhu, & Chen, 2023; Zhu & Rahman, 2024).

In summary, while scholars have yet to reach a definitive conclusion on the ESG–financial performance relationship, the majority lean towards a positive correlation. Based on these findings, this study proposes the following hypothesis:

H1: There is a positive relationship between ESG and enterprise's financial performance.

3.2 The Moderating Effect of Different Enterprise Natures on the Relationship between ESG and Financial Performance

Corporate social responsibility (CSR) is shaped by external pressures and constraints (Campopiano & De, 2015). Song (2018) noted that during China's planned economy era, enterprises were government-owned and primarily focused on fulfilling state-directed tasks, with little awareness of CSR. Over decades of economic reform, SOEs have shifted towards profit orientation but still retain ties to the national political system. In contrast, non- SOEs are established primarily for profit and may leverage CSR as a business strategy to gain political legitimacy and access to government resources (Long et al., 2020).

Peterson and Pfitzer (2008) argued that companies utilizing government relations for CSR activities can effectively address social issues due to better access to information and special advantages from these political connections. Cheung (2015) found that SOEs are generally less active in ESG practices compared to non-SOEs, as they often rely on continuous government financial support and are less influenced by external stakeholders. For private enterprises, the costs associated with fulfilling CSR obligations tend to be lower than for SOEs, making them more conscious of the benefits of these actions (Huang & Zhao, 2016). Recent studies suggest that as private enterprises increasingly focus on CSR, their sense of responsibility and performance improve (Fontaine, 2013).

In summary, existing research indicates that non-SOEs are more attentive to ESG performance than SOEs. Therefore, this study proposes the following hypothesis:

H2: Compared with SOEs, non-SOEs have a more positive moderating effect on the relationship between ESG and financial performance.

4.   Research Design

4.1   Sample Selection

This study takes all A-share listed companies in China as research samples. In order to ensure data consistency, this study uses the annual reports disclosed by enterprises from 2014 to 2020 and continuous ESG rating data for analysis. Relevant financial data of the listed companies were selected from the CSMAR database, and ESG scoring data were selected from the Bloomberg database. To make the sample more standardized and the research results more authentic, companies with incomplete data on relevant dependent variables, independent variables, and control variables were deleted, as well as the data of special treatment (ST) and ST*[1] companies. Finally, relevant data from 704 sample companies, with a total of 4,928 firm-year observations, were collected. This study uses Microsoft Excel to filter, categorize, and match the data, and the data are sorted into complete sectional and panel data. In the end, regression analysis is conducted with Stata17 to test the research hypothesis. Table 7.1 shows the sample selection process.

4.2   Model Specification

To test the relationship between ESG and financial performance in H1, this study uses the following model:

$$\begin{aligned}
EVA_{i,t} = {} & \alpha_0 + \alpha_1 ESG_{i,t} + \alpha_2 Size_{i,t} + \alpha_3 Lev_{i,t} + \alpha_4 Growth_{i,t} + \alpha_5 Age_{i,t} \\
& + \alpha_6 Big4_{i,t} + \alpha_7 Top1_{i,t} + \alpha_8 BOD_Ind_{i,t} + \alpha_9 Salary_{i,t} \\
& + \alpha_{10} BOD_Gender_{i,t} + \alpha_{11} Industry_{i,t} + \alpha_{12} Year_{i,t} + \varepsilon_1
\end{aligned} \tag{1}$$

Table 7.1 Sample Selection

Total A-Share Listed Firms from 2014 to 2020 in China	*5,222*
Exclusions:	
ST and ST* share companies	190
Total number excluded	190
Sample pool	5,032
Less:	
Missing ESG data	4,188
Missing financial data	137
Missing SOE data	3
Final sample firms	704

To further test the moderating effect of enterprises with different natures on the relationship between ESG and financial performance in H2, model 2 adds the moderating variable ESG*SOE:

$$\begin{aligned}
EVA_{i,t} = {} & \alpha_0 + \alpha_1 ESG_{i,t} + \alpha_2 SOE_{i,t} + \alpha_3 ESG\text{*}SOE_{i,t} + \alpha_4 Size_{i,t} \\
& + \alpha_5 Lev_{i,t} + \alpha_6\, Growth_{i,t} + \alpha_7 Age_{i,t} + \alpha_8 Big4_{i,t} + \alpha_9 Top1_{i,t} \\
& + \alpha_{10} BOD_Ind_{i,t} + \alpha_{11} Salary_{i,t} + \alpha_{12} BOD_Gender_{i,t} \\
& + \alpha_{13} Industry_{i,t} + \alpha_{14} Year_{i,t} + \varepsilon 1
\end{aligned} \tag{2}$$

4.2.1 Dependent Variable

The dependent variable in this study is the company's financial performance measured by EVA. Many previous scholars used ROA and Tobin's Q to measure the financial performance of enterprises and study the relationship between them and ESG. However, Gregory (2021) claimed that because E, S, and G standards affect productivity and debt costs, this has caused measurement errors in the dependent variables of regression ROA and Tobin's Q on E, S, and G standards, which has led to biased estimates and exaggerated standard errors. According to Dumitrascu (2014), to examine the actual value or profit that shareholders should obtain, EVA measurement is the best method to measure the overall efficiency of enterprise assets. EVA is the assessment of the company's annual profit, which represents the residual income after covering all the opportunity costs of capital (Niresh & Alfred, 2014). ESG disclosure practices need to be reviewed to determine whether they bring value to a company and its shareholders.

4.2.2 Independent Variables

ESG indicators assess the sustainability of enterprise operations and their impact on social value from three dimensions of environment, society, and corporate governance (Hoang, 2018). Based on Saini et al.'s (2022) study, the enterprise ESG score in this study is selected from the Bloomberg database, with a full score of 100. A higher score indicates a better enterprise ESG performance (Zhang & Jin, 2022).

In model 2, the new independent variable SOE and the cross-multiplication term ESG*SOE are added. SOE is a dummy variable, where 1 represents SOEs and 0 represents non-SOEs (Pan et al., 2022). ESG*SOE can measure the impact of different enterprise natures on ESG performance (Bing & Li, 2019).

4.2.3 Control Variables

As the dependent variables in this study reflect the financial performance of enterprises, the relevant factors that may affect the financial performance are controlled in the study. Enterprise size (Size) has a significant impact on the financial performance of enterprises. Larger enterprises have better financial performance than small and medium-sized enterprises (Garay & Font, 2012). Financial leverage (Lev) is a means by which an enterprise uses liabilities to adjust its equity capital gains (Akhtar et al.,

2012). And a high financial leverage indicates the significate pressure on an enterprise to repay its creditors and ensure good economic performance, which will have a negative impact on its financial performance. Whether the company is audited by the Big Four accounting firms (Big4) is also one of the critical factors that affect the financial performance of enterprises. Generally, the enterprises audited by the Big Four accounting firms will have more satisfactory financial performance than other enterprises (Palepu et al., 2020). In addition, the shareholding ratio of the largest shareholder (Top1) has a specific positive impact on corporate governance. Companies with more concentrated equity have a positive role in promoting financial performance (Chen & Zhang, 2014). Table 7.2 shows the variable measurements.

Table 7.2 Variable Measurements

Variable	Measurement	Reference	Data Source
EVA	Used to measure the enterprise's financial performance, calculated as (after-tax operating profit minus debt and equity costs)/100,000,000; it is the residual income after all costs are deducted	Atan et al. (2016)	CSMAR
ESG	Firm performance of environmental, social and governance, measured by ESG score from the Bloomberg database	Zhang & Jin (2022)	Bloomberg
SOE	Nature of corporate property rights as dummy variables to deal with and sets the value of SOEs as 1 and non-SOEs as 0	Khalid et al. (2021), Zhu & Rahman (2024)	CSMAR
Size	Enterprise size, measured by the amount of total assets/1,000,000	Zhang & Jin (2020), Rahman & Zhu (2024)	CSMAR
Lev	Financial leverage, calculated as total liabilities divided by the total assets	Zhang & Jin (2022), Rahman & Zhu (2023)	CSMAR
Growth	Annual change in total sales, calculated as (the sales amount of the enterprise in the current year – the sales amount of the previous year)/the sales amount of the previous year	Khalid et al. (2021)	CSMAR
Age	Value obtained by the subtraction of the current year from the year of the company's listing	Khalid et al. (2021), Rahman, Zhu, Zhang et al. (2024)	CSMAR
Big4	Dummy variable, 1 if the company is audited by Big Four accounting firms and 0 if otherwise	Zhou et al. (2022), Rahman, Zhu, & Yue (2024), Rahman, Zhu, & Jiang (2024)	CSMAR

(Continued)

Table 7.2 (Continued)

Variable	Measurement	Reference	Data Source
Top1	Shareholding ratio of the largest shareholder, calculated as (number of shares held by the largest shareholder/total number of shares) * 100	Zhou et al. (2022)	CSMAR
BOD_Ind	Independence of board, calculated as the number of independent directors divided by the number of directors	Zhou et al. (2022)	CSMAR
Salary	Amount of executive compensation, measured by executive compensation divided by 1,000,000	Deng & Cheng (2019)	CSMAR
BOD_Gender	Percentage of female on the board, calculated as the number of female directors divided by the number of directors.	Zhou et al. (2022)	CSMAR
Industry	Dummy variable, 1 if the company operates in industry i and 0 if otherwise	Zhang & Jin (2022)	CSMAR
Year	Dummy variable, 1 if in year i and 0 if otherwise	Zhang & Jin (2022)	CSMAR

5. Results

5.1 Descriptive Statistics

Table 7.3 presents the descriptive statistics of each variable. Among the 4,928 observations, the average value of financial performance (EVA) is 6.132, while the standard deviation is 68.405. The maximum value of EVA is 1,194.803, and the minimum is −705.390. The median value of EVA is 0.312, showing that the financial performance of most sample enterprises is good. The average value of ESG is 22.706, of which the minimum value is 1.240, the maximum value is 60.744, and the standard deviation is 7.449. This shows that different enterprises have significant differences in the performance of ESG, and the vast majority have poor performance and low ESG scores. Enterprises still need to explore more strategies to improve their ESG performance. The average value of SOE is 0.602, with a median of 1, indicating that the number of SOEs in the sample is more than that of non-SOEs.

The maximum value of Size is 33,345,058, while the minimum value is 279.122. The standard deviation of Size is 1,852,907.7, showing that there is a large gap between the total assets of the sample enterprises, and the scale of enterprises is relatively unbalanced. The standard deviation of financial leverage (Lev) is 0.212, which indicates that the dispersion of financial leverage of sample enterprises is small. The maximum value of Lev is 1.698, and the minimum value is 0.009, indicating that different enterprises have different risk preferences and risk resistance

Table 7.3 Summary Statistics

Variable	Obs	Mean	SD	Min	Median	Max
EVA	4,928	6.132	68.405	−705.390	0.312	1,194.803
ESG	4,928	22.706	7.449	1.240	21.488	60.744
SOE	4,928	0.602	0.489	0.000	1.000	1.000
Size	4,928	271,725.7	1,852,907.7	279.122	14,488.065	33,345,058
Lev	4,928	0.505	0.212	0.009	0.513	1.698
Growth	4,928	0.145	1.568	−0.875	0.075	102.635
Age	4,928	14.868	6.126	3.000	15.000	30.000
Big4	4,928	0.166	0.372	0.000	0.000	1.000
Top1	4,928	36.939	15.902	3.000	35.660	87.460
BOD_Ind	4,928	0.378	0.060	0.200	0.364	0.800
Salary	4,928	7.360	9.456	0.000	4.500	142.122
BOD_Gender	4,928	0.136	0.126	0.000	0.111	0.778

capabilities. The mean value of Growth is 0.145, the maximum value is 102.635, and the minimum value is −0.875. This means a significant gap in the annual sales growth rate of the sample enterprises. Some sample companies may be affected by the market and other factors to reduce the sales increment. The average value of Big4 is 0.166, with a median of 0, showing that most enterprises are not audited by the Big Four accounting firms but by other accounting firms.

5.2　Correlation Analysis

Table 7.4 shows the correlation analysis of variables. Through correlation analysis, this study found that there is a significant positive relationship of 0.103 between ESG and financial performance (EVA) at the 1% level. Enterprise nature (SOE), the scale of the enterprise (Size), financial leverage (Lev), whether it is audited by the Big Four accounting firms (Big4), the shareholding ratio of the largest shareholder (Top1), the independence of board (BOD_Ind), and executive compensation (Salary) are also significantly positively correlated with EVA at different levels. The firm listing year (Age) is negatively related to financial performance at the 1% level.

5.3　Main Results

In this study, the individual effects of the two hypotheses are determined by the Hausman test, and the fixed-effect (FE) model and the random effect (RE) model are compared. According to the Hausman test, the regression result of the FE model is better, so for this study, we selected the FE model to test the hypothesis. The regression results are shown in Table 7.5.

5.3.1　Result for Hypothesis 1

Model 1 shows the impact of ESG on financial performance (EVA). According to the empirical results of model 1, the ESG performance in the sample companies is significantly positively correlated with financial performance (EVA), with a regression

Table 7.4 Correlation Matrix

Variable	EVA	ESG	SOE	Size	Lev	Growth	Age
EVA	1						
ESG	0.103***	1					
SOE	0.046***	0.191***	1				
Size	0.595***	0.207***	0.077***	1			
Lev	0.059***	0.195***	0.179***	0.249***	1		
Growth	0.00900	−0.00600	0	−0.00600	0.0220	1	
Age	−0.060***	0.100***	0.307***	−0.068***	0.108***	0.00200	1
Big4	0.148***	0.427***	0.155***	0.294***	0.248***	−0.0120	−0.00600
Top1	0.062***	0.090***	0.286***	0.041***	0.00700	0.00800	−0.0130
BOD_Ind	0.029**	0.096***	0.076***	0.0200	0.054***	−0.00300	−0.028**
Salary	0.090***	0.286***	−0.082***	0.086***	0.215***	0.00300	0.068***
BOD_Gender	0.00600	−0.094***	−0.213***	−0.00400	−0.073***	−0.00200	0.031**

Variable	Big4	Top1	BOD_Ind	Salary	BOD_Gender
Big4	1				
Top1	0.102***	1			
BOD_Ind	0.114***	0.094***	1		
Salary	0.311***	−0.144***	0.00800	1	
BOD_Gender	−0.091***	−0.122***	−0.0130	−0.0220	1

Note: *t* statistics in parentheses
*$p < 0.1$, **$p < 0.05$, and ***$p < 0.01$.

coefficient of 0.2933 and significance at the 10% level, which supports hypothesis 1; that is, there is a positive relationship between ESG and financial performance.

5.3.2   Result for Hypothesis 2

Model 2 shows the impact of different enterprise natures on the relationship between ESG and financial performance. In the regression result of model 2, the regression coefficient of ESG and EVA is 0.9024 and significant at the 1% level, while the regression coefficient of ESG*SOE and EVA is −0.9163 and significant at the 1% level. This shows that the nature of enterprises has a negative moderating effect on the relationship between ESG and financial performance, which means that compared with SOEs, non-SOEs have a more positive moderating effect on the relationship between ESG and financial performance. This verifies the correctness of hypothesis 2.

5.4   Endogeneity Test

Endogeneity problem refers to the problem that one or several explanatory variables in the model interact with the disturbance term, which is a common problem with panel data (Duncan et al., 2004). In order to reduce the influence of endogeneity

Table 7.5 Regression Analysis Results

VARIABLE	(1) EVA FE Coeff. (t-value)	(2) EVA FE Coeff. (t-value)
ESG	0.2933* (1.8907)	0.9024*** (3.6395)
ESG*SOE		−0.9163*** (−3.1470)
Size	−0.0001*** (−47.4533)	−0.0001*** (−47.3510)
Lev	−27.1267*** (−4.0095)	−27.5859*** (−4.0807)
Growth	0.6237** (1.9759)	0.6391** (2.0265)
Age	−1.1011*** (−3.4257)	−1.0978*** (−3.4189)
Big4	4.2129 (0.8644)	4.0938 (0.8408)
Top1	0.0187 (0.1655)	0.0145 (0.1286)
BOD_Ind	25.9662* (1.8759)	28.5814** (2.0633)
Salary	0.5942*** (5.5734)	0.5647*** (5.2810)
BOD_Gender	−1.6901 (−0.2424)	−2.2862 (−0.3281)
Industry-fixed effects	Yes	Yes
Year-fixed effects	Yes	Yes
_cons	33.3802*** (3.6396)	32.3575*** (3.5296)
N	4,928	4,928
Adj. R^2	0.254	0.256

t-statistics in parentheses.
* $p < 0.1$, ** $p < 0.05$, *** $p < 0.01$.

on the model variables in this study as much as possible, this study employs an additional endogeneity test.

The Heckman two-stage model is suitable for solving the endogenous problem caused by sample selection bias (Certo et al., 2016). Thus, the Heckman two-stage model is used for endogeneity test in this study. In the first stage, this study uses median_EVA to represent the median of EVA and sets a dummy variable EVA_dummy. Then, this study distinguishes EVA_dummy that is greater than median_EVA, less than median_EVA, and null. Next, based on EVA_dummy, this study uses the Probit model to calculate the inverse Mills ratio (IMR), the value used to correct the sample selection deviation of each sample. In the second stage of the Heckman test, these predicted individual probabilities IMR is combined with an additional variable, and regression analysis is performed with other control variables. The models of the second stage are as follows.

For Hypothesis 1:

$$\begin{aligned}
EVA_{i,t} = {} & \alpha_0 + \alpha_1 ESG_{i,t} + \alpha_2 Size_{i,t} + \alpha_3 Lev_{i,t} + \alpha_4 Growth_{i,t} + \alpha_5 Age_{i,t} \\
& + \alpha_6 Big4_{i,t} + \alpha_7 Top1_{i,t} + \alpha_8 BOD_Ind_{i,t} + \alpha_9 Salary_{i,t} \\
& + \alpha_{10} BOD_Gender_{i,t} + \alpha_{11} Industry_{i,t} + \alpha_{12} Year_{i,t} + \varepsilon_1 imr
\end{aligned} \tag{3}$$

For Hypothesis 2:

$$EVA_{i,t} = \alpha_0 + \alpha_1 ESG_{i,t} + \alpha_2 SOE_{i,t} + \alpha_3 ESG*SOE_{i,t} + \alpha_4 Size_{i,t} + \alpha_5 Lev_{i,t}$$
$$+ \alpha_6 Growth_{i,t} + \alpha_7 Age_{i,t} + \alpha_8 Big4_{i,t} + \alpha_9 Top1_{i,t} \quad (4)$$
$$+ \alpha_{10} BOD\_Ind_{i,t} + \alpha_{11} Salary_{i,t} + \alpha_{12} BOD\_Gender_{i,t}$$
$$+ \alpha_{13} Industry_{i,t} + \alpha_{14} Year_{i,t} + \varepsilon 1 imr$$

Results of the Heckman two-stage regression tests are shown in Table 7.6. The first column is the first-stage regression result of EVA_dummy and the second and third columns are the second-stage regression results of model 1 and model 2 Heckman tests, respectively. From the regression results, the correlation coefficients between IMR and EVA are significant, indicating that there is a certain bias in the selection of the original sample, which shows the necessity of using the Heckman two-step model to correct the sample selection bias. In the second-stage regression test of model 1, the regression coefficient of ESG and EVA is 0.306, and there is a significant positive correlation at the 5% level. This proves that the previous conclusion on hypothesis 1 is robust. In the second-stage regression test of model 2, the regression coefficient of ESG and EVA is 0.967 and significantly positively correlated at the 1% level, while the regression coefficient between ESG * SOE and EVA

Table 7.6 Endogeneity Test Results

	(1)	(2)	(3)
	EVA_dummy	*EVA*	*EVA*
	Coeff. (t-value)	*Coeff. (t-value)*	*Coeff. (t-value)*
ESG		0.306** (1.97)	0.967*** (3.89)
ESG * SOE			−0.993*** (−3.41)
Size	0.000*** (4.33)	−0.000*** (−47.61)	−0.000*** (−47.53)
Lev	−1.664*** (−13.29)	1.356 (0.14)	1.760 (0.18)
Growth	0.277** (2.34)	−0.206 (−0.54)	−0.215 (−0.57)
Age	−0.008** (−2.02)	−0.950*** (−2.93)	−0.942*** (−2.91)
Big4	0.256*** (4.00)	0.137 (0.03)	−0.120 (−0.02)
Top1	0.007*** (5.22)	−0.163 (−1.34)	−0.173 (−1.43)
BOD_Ind	−0.284 (−0.82)	31.567** (2.27)	34.578** (2.48)
Salary	0.030*** (6.49)	0.226 (1.60)	0.183 (1.28)
BOD_Gender	0.010 (0.06)	−1.384 (−0.20)	−2.016 (−0.29)
Imr		−28.029*** (−3.88)	−28.910*** (−4.01)
Industry-fixed effects	Yes	Yes	Yes
Year-fixed effects	Yes	Yes	Yes
_cons	0.258 (1.00)	48.557*** (4.88)	47.939*** (4.82)
N	4,900	4,900	4,900
Adj. R^2		0.257	0.259
r2_p	0.149		

t-statistics in parentheses.
* $p < 0.1$, ** $p < 0.05$, *** $p < 0.01$.

is −0.993 and significantly negatively correlated at the 1% level. The result is the same as the previous main regression result, proving the stability of hypothesis 2 and the original regression result after alleviating the influence of endogenous problems.

5.5 *Robustness Test*

According to Gelenbe and Caseau (2015), the energy consumption of computers and information and communication technology (ICT) is still a growing problem in terms of carbon dioxide emissions and economic costs. At the same time, with the advent of the digital economy, data centres, telecommunication base stations, and other scientific and technological fields may become the main sources of energy consumption in the future (Kagermann, 2015). Therefore, in this study, the computer, communication, and electronic equipment manufacturing industry was selected, instead of all industries, to do the robustness test. ROA (Alareeni & Hamdan, 2020) and IMR were added as the control variables, and FE regression analysis of two models was carried out. The regression results of the robustness test are shown in Table 7.7.

In the robustness test for model 1, the regression coefficient of ESG and EVA is 0.9036, and there is a significant positive relationship at the 1% level. This is consistent with the main regression result of model 1, so it supports the correctness of hypothesis 1; that is, there is a positive relationship between ESG and

Table 7.7 Robustness Test Results

Variable	(1)	(2)
	EVA	*EVA*
	Coeff. (t-value)	*Coeff. (t-value)*
ESG	0.9036*** (3.6443)	1.5556*** (4.5407)
ESG * SOE		−1.1826*** (−2.7157)
Size	−0.0002*** (−3.3112)	−0.0002*** (−3.5357)
Lev	0.5785 (0.0434)	−2.3714 (−0.1797)
Growth	−36.0640*** (−2.7049)	−38.9689*** (−2.9556)
Age	−0.1318 (−0.2376)	−0.0271 (−0.0494)
Big4	33.3770*** (4.1451)	33.6646*** (4.2414)
Top1	−0.4727* (−1.9312)	−0.4146* (−1.7119)
ROA	54.2829*** (4.2614)	52.3064*** (4.1591)
BOD_Ind	30.3754 (1.3938)	37.7341* (1.7429)
Salary	−0.4316 (−1.0282)	−0.4485 (−1.0841)
BOD_Gender	−4.6929 (−0.3330)	−11.8478 (−0.8381)
IMR	−29.2795*** (−2.8665)	−31.6108*** (−3.1285)
Year-fixed effects	Yes	Yes
_cons	24.0741 (1.0276)	24.6205 (1.0662)
Observations	273	273
R^2	0.122	0.147

t-statistics in parentheses.
*** $p < 0.01$, ** $p < 0.05$, * $p < 0.1$.

financial performance. In the robustness test for model 2, the regression coefficient of ESG and EVA is 1.5556, and there is a significant positive correlation at the 1% level. Multiplication of ESG and SOE, ESG * SOE, has a regression coefficient of −1.1826, and they are significantly negatively correlated at the 1% level, which supports the original regression result of model 2, proving that non-SOEs have better moderating effects on ESG and financial performance than SOEs.

6. Discussion

6.1 Relationship/Comparison with Prior Literature

Many scholars have explored the relationship between ESG and corporate financial performance. Consistent with this study's findings, Rose (2016) identified a positive correlation between corporate financial performance and governance. Auer (2016) found that in the Asia Pacific region and the United States, investors achieve better returns by considering ESG performance. Additionally, some researchers argue that embracing ESG responsibilities allows companies to identify strengths and weaknesses in environmental and social governance, enhancing their reputation and attracting more investors, which in turn boosts financial performance and supports sustainable development (Karwowski & Raulinajtys-Grzybek, 2021). This aligns with the results of this study.

However, some findings contradict this perspective. For instance, Reverte (2012) contended that ESG performance does not lead to real financial growth, suggesting that companies use social responsibility disclosures to obscure negative impacts. Other scholars argue that investing in ESG projects entails higher risks and uncertainties, diverting managers from core development goals, leading them to focus on more valuable areas instead (Settembre-Blundo et al., 2021). This study also found that non-SOEs are more significantly influenced by ESG in relation to financial performance. Dong and Baek (2022) observed that for companies listed on the Shanghai Composite Index from 2017 to 2020, ESG ratings were more closely linked to enterprise value in non-SOEs. Ruan and Liu (2021) found that ESG activities negatively impacted corporate performance, with non-SOEs providing stronger evidence. Wu et al. (2024) noted that state ownership diminishes the value of ESG activities, while Liu and Lin (2022) found that ESG information disclosure more effectively controls financial risk in non-SOEs. These studies support the second finding of this study.

6.2 Theoretical Contributions

An empirical analysis of the relationship between ESG and financial performance was conducted in this study, and the empirical results provided meaningful evidence and theoretical support for the promotion of ESG on corporate financial performance. At the same time, the use of EVA to measure corporate financial performance was also found to expand the common indicators of corporate financial performance.

Secondly, this study makes up for the gap in the domestic research on the moderating influence of the nature of enterprises on ESG and financial performance and provides theoretical evidence for later scholars to study the impact of SOEs or non-SOEs on ESG or financial performance.

6.3   *Practical Implications*

This study offers valuable decision-making and investment data for company managers and external investors. Generally, strong corporate governance, social responsibility, and environmental protection enhance enterprise value, attracting more external investment. Conversely, capital from external investors can improve a company's financial performance. Additionally, this study provides a foundation for government decisions aimed at promoting the coordinated development of SOEs and non-SOEs. As noted in hypothesis 2, non-SOEs tend to prioritize sustainable development more than their state-owned counterparts. However, challenges remain, including unequal development conditions and inadequate policies (Capalbo & Palumbo, 2012). By enhancing ESG performance in both types of enterprises, significant scale benefits can be achieved, effectively fostering sustainable societal development.

7.   Conclusion

This study examines the relationship between ESG and corporate financial performance, including the moderating effect of firm type. Using data from Chinese A-share listed companies between 2014 and 2020, we conducted empirical tests with a fixed-effect model, analysing 4,928 observations from 704 companies. Financial performance is measured by EVA, with ESG scores sourced from the Bloomberg database. To address potential endogeneity, we employed the Heckman two-stage model. In the first stage, we identified the median EVA and created a dummy variable, EVA_dummy. We also calculated an IMR for each observation for the second-stage regression analysis. Additionally, we performed robustness tests using data from the computer, communication, and electronic equipment manufacturing industry.

The empirical results indicate a positive relationship between ESG and financial performance, suggesting that ESG activities enhance financial outcomes. Notably, ESG has a more substantial positive impact on the financial performance of non-SOEs compared to SOEs. This disparity may stem from SOEs receiving government support, leading to lower expectations regarding their ESG responsibilities. In contrast, non-SOEs, facing less governmental backing, are more motivated to enhance their reputation and fulfil social responsibilities. The results of our endogeneity and robustness tests align with the primary regression outcomes, confirming the stability of our models and conclusions.

The theoretical contribution of this study is twofold: it enriches the literature on the positive impact of ESG on financial performance and addresses the research gap regarding the moderating effect of firm type on this relationship. Practically, the findings provide robust data to support government decision-making and regulations for different types of enterprises, facilitating the sustainable development of businesses and improving governmental decision-making efficiency.

Note

1 ST stands for Special Treatment in the context of Chinese A-share listed companies. Companies labeled as "ST" are typically those facing financial difficulties, such as continuous losses, and are subject to stricter regulatory scrutiny and potential delisting.

References

Akhtar, S., Javed, B., Maryam, A., & Sadia, H. (2012). Relationship between financial leverage and financial performance: Evidence from fuel & energy sector of Pakistan. *European Journal of Business and management, 4*(11), 7–17.

Alareeni, B. A., & Hamdan, A. (2020). ESG impact on performance of US S&P 500-listed firms. *Corporate Governance: The International Journal of Business in Society, 20*(7), 1409–1428.

Amel-Zadeh, A., & Serafeim, G. (2018). Why and how investors use ESG information: Evidence from a global survey. *Financial Analysts Journal, 74*(3), 87–103.

Atan, R., Alam, M. M., Said, J., & Zamri, M. (2018). The impacts of environmental, social, and governance factors on firm performance: Panel study of Malaysian companies. *Management of Environmental Quality: An International Journal, 29*(2), 182–194.

Atan, R., Razali, F. A., Said, J. A. M. A. L. I. A. H., & Zainun, S. (2016). Environmental, social and governance (ESG) disclosure and its effect on firm's performance: A comparative study. *International Journal of Economics and Management, 10*(2), 355–375.

Auer, B. R. (2016). Do socially responsible investment policies add or destroy European stock portfolio value? *Journal of Business Ethics, 135*, 381–397.

Auer, B. R., & Schuhmacher, F. (2016). Do socially (ir)responsible investments pay? New evidence from international ESG data. *The Quarterly Review of Economics and Finance, 59*, 51–62.

Bing, T., & Li, M. (2019). Does CSR signal the firm value? Evidence from China. *Sustainability, 11*(15), 4255.

Breuer, N., & Nau, C. (2014). ESG performance and corporate financial performance. *Department of Business Administration*.

Busco, C., Consolandi, C., Eccles, R. G., & Sofra, E. (2020). A preliminary analysis of SASB reporting: Disclosure topics, financial relevance, and the financial intensity of ESG materiality. *Journal of Applied Corporate Finance, 32*(2), 117–125.

Campopiano, G., & De Massis, A. (2015). Corporate social responsibility reporting: A content analysis in family and non-family firms. *Journal of Business Ethics, 129*(3), 511–534.

Capalbo, F., & Palumbo, R. (2012). The imperfect match of public accountability of state-owned enterprises and private-sector-type financial reporting: The case of Italy. *Capalbo, Francesco and Palumbo, Riccardo, The Imperfect Match of Public Accountability of State-Owned Enterprises and Private-Sector-Type Financial Reporting: The Case of Italy, Australasian Accounting Business and Finance Journal, 7*(4), 2013.

Certo, S. T., Busenbark, J. R., Woo, H. S., & Semadeni, M. (2016). Sample selection bias and Heckman models in strategic management research. *Strategic Management Journal, 37*(13), 2639–2657.

Chen, J. J., & Zhang, H. (2014). The impact of the corporate governance code on earnings management: Evidence from Chinese listed companies. *European Financial Management, 20*(3), 596–632.

Cheung, S. (2015). Asia starts to embrace ESG amid investor pressure. *The Private Equity Analyst*.

Chouaibi, S., Rossi, M., Siggia, D., & Chouaibi, J. (2021). Exploring the moderating role of social and ethical practices in the relationship between environmental disclosure and financial performance: Evidence from ESG companies. *Sustainability, 14*(1), 209.

Cole, L. (2020). *Who cares wins: Reasons for optimism in a changing world*. Rizzoli Publications.

Dalal, K. K., & Thaker, N. (2019). ESG and corporate financial performance: A panel study of Indian companies. *IUP Journal of Corporate Governance, 18*(1), 44–59.

Deng, X., & Cheng, X. (2019). Can ESG indices improve the enterprises' stock market performance? – An empirical study from China. *Sustainability, 11*(17), 4765.

Dong, M., & Baek, K. (2022). The effect of ESG ratings on the value of Chinese listed companies. *Asia-Pacific Journal of Business, 13*(1), 153–166.

Du, S., Bhattacharya, C. B., & Sen, S. (2010). Maximizing business returns to corporate social responsibility (CSR): The role of CSR communication. *International Journal of Management Reviews, 12*(1), 8–19.

Dumitrascu, R. A. (2014). The process of creating economic value added: Causes, factors and implications. *Knowledge Horizons Economics, 6*(1), 30.

Duncan, G. J., Magnuson, K. A., & Ludwig, J. (2004). The endogeneity problem in developmental studies. *Research in Human Development, 1*(1–2), 59–80.

Dyck, A., Lins, K. V., Roth, L., Towner, M., & Wagner, H. F. (2018). *Entrenched insiders and corporate sustainability (ESG): How much does the "G" matter for "E" and "S" performance around the world?*. Working paper. University of Toronto.

Fatemi, A., Glaum, M., & Kaiser, S. (2018). ESG performance and firm value: The moderating role of disclosure. *Global Finance Journal, 38*, 45–64.

Fontaine, M. (2013). Corporate social responsibility and sustainability: The new bottom line? *International Journal of Business and Social Science, 4*(4).

Friede, G., Busch, T., & Bassen, A. (2015). ESG and financial performance: Aggregated evidence from more than 2000 empirical studies. *Journal of Sustainable Finance & Investment, 5*(4), 210–233.

Gadinis, S., & Miazad, A. (2020). Corporate law and social risk. *Vanderbilt Law Review, 73*, 1401.

Garay, L., & Font, X. (2012). Doing good to do well? Corporate social responsibility reasons, practices and impacts in small and medium accommodation enterprises. *International Journal of Hospitality Management, 31*(2), 329–337.

Garcia, A. S., Mendes-Da-Silva, W., & Orsato, R. J. (2017). Sensitive industries produce better ESG performance: Evidence from emerging markets. *Journal of Cleaner Production, 150*, 135–147.

Gelenbe, E., & Caseau, Y. (2015). The impact of information technology on energy consumption and carbon emissions. *Ubiquity, 2015*(June), 1–15.

Gregory, R. P. (2021). Why ROE, ROA, and Tobin's Q in regressions aren't good measures of corporate financial performance for ESG criteria. *SSRN 3775789*.

Held, N. A., Webb, E. A., McIlvin, M. M., Hutchins, D. A., Cohen, N. R., Moran, D. M., . . . & Saito, M. A. (2020). Co-occurrence of Fe and P stress in natural populations of the marine diazotroph Trichodesmium. *Biogeosciences, 17*(9), 2537–2551.

Hoang, T. (2018). The role of the integrated reporting in raising awareness of environmental, social and corporate governance (ESG) performance. In *Stakeholders, governance and responsibility*. Emerald Publishing Limited.

Huang, H., & Zhao, Z. (2016). The influence of political connection on corporate social responsibility – Evidence from listed private companies in China. *International Journal of Corporate Social Responsibility, 1*(1), 1–19.

Huang, L., Liu, S., Han, Y., & Peng, K. (2020). The nature of state-owned enterprises and collection of pollutant discharge fees: A study based on Chinese industrial enterprises. *Journal of Cleaner Production, 271*, 122420.

Kagermann, H. (2015). Change through digitization – Value creation in the age of Industry 4.0. In *Management of permanent change* (pp. 23–45). Springer Gabler.

Karwowski, M., & Raulinajtys-Grzybek, M. (2021). The application of corporate social responsibility (CSR) actions for mitigation of environmental, social, corporate governance (ESG) and reputational risk in integrated reports. *Corporate Social Responsibility and Environmental Management, 28*(4), 1270–1284.

Khalid, F., Sun, J., Huang, G., & Su, C. Y. (2021). Environmental, social and governance performance of Chinese multinationals: A comparison of state-and non-state-owned enterprises. *Sustainability, 13*(7), 4020.

Kim, J., Chung, S., & Park, C. (2013). Corporate social responsibility and financial performance: The impact of the MSCI ESG ratings on Korean firms. *Journal of the Korea Academia-Industrial Cooperation Society, 14*(11), 5586–5593.

Kot, S. (2018). Sustainable supply chain management in small and medium enterprises. *Sustainability, 10*(4), 1143.

Leins, S. (2020). "Responsible investment": ESG and the post-crisis ethical order. *Economy and Society, 49*(1), 71–91.

Liang, H., Fernandez, D., & Larsen, M. (2022). Impact assessment and measurement with sustainable development goals. In *Handbook on the business of sustainability* (pp. 424–438). Edward Elgar Publishing.

Limkriangkrai, M., Koh, S., & Durand, R. B. (2017). Environmental, social, and governance (ESG) profiles, stock returns, and financial policy: Australian evidence. *International Review of Finance, 17*(3), 461–471.

Liu, Y., & Lin, M. (2022). Research on the influence of ESG information disclosure on enterprise financial risk: Taking pharmaceutical industry as an example. *Frontiers in Business, Economics and Management, 5*(3), 264–271.

Long, W., Li, S., Wu, H., & Song, X. (2020). Corporate social responsibility and financial performance: The roles of government intervention and market competition. *Corporate Social Responsibility and Environmental Management, 27*(2), 525–541.

Martins, H. C. (2022). Competition and ESG practices in emerging markets: Evidence from a difference-in-differences model. *Finance Research Letters, 46*, 102371.

Morales, L., Soler-Domínguez, A., & Hanly, J. (2019). The power of ethical investment in the context of political uncertainty. *Journal of Applied Economics, 22*(1), 554–580.

Niresh, A. J., & Alfred, M. (2014). The association between economic value added, market value added and leverage. *International Journal of Business and Management, 9*(10), 126.

Nollet, J., Filis, G., & Mitrokostas, E. (2016). Corporate social responsibility and financial performance: A non-linear and disaggregated approach. *Economic Modelling, 52*, 400–407.

Oehmke, M., & Opp, M. M. (2024). A theory of socially responsible investment. *Review of Economic Studies*, rdae048.

Ortas, E., Álvarez, I., Jaussaud, J., & Garayar, A. (2015). The impact of institutional and social context on corporate environmental, social and governance performance of companies committed to voluntary corporate social responsibility initiatives. *Journal of Cleaner Production, 108*, 673–684.

Palepu, K. G., Healy, P. M., Wright, S., Bradbury, M., & Coulton, J. (2020). *Business analysis and valuation: Using financial statements*. Cengage AU.

Pan, X., Cheng, W., & Gao, Y. (2022). The impact of privatization of state-owned enterprises on innovation in China: A tale of privatization degree. *Technovation, 118*, 102587.

Peterson, K., & Pfitzer, M. (2008). *Lobbying for good*. FSG.

Rahman, M. J., & Wu, J. (2023). M&A activity and ESG performance: Evidence from China. *Managerial Finance, 50*(1), 179–197.

Rahman, M. J., Wu, Q., & Zhu, H. (2024). Corporate social responsibility in times of social distancing: Evidence from China. *Business Ethics, the Environment & Responsibility*.

Rahman, M. J., & Zhu, H. (2023). Predicting accounting fraud using imbalanced ensemble learning classifiers – Evidence from China. *Accounting & Finance, 63*(3), 3455–3486.

Rahman, M. J., & Zhu, H. (2024). Detecting accounting fraud in family firms: Evidence from machine learning approaches. *Advances in Accounting, 64*, 100722.

Rahman, M. J., Zhu, H., & Chen, S. (2023). Does CSR reduce financial distress? Moderating effect of firm characteristics, auditor characteristics, and Covid-19. *International Journal of Accounting & Information Management, 31*(5), 756–784.

Rahman, M. J., Zhu, H., & Jiang, X. (2024). Family firms, client importance, and auditor reporting behavior: Evidence from China. *Meditari Accountancy Research, 32*(2), 543–578.

Rahman, M. J., Zhu, H., & Yue, L. (2024). Does the adoption of artificial intelligence by audit firms and their clients affect audit quality and efficiency? Evidence from China. *Managerial Auditing Journal.*

Rahman, M. J., Zhu, H., Zhang, Y., & Hossain, M. M. (2024). Effect of female representation in audit committees on non-audit fees: Evidence from China. *Meditari Accountancy Research.*

Reverte, C. (2012). The impact of better corporate social responsibility disclosure on the cost of equity capital. *Corporate Social Responsibility and Environmental Management, 19*(5), 253–272.

Rose, C. (2016). Firm performance and comply or explain disclosure in corporate governance. *European Management Journal, 34*(3), 202–222.

Ruan, L., & Liu, H. (2021). Environmental, social, governance activities and firm performance: Evidence from China. *Sustainability, 13*(2), 767.

Ruggiero, P., & Cupertino, S. (2018). CSR strategic approach, financial resources and corporate social performance: The mediating effect of innovation. *Sustainability, 10*(10), 3611.

Saini, N., Antil, A., Gunasekaran, A., Malik, K., & Balakumar, S. (2022). Environment-social-governance disclosures nexus between financial performance: A sustainable value chain approach. *Resources, Conservation and Recycling, 186*, 106571.

Sassen, R., Hinze, A. K., & Hardeck, I. (2016). Impact of ESG factors on firm risk in Europe. *Journal of business economics, 86*(8), 867–904.

Sethi, S. P., & Schepers, D. H. (2014). United Nations global compact: The promise–performance gap. *Journal of Business Ethics, 122*(2), 193–208.

Settembre-Blundo, D., González-Sánchez, R., Medina-Salgado, S., & García-Muiña, F. E. (2021). Flexibility and resilience in corporate decision making: A new sustainability-based risk management system in uncertain times. *Global Journal of Flexible Systems Management, 22*(2), 107–132.

Sharma, P., Panday, P., & Dangwal, R. C. (2020). Determinants of environmental, social and corporate governance (ESG) disclosure: A study of Indian companies. *International Journal of Disclosure and Governance, 17*(4), 208–217.

Sinha, R., Datta, M., & Ziolo, M. (2019). Inclusion of ESG factors in investments and value addition: A meta-analysis of the relationship. In *Effective investments on capital markets* (pp. 93–109). Springer.

Song, L. (2018). 19. State-owned enterprise reform in China: Past, present and prospects. *China's 40 Years of Reform and Development, 345.*

Stan, C. V., Peng, M. W., & Bruton, G. D. (2014). Slack and the performance of state-owned enterprises. *Asia Pacific Journal of Management, 31*(2), 473–495.

Tian, L. (2021). Unraveling the relationship between ESG and corporate financial performance-logistic regression model with evidence from China. *SSRN 3897207.*

Tian, L., & Ma, W. (2009). Government intervention in city development of China: A tool of land supply. *Land Use Policy, 26*(3), 599–609.

Wang, F., & Sun, Z. (2022). Does the environmental regulation intensity and ESG performance have a substitution effect on the impact of enterprise green innovation: Evidence from China. *International Journal of Environmental Research and Public Health, 19*(14), 8558.

Willard, B. (2012). *The new sustainability advantage: Seven business case benefits of a triple bottom line*. New Society Publishers.

Wu, Z., Pownall, R. A., Shih, Y. C., & Zhang, C. (2024, March 4). *ESG, State Ownership, and Stock Returns: Evidence from China* (March 4, 2024). Available at SSRN: https://ssrn.com/abstract=4251519 or http://dx.doi.org/10.2139/ssrn.4251519

Yilmaz, I. (2021). Sustainability and financial performance relationship: International evidence. *World Journal of Entrepreneurship, Management and Sustainable Development*, *17*(3), 537–549.

Zhang, C., & Jin, S. (2022). What drives sustainable development of enterprises? Focusing on ESG management and green technology innovation. *Sustainability*, *14*(18), 11695.

Zhou, G., Liu, L., & Luo, S. (2022). Sustainable development, ESG performance and company market value: Mediating effect of financial performance. *Business Strategy and the Environment, 31*(7), 3371–3387.

Zhu, H., & Rahman, M. J. (2024). Ex-ante expected changes in ESG and future stock returns based on machine learning. *The British Accounting Review*, 101457.

Part 3

Sustainability Accounting and Disclosure

8 Mandatory or Voluntary? Explore the Effectiveness of Chinese Environmental Information Disclosure Policy

Md Jahidur Rahman, Tarek Rana, Hongtao Zhu, and YE Xiaowei

1. Introduction

With the global attention to ecologically sustainable development, environmental information disclosure (EID) has become a major discussion, especially for manufacturing firms. Environmental investment and EID have contributed to building a transparent and sustainable environment. Many countries have taken steps to deal with this issue. America applied the legislative activity to regulate the firms to obey the environmental policy, such as the Clean Air Act. Securities and Exchange Commission (SEC) has updated the environment disclosure requirements for US firms, which refer to the environmental penalties and other environmental impacts (Duvall & Stern, 2020). The other parties in the United States also contribute to the regulations on environmental accounting information disclosure, such as the Financial Accounting Standards Board, American Institute of Certified Public Accountants, and American Accounting Association. To sum up, the United States has formed a normative system of mandatory EID through environmental laws.

Meng et al. (2012) stated that there is no enforced requirement for Chinese companies to disclose environmental information until the release of *Measures for the Disclosure of Environmental Information* (MDEI) in 2007 and the *Guidelines on Environmental Information Disclosure by Companies Listed on the Shanghai Stock Exchange* (SSE Guideline) in 2008. The listed companies, classified as heavy polluters by the Ministry of Environmental Protection (MEP), should disclose the primary environmental information under the provisions in the SSE Guideline. It can be seen that China's requirements for EID are mainly focused on specific public institutions. In the research from Meng et al. (2012), the practice of EID in China is in its infancy, and the regulations need to be strengthened. And the Chinese government has recognized the challenges of environmental issues and has gradually issued guidelines and regulations related to EID (Situ & Tilt, 2018). *Environmental Information Disclosure Guidelines for Listed Companies in China* were presented in 2010 to develop the index of EID (Meng et al., 2012).

A series of guidelines and regulations to encourage enterprises to disclose environmental information have been issued (Meng et al., 2012). However, since the mandatory disclosure requirement only focuses on companies with heavy

DOI: 10.4324/9781003488965-11

pollutants, China has no universally mandatory corporate environmental disclosure (Situ & Tilt, 2018). Enterprises in China are divided into state-owned enterprises and non-state-owned enterprises, but prior research about exploring the relationship between ownership and environmental disclosure quality presented a mixed result. The stakeholder theory suggested that the environmental disclosure performance of state-owned enterprises will be higher than that of non-state-owned enterprises as they are concerned by the public (Zeng et al., 2012; Meng et al., 2012; Acar et al., 2021). In contrast, the rent-seeking behaviour theory believed that state-owned enterprises would cover some adverse environmental problems and cause a low environmental disclosure performance (Li et al., 2021). The mixed results suggested that the managers consider environmental disclosure's motivation objectively when designing the environmental disclosure strategy. New guidelines require the state-owned enterprises to assume social and environmental responsibilities more actively (Li et al., 2018), but the performance has not been evaluated yet.

Since China's existing disclosure guidelines mainly focus on specific heavy-polluting enterprises, the classification of heavily-polluting enterprises listed by the MEP provides a good classification standard for studying the effect of China's disclosure guidelines. This study, therefore, seeks to (1) explore the response of Chinese enterprises to the existing disclosure policy by studying the difference in environmental disclosure quality between Chinese heavy-polluting enterprises and other enterprises and (2) confirm the moderate effect of ownership type on environmental disclosure quality in China.

The previous study involved lots of variables to measure EID. However, when the regulation was introduced as a variable, the existing literature was studied from the perspective of different countries (Alshbili et al., 2019). When evaluating the relationship between potential variables and level of EID, the previous study also suggested performing a cross-culture analysis between different countries (Meng et al., 2012). However, Cormier and Gomez-Gutierrez (2018) found that the cultural background of different countries will present the main barriers to the worldwide institutionalization of social and environmental disclosure. Also, China's economic environment is very different from that of other countries, and the EID levels in different regions perform differently due to the different regional economic development (Zhang et al., 2010). Therefore, cross-cultural analysis in different cultural backgrounds may not be very effective and may not provide good suggestions for China's environmental disclosure strategy. Thus, I believe it is better to test the implementation effectiveness of the environmental disclosure policy in China before conducting cross-cultural research. Therefore, we wanted to explore the factors associated with EID in the sample Chinese companies only and find any recommendations generated within China.

In this study, we analysed how the nature of environmental disclosure requirements would affect corporate EID in China regarding the transmission from voluntarism to regulation under the current Chinese context (Meng et al., 2012). However, it was not objective to consider China's transfer to mandatory environmental disclosure, and we selected all of the listed Chinese firms as a sample to evaluate the regulation effect. According to the SSE Guideline, only the heavily polluted companies listed by the MEP are required to disclose their environmental information. It means that Chinese environmental disclosure requirements were mandatory only for polluting companies, while others were encouraged to make voluntary disclosures (Situ & Tilt, 2018). Thus, this study divides the listed Chinese companies into heavy-polluting enterprises and others. Based on the SSE Guideline, these two groups can be referred to as the mandatory disclosure group and the voluntary group, which are the sample group in our research to explore the role of environmental disclosure policy in China.

This study uses empirical analysis to investigate the impact of disclosure requirements on Chinese EID quality. There are 248 Chinese listed companies selected as a sample from the CSMAR database. The EID score is quantitatively summed up by the six components of EID. The regression model and fixed effect are introduced to analyse the relationship between the panel data. The results show that China's heavily polluted companies, which are objective to mandatory disclosure, have a higher level of EID than other companies. And the proportion of state-owned shares will moderate the relationship between the type of environmental disclosure policy and EID level.

Given China's different economic and cultural backgrounds, it is inappropriate to apply other countries' disclosure requirements to China directly. This study contributes to exploring the factors associated with EID in the sample of Chinese companies only and finding any recommendations that can be generated within China. Dividing China's listed companies into groups that are required to make compulsory disclosure and groups that are encouraged to make a voluntary disclosure can help China have a deeper understanding of what kind of disclosure requirements are suitable for China. The analysis results will give regulators or managers a timely insight into EID. And this study has practical implications for regulators to the growing body of research on regulation.

In the research from Deswanto and Siregar (2018), environmental disclosure has a positive relationship with investment recommendation and firms' value with the transparent information. However, the primary purpose of regulating EID is for business sustainability and reminding businesses of their responsibility for environmental and economic sustainability. It is convincing that the results of this research can provide a guideline to strengthen the body of Chinese environmental legislation and supervision.

The remainder of this chapter is organized as follows. Section 2 presents the literature review and hypothesis development. The methodology followed is discussed in Section 3. Results and discussion are analysed in Section 4. And Section 5 summarizes the key findings of this study.

2.   Literature Review and Hypothesis Development

2.1   *Literature Review*

2.1.1   *Environmental Disclosure, Environmental Performance, and Economic Performance*

Green or environmental disclosure, defined by Ramdhony (2015), is the evidence about the relationship between organizations and society. Specifically, it is the process of releasing information to fulfil their environmental accountability. Previous studies on EID indicate that the primary function of EID is showing a company's effort towards environmental responsibility (Patten, 2005; Schneider et al., 2017). A study showed no significant relationship between environmental performance and environmental disclosure, even though a poor performer would report more content than a good performer (Hughes et al., 2001). However, the research performed by Pattern (2002) indicated the negative correlation between environmental disclosure and environmental performance with the proxy of toxics release inventory (TRI) data. Al-Tuwaijri et al. (2004) applied a simultaneous equations approach to assessing the relationship among environmental disclosure, environmental performance, and economic performance. The results revealed that environmental disclosure is related positively to environmental performance, which consisted of a statement from Ahmadi and Bouri (2017). To overcome the mixed results of environmental disclosure and environmental performance, Clarkson et al. (2008) revisited this relation based on the economics-based and socio-political theories of voluntary disclosure. Al-Tuwaijri et al. (2004) used the market-based proxies to represent economic performance and found a considerably positive correlation between environmental performance and economic performance, and they were closely related. Several studies in China also investigate the relationship between environmental accounting and firm performance (Rahman & Wu, 2023; Rahman et al., 2023, 2024; Zhu & Rahman, 2024).

2.1.2 Factors Affecting Environmental Disclosure

Stakeholders' influence. Based on the stakeholder theory, firms should consider the interest of stakeholders since they will be affected by the actions and decisions of firms. Maama and Appiah (2019) suggested that companies include stakeholders' views on the reportable environmental and social information in their annual reports. To improve the relationship with the stakeholder groups, managers can show their better environmental performance by providing complete information about their environmental performance or disclosing the environmental information selectively (Gray et al., 1995). The latter choice indicates that EID is viewed as a tool to manage stakeholder impressions of environmental performance (Patten, 2005). The influence of stakeholders on EID is not only in a firm's external disclosure but also in the environmental performance management processes within the company. Rodrigue et al. (2013) applied internal environmental performance indicators (EPIs) to illustrate the impact of stakeholders on the company's environmental strategy. The pressure generated by stakeholders' concern for environmental issues will directly affect the company's environmental strategy. In contrast, EPI will be indirectly affected as an indicator for formulating and monitoring environmental strategy (Rodrigue et al., 2013). Due to the interaction between EPI and environmental strategy, it is worth considering the impact of stakeholders on environmental performance management processes.

Company characteristics. Zeng et al. (2012), Meng et al. (2013), and Acar et al. (2021) stated that companies with higher state ownership have a positive correlation with voluntary disclosure and higher environmental disclosure scores, while companies with higher institutional ownership have a negative correlation with voluntary disclosure and lower disclosure scores. However, the rent-seeking theory suggests that the government's political connection will provide poor EID quality in China, since top executives might cover up some environmental information to maintain the relationship with the government (Li et al., 2021). Wang and Zhang (2019) stated that the balance of equity (proportion of state-owned shares) had no significant correlation with EID quality. Company size has been viewed as the most relevant company characteristic with the quality of EID, and it embodies a positive relationship between company size and the quality of EID (Wang & Zhang, 2019), and the driving factors can be referred to the economic development of a company. As for the board characteristics, the research from Khaireddine et al. (2020) indicated that companies with a large board size had a high quality of EID.

Economic and legitimacy incentives. Based on the economic theory, high-performing firms will have higher benefits and lower costs. Considering cost and benefit will encourage the high-performing firms to disclose more negative information (Al-Tuwaijri et al., 2004) and more confirmable information (Clarkson et al., 2008). Aragón-Correa et al. (2016) compared the information disclosed by the most significant 100 international companies in different industries. They concluded that those companies tried to make their operations legitimate and conform to social norms by disclosing environmental information. It means poor performers will disclose environmental information to maintain legitimacy. Tadros and Magnan (2019) reconciled these two opposing viewpoints. Both legitimacy and economic factors significantly influence EID, and the low- and high-performance groups have different responses to these two factors. The results show that the impact of economic factors on EID is more significant for low-performance groups than for high-performance groups. Although both high and low performers maintained their positive image by disclosing less negative information, low performers disclosed more negative information than high performers. The results did not prove that low performers used disclosure to legitimate their behaviour.

2.1.3   *Benefit/Motivation of High-Level Environmental Disclosure*

Contributes to obtaining investment. The high EID level is considered the strategy to improve public image and gain public acceptance (Maama & Appiah, 2019). Previous researches on the relationship between environmental disclosure and financial performance showed that high EID is valuable to both shareholders and stakeholders. Substantive environmental disclosures allow a company to gain the trust and expectations of investors, which can be referred to as a form of "green goodwill" (M. Acar & Temiz, 2020). The research from Griffin et al. (2017) suggested that the stockholders respond positively to firms' voluntary disclosures about greenhouse gas emissions. This impression of good environmental disclosure helps a company obtain financing quickly, increasing the firm value. Ahmadi and Bouri (2017) also stated that high-quality environmental disclosure will reflect the effectiveness of corporate governance, and the company will often face fewer difficulties in accessing capital markets. Those statements were consistent with investors' preferences for equities of environmentally responsible firms (Al-Tuwaijri et al., 2004).

Make a better forecast for financial health. By disclosing environmental information, enterprises can better understand their environmental behaviour and even increase their ability to predict the company's operation. In the research from Cormier and Magnan (2013), poor relative environmental disclosure would weaken financial analysts' ability to make earning forecasts. Thus, better EID will help improve financial analysts' ability to predict a company's earning situation, which is one of the indicators of company operation.

Increase firm value. Previous studies indicate the investors preferred the company with high-quality EID (Al-Tuwaijri et al., 2004; Griffin et al., 2017), which led to increasing the firm value. The higher level of environmental disclosure can also contribute to company value by reducing the cost (Griffin et al., 2017). It was achieved by reducing information asymmetry in the company's non-financial disclosures, thus reducing the cost of information acquisition for the company (Rezaee et al., 2020).

2.1.4   Institution Development for EID Requirement in China

Environmental governance. Environmental protection was included in the Constitution, and the Environmental Protection Law was enacted at the end of the 1970s in China (Wang & Bernell, 2013). A new Environmental Protection Law (EPL) was enacted in 1989, and the revised version took effect in 2015, which became the fundamental law governing environmental protection in China (Situ & Tilt, 2018). The enactment of the Environmental Impact Assessment Law in 2003 and the MDEI issued by the MEP in 2007 marked the new progress of the Chinese government in environmental governance (Situ & Tilt, 2018).

Environmental Disclosure Guidelines and Regulations. The related Chinese guidelines for firms to disclose social and environmental information have been issued since 2006, such as the SSE *Notice*, the *Guidelines of State-owned Enterprises Performing Social Responsibilities* (the SSE Guideline), and so on (Situ & Tilt, 2018). The first regulation on EID was issued in 2007 by China State Environmental Protection Administration (SEPA), requesting companies to disclose the environmental information related to pollution and protection in their annual reports (Yekini et al., 2019). The *Guidelines on Environmental Information Disclosure by Companies Listed on the Shanghai Stock Exchange* (SSE Guideline) was issued in 2008. And SEPA further released EID Guidelines in 2010, which specify further details on mandatory and voluntary disclosure for listed companies (Yekini et al., 2019). China Securities Regulatory Commission revised the *Standards for the Contents and Formats of Information Disclosure by Companies Offering Securities to the Public No. 3* in 2021. There is no significant modification of the environmental disclosure requirement. Based on this document, the listed companies of key pollutant emission should disclose the primary environmental information following the provisions. A full explanation should be given if there is no related disclosure (China Securities Regulatory Commission, 2021). And the other enterprises can refer to the provisions to make the EID. Environmental disclosure requirements presented are the same as the previous versions.

2.2 Hypothesis Development

The nature of environmental disclosure requirement in China transferred from voluntarism to mandatory for the heavy-polluted listed companies after the releasing of a legal document MDEI issued by the MEP (Situ & Tilt, 2018), and SSE Guidelines issued by the SSE in response to MDEI, which required the heavy-polluted listed companies to disclose the following information (Situ & Tilt, 2018): type, volume, and content of pollutants discharged; information on the construction and operation of their environmental protection facilities; plans that deal with environmental accidents; and strategies and plans to reduce pollutants discharged.

Meng et al. (2012) suggested that the pressure–legitimacy theory could explain the performance of mandatory environmental disclosure, while the performance–impression theory can explain the disclosure behaviours during the period of voluntary disclosure since the Chinese environmental disclosure requirements were mandatory only for polluting companies, while others were encouraged to make voluntary disclosures (Situ & Tilt, 2018). A recent study presented that mandatory corporate social responsibility (CSR) disclosure has a significantly positive impact on corporate environmental responsibility (CER), an essential part of CSR (Liu et al., 2021). Also, the results indicated the CER score of mandatory disclosure enterprises is higher than those of voluntary disclosure enterprises. With the strengthening of the policy, voluntary disclosure enterprises are also gradually strengthening their environmental responsibility. To identify the effectiveness of the mandatory disclosure policy, the research proposes that the following hypothesis:

Hypothesis H1: Firms with mandatory environmental disclosure requirements are positively associated with EID.

Prior studies related to the effect of ownership on the environmental disclosure level had mixed results, based on the stakeholder theory and rent-seeking behaviour theory. These researches mainly focused on the motivation of the state-owned enterprises to analyse their EID. With the implementation of mandatory disclosure requirements of CSR, shareholders of non-state-owned enterprises started to pay more attention to the fulfilment of environmental responsibility. The CER promotion level of non-state-owned enterprises is higher than that of state-owned enterprises (Liu et al., 2021). Hence, in the light of the preceding analysis, it is hypothesized that

Hypothesis H2: The proportion of states shares will moderate the relationship between the mandatory environmental disclosure policy and EID level.

3. Methodology

3.1 Data and Sample

In this research, the time interval is set from 2011 to 2019. This period is timely to examine the level of EID regarding the regulation on EID in 2010. The sampled companies are all Chinese listed firms in China's CSMAR database from 2011 to 2019 to evaluate the hypothesis. This multi-level capital market information database includes all the China Securities Regulatory Commission (Yekini et al., 2019). And the Chinese marketization index is captured in the Wind Economic Database, which presents the most comprehensive insights into China's macroeconomy and industry. Since the type of industry might significantly relate to the nature of disclosure requirement, the sample was not screened by industry type. In total, 2,232 observations were collected from CSMAR after screening the above conditions.

3.2 Variables

3.2.1 Dependent Variables

Corporate EID is measured using the disclosure-scoring method, a coding instrument developed from context analysis (Meng et al., 2012). Considering the provided information in CSMAR, this study will measure the EID scores based on the six items from the evaluation components of EID, which was developed by Zeng et al. (2010): (1) information related to ISO environmental system authentication; (2) firms' environmental protection policies, strategies, and goals; (3) loans related to environmental protection; (4) disposal and treatment of generated waste, recycling, and integrated utilization of waste products; (5) construction and operation of environmental improvement; and (6) other environment-related information. Each component is scored according to its level of disclosure, ranging between 0 and 3: 3 if the component is described in monetary and quantitative terms, 2 if it is explicitly described, 1 if it is discussed in general, and 0 if there is no information (e.g., Al-Tuwaijri et al., 2004; Hughes et al., 2001; Zeng et al., 2010). The data downloaded from the CSMAR presented the scores based on these scoring criteria.

Then the sum of the EID score was evaluated based on Equation (1)

$$SEID_i = \sum_{j=i}^{n} SCID_i \tag{1}$$

where social and environmental impact disclosure ($SEID_i$) is the total score of EID for the firm i, and social and corporate impact disclosure ($SCID_i$) is the score of the jth component for the firm i, in which $j = 1, 2, \ldots, 10$.

3.2.2 Independent Variables

Type of environmental information disclosure policy. Based on the SSE Guideline, the heavily polluted enterprises and the other enterprises can be referred to as mandatory disclosure groups and voluntary groups, respectively. In this study, the dummy variable is introduced for the type of policy. The value of "1" will be assigned for heavily polluted enterprises (refer to mandatory disclosure policy); otherwise (refer to voluntary disclosure policy), it is "0".

3.2.3   Moderating Variables

State-owned share is based on the proportion of shareholding owned by state-own legal-person shareholder and state shareholder, which can be generated from CSMAR.

3.2.4   Control Variables

Following previous research, this study mainly uses the following five variables as controls: (1) Firm size: measured by the natural logarithm of total year-end assets as a proxy variable (e.g., Gray et al., 2001); (2) firm age: this study uses the number of years the company has been listed in the Chinese Stock Market at the end of the reporting year (Zeng et al., 2010, 2012). (3) Return to equity (ROE): compared to return to assets (ROA), ROE is widely used to measure a company's social and environmental disclosure (Meng et al., 2013). (4) Marketization: the EID levels in different regions perform differently due to regional economic development (Zhang et al., 2010). And marketization is measured by the marketization index of the region (Meng et al., 2012). (5) Leverage: this is an essential indicator of a firm's financial risk and availability of financial resources. Since extreme leverage will also affect EID, it should be treated as a control variable. Leverage is measured as the ratio of a firm's total debt to total assets (Meng et al., 2013).

3.2.5   Empirical Model

To evaluate the impact of environmental disclosure requirements on EID, this research builds Equation (2) based on Hypothesis H1.

$$EID_{it} = \alpha + \beta_1 PL_i + \sum_{j=1}^{n} \gamma_j \, X_{jit} + Fixed\ effext + \varepsilon_{it} \tag{2}$$

Here, EID_{it} is the score of EID of company i in year t; PL_i is a dummy variable, denoting whether company i is required to disclose environmental information mandatorily – if yes, it is 1, otherwise it is 0; β is the coefficient, and there will be a positive relationship between the mandatory disclosure requirement and EID if β has a positive value. X_{jit} donates the j-th control variable in the t-th year of company i, including firm size, firm age, ROE, marketization index, and leverage. Since the sample involves panel data, and the EID might be affected by different firms and time, firm-fixed effect and time-fixed effect are included in the study. ε_{it} is the random error.

Based on hypothesis H2, state-owned shares are introduced as moderating variables in this study, ad the moderating effect is evaluated by Equation (3):

$$\begin{aligned} EID_{it} = {} & \alpha + \beta_1 PL_{it} + \beta_2\, State-owned\ shares_{it} \\ & + \beta_3\, State-owned\ shares_{it} \times PL_{it} + \sum_{j=1}^{n} \gamma_j X_{jit} \\ & + Fixed\ effext + \varepsilon_{it} \end{aligned} \tag{3}$$

Here, the state-owned share is the moderating variable, and the interaction of state-owned shares and PL denotes the moderating effect on EID.

4. Results

4.1 Descriptive Statistical Analysis

Table 8.1 presents the descriptive statistics of EID for the 248 year-firm observations from 2011 to 2019. Because of the different effects of different disclosure requirements and property rights, the samples are divided into mandatory requirement firms, voluntary requirement firms, state-owned firms, and non-state-owned firms. As shown in Table 8.1, the average level of EID of all samples increased from 2011 to 2019. On an annual basis, enterprises required to make mandatory disclosure have higher EID scores than the voluntary disclosure enterprises, and state-owned enterprises have higher EID scores than non-state-owned enterprises. However, the average EID score enterprises with mandatory disclosure policy started to decline in 2017, while the voluntary disclosure firms remain increasing. Due to the availability of the marketization index, the sample in this study only covers the period from 2011 to 2019. Also, the impact of COVID-19 EID has not been evaluated.

The mean, standard deviation, minimum, maximum, and Pearson's correlation coefficients of each variable are shown in Tables 8.2 and 8.3, separately. The correlation coefficient between EID and policy is positive and significant ($p < 0.05$). Also, the correlation is <0.5, demonstrating the multiple collinearities will not affect the study (Meng et al., 2012; Cailou et al., 2021).

4.2 Baseline Regression

To make the results more convincing, the EID score applied in the regression analysis excluded the censored data with a zero value, and the results are reported in

Table 8.1 Descriptive Statistics of EID

Year	Full Sample		Mandatory Policy Firms		Voluntary Policy Firms		Stated-Owned Firms		Non-Stated-Owned Firms	
	Mean	*SD*	*Mean*	*SD*	*Mean*	*SD*	*Mean*	*SD*	*Mean*	*SD*
2011	6.42	6.52	14.00	–	6.42	6.53	10.00	6.12	6.02	6.45
2012	6.93	6.26	14.33	5.38	6.46	6.03	10.60	6.39	6.52	6.13
2013	7.13	6.55	15.29	6.08	6.44	6.12	9.76	7.52	6.84	6.38
2014	7.38	6.40	15.63	6.82	6.86	6.12	10.36	7.61	7.04	6.18
2015	7.50	6.45	15.80	7.45	7.00	5.96	9.76	8.00	7.24	6.23
2016	8.38	6.72	14.60	7.16	7.02	5.96	10.84	8.20	8.10	6.50
2017	9.09	7.27	15.68	6.70	6.56	5.81	11.64	7.97	8.81	7.15
2018	10.36	7.09	13.82	7.39	7.10	5.50	11.68	7.74	10.22	7.02
2019	11.04	7.52	13.09	6.67	8.50	6.66	13.24	7.19	10.79	7.53
Total	8.26	6.75	14.66	6.65	6.95	6.08	10.86	7.40	7.97	6.62

Table 8.2 Mean, Standard Deviation, Min, and Max of Each Variable

Variable	Obs	Mean	SD	Min	Max
EID	2232	8.248	6.923	0.000	32.000
Policy	2232	0.195	0.397	0.000	1.000
State-owned shares	2232	0.302	0.249	0.000	0.877
Firm size	2232	23.326	1.835	18.811	31.036
Firm age	2232	13.468	5.975	0.000	28.000
ROE	2232	−0.002	4.576	−213.809	17.898
Marketization	2232	8.111	1.926	−1.420	11.400
Leverage	2232	0.520	0.229	0.026	2.290

Table 8.3 Correlation of Each Variable

	EID	Policy	State-Owned Shares	Firm Size	Firm Age	ROE	Marketization	Leverage
1	1.000							
2	0.414**	1.000						
3	0.148**	−0.008	1.000					
4	0.248**	0.057**	0.352**	1.000				
5	0.098**	0.221**	0.125**	0.104**	1.000			
6	0.021	0.006	0.009	0.005	−0.207	1.000		
7	0.057**	0.022	−0.027	0.138**	0.070**	0.019	1.000	
8	0.023	0.007	0.170**	0.526**	0.126**	−0.042*	−0.330	1.000

$** p < 0.05, * p < 0.1.$

Table 8.4. In Table 8.4, Column (1) presents the regression results of the type of policy and EID, column (2) presents the control variables in the regression process, and column (3) presents the regression results after controlling the firm and time effect. After adding the control variables, firm- and time-fixed effects, the coefficient of policy is positive and significant at the 1% level, indicating that firms with mandatory disclosure policy pose a significant positive relationship with EID score. The regression results verify hypothesis H1, which states that mandatory environmental disclosure policy can promote a firm's EID level, and are consistent with the existing literature, which applied the pressure–legitimacy theory to explain the EID score with mandatory disclosure policy (Meng et al., 2012). EID is defined as releasing information to fulfil firms' environmental accountability (Ramdhony, 2015). Besides the environmental liability and protection expense, EID score also includes the environmental management terms, such as protection policy, and the relevant education training. Thus, firms with higher EID indicate they have better environmental management. Since the mandatory environmental disclosure policy can lead to a significant positive increase in EID, regulators are required to expand the scope of companies mandated to disclose environmental information. It can not only meet the requirement of a transparent disclosure environment but also drive the companies to promote their environmental management.

Table 8.4 Results of Baseline Regression

Variable	(1) EID Coeff. (t-value)	(2) EID Coeff. (t-value)	(3) EID Coeff. (t-value)	(4) EID Coeff. (t-value)
Policy	6.188*** (18.73)	6.053*** (18.30)	2.266*** (7.65)	2.212*** (7.44)
State-owned shares				−2.557 (−1.42)
Policy × State-owned shares[a]				−1.808* (−1.71)
Firm size	−	0.885*** (9.43)	0.457 (1.60)	0.474* (1.65)
Firm age	−	0.029 (1.22)	0.405*** (5.19)	0.395*** (5.01)
ROE	−	−0.127 (−0.52)	−0.127 (−0.72)	−0.141 (−0.81)
Marketization	−	−0.474*** (−6.66)	−0.368 (−1.45)	−0.350 (−1.37)
Leverage	−	−4.763*** (−6.16)	−1.949** (−2.03)	−1.825* (−1.90)
Control	No	Yes	Yes	Yes
Firm FE	No	No	Yes	Yes
Year FE	No	No	Yes	Yes
Cons	8.249*** (52.96)	−6.558*** (−3.33)	−3.179 (−0.51)	−2.853 (−0.45)
Log likelihood	−6,143.663	−6,086.6391	−5,107.800	−5,104.9641
R^2	0.155	0.204	0.153	0.156
N	1,913	1,913	1,913	1,913

Note

[a]The variables (i.e., policy and state-owned shares) involved in the interaction term (Policy * State-owned shares) are centred (converted to zero mean).

*** $p < 0.01$, ** $p < 0.05$, * $p < 0.1$.

Column (4) in Table 8.4 presents the moderating effects of state-owned shares on EID. And the variables involved in the interaction term (policy and state-owned shares) are centralized (Meng et al., 2012). From column (4), as a whole sample, we find that state-owned shares have negative coefficients and are not statistically significant. However, the interaction term (policy * state-owned shares) is negatively and statistically significant with EID, which indicates that the proportion of state-owned shares in a firm weakens the promotion of mandatory environmental disclosure policy on EID. And the finding supports hypothesis H2, which states that the state-owned shares moderate the relationship between the mandatory environmental disclosure policy and EID level. The results support the previous studies, indicating the impact of rent-seeking behaviour on EID score (Li et al., 2021). This is because the state-owned shares of firms can refer to their relationship with the government. To maintain a good image and obtain the benefits from a close relationship with the government, managers might cover up some adverse environmental information. Thus, state-owned shares will impair the effectiveness of mandatory policy to promote EID. It is also consistent with the existing study that managers will use EID as an impressive tool to show their environmental performance to their stakeholders (Patten, 2005). Patten (2005) also proved the stakeholders' influence on companies' environmental strategy. Thus, there should be stricter supervision for companies that hold more state-owned shares to ensure the authenticity of the information disclosed and environmental strategy.

4.3   *Robustness Testing*

In this study, further robustness tests were performed to evaluate the robustness of the results. Firstly, the variance inflation factors (VIFs) of each variable are presented in Table 8.5. All VIFs were close to 1, much less than the recommended maximum threshold of 10, indicating Equations (2) and (3) are robust. Since the EID is a count data and the standard deviation of EID (shown in Table 8.2) shows the dispersion of data, the negative binomial regression is applied to test models (1) and (2) (Cailou et al., 2021). And the results are presented in columns (1) and (2) in Table 8.6. It can be seen from Table 8.6 that the regression coefficient of policy is 0.235 and significantly positive, indicating the mandatory environmental disclosure requirement does lead to an increase in the EID score. Column (2) presents the moderating effect of state-owned shares on EID. It shows that the impact coefficient of the interaction term on EID is significantly negative ($p < 0.05$), indicating the state-owned shares play a moderating role in the effect of environmental disclosure policy on EID. The robustness results support and prove the previous regression results are robust.

5.   Limitations and Future Research

This study has several limitations. The marketization index of Chinese regions issued by the Wind Economic Database was only updated through 2019; therefore, the effect of COVID-19 on EID cannot be tested. Also, the EID score will only be affected by whether certain components are presented because of the code measurement, so the authenticity of the disclosed content cannot be further tested. And the EID score might be biased since we only selected six items from the ten components of EID, developed by Zeng et al. (2010).

Future research is suggested with adding data after 2019 to provide insights on the COVID-19 effect. The average EID score enterprises with mandatory disclosure policy started to decline in 2017, while the EID scores of firms with voluntary disclosure policy remain increasing. Thus, future research are recommended to explore the reason for decreased EID with mandatory disclosure policy and confirm

Table 8.5 Variance Inflation Factors

	Policy	State-Owned Shares	Firm Size	Firm Age	ROE	Marketization	Leverage
VIFs	1.058	1.166	1.605	1.090	1.004	1.049	1.426

Table 8.6 Results of Negative Binomial Regression

Variable	EID	EID
	Coeff. (t-value)	*Coeff. (t-value)*
Policy	0.235*** (8.11)	0.232*** (8.02)
State-owned shares		−0.095 (−0.77)
Policy × State-owned shares[a]		−0.211** (−2.07)
Firm size	0.098*** (4.88)	0.228*** (4.77)
Firm age	0.022*** (4.75)	0.099*** (4.77)
ROE	−0.019 (−0.89)	−0.02 (−0.99)
Marketization	−0.009 (−0.62)	−0.011 (−0.72)
Leverage	−0.344*** (−3.40)	−0.331*** (−3.28)
Control	Yes	Yes
Firm FE	Yes	Yes
Year FE	Yes	Yes
Cons	0.542 (1.23)	0.573 (1.28)
Log likelihood	−5,402.9893	−5,400.4674
N	1,913	1,913

Note

[a]The variables (i.e., policy and state-owned shares) involved in the interaction term (Policy * State-owned shares) are centred (converted to zero mean).

*** $p < 0.01$, ** $p < 0.05$, * $p < 0.1$.

whether it will decline with the extended period. Also, it is better to further measure EID score, as an explained variable, using more comprehensive indicators and analysis methods.

6.   Conclusion

As national and local governments gradually pay attention to environmental information laws and policies, more and more research on EID is conducted. Since different cultural backgrounds will present barriers to the global institutionalization of environmental disclosure (Cormier & Gomez-Gutierrez, 2018), this study only involves Chinese samples. There are lots of factors that can influence the EID score, such as stakeholder's influence, characteristics of a company, and economic and legitimate incentives. When the type of disclosure policy was studied as a variable, Meng et al. (2012) indicated that China's policy on EID has changed from voluntary to mandatory since the first regulation on EID was issued in 2007 by China SEPA. However, China does not have a comprehensive system of mandatory disclosure policy because only high-polluting enterprises published by the MEP are required to make mandatory EID. Therefore, this study was driven to divide Chinese samples into mandatory and voluntary disclosure groups and then compare these two groups' environmental disclosure quality in the same period.

Based on the regulation from MEP, heavily polluted firms have been objected to a mandatory disclosure group. To test the effectiveness of mandatory disclosure policy and generate valuable recommendations, this study evaluated the relationship between EID policy and EID score and examined the moderating effect of state-owned share proportion in China. After controlling the firm-fixed and time-fixed effect, the regression results of 1,913 observations show that the mandatory policy has a significant positive relationship with the EID level. And the moderating effect of state-owned shares on the relationship between the type of policy and EID score has been proved in this research. Specifically, the higher proportion of state-owned shares will weaken the positive relationship between mandatory disclosure policy and EID score.

Based on the above research, this study proposes the following recommendations for regulators: firstly, mandatory disclosure policy is an effective method to promote the EID score, and regulators are required to establish a comprehensive system of the mandatory disclosure policy, which can encourage the enterprises to realize their environmental accountability. Secondly, it is crucial to enhance the external supervision of EID, especially for the company with a high proportion of state-owned shares. There are many incentives for enterprises to make environmental disclosure, including stakeholder's influence, economic benefit, and impression tools. Those motivations might affect the quality of EID. Furthermore, the government should encourage enterprises to create a transparent disclosure environment and then continuously fulfil their environmental and economic sustainability responsibility.

References

Acar, E., Tunca Çalıyurt, K., & Zengin-Karaibrahimoglu, Y. (2021). Does ownership type affect environmental disclosure? *International Journal of Climate Change Strategies and Management, 13*(2), 120–141.

Acar, M., & Temiz, H. (2020). The relationship between environmental disclosure and financial performance: An empirical study on BIST-100 companies. *Environmental Management and Sustainability, 4*(2), 67–85.

Ahmadi, A., & Bouri, A. (2017). The relationship between environmental disclosure, environmental performance, and economic performance: Evidence from French firms. *Journal of Business Ethics, 149*(2), 343–364.

Alshbili, I., Elamer, A. A., & Beddewela, E. (2019). The influence of institutional context on corporate social responsibility disclosure: A comparative analysis of the UK and Egypt. *International Journal of Corporate Social Responsibility, 4*(1), 1–25.

Al-Tuwaijri, S. A., Christensen, T. E., & Hughes, K. E. (2004). The relations among environmental disclosure, environmental performance, and economic performance: A simultaneous equations approach. *Accounting, Organizations and Society, 29*(5–6), 447–471.

Aragón-Correa, J. A., Marcus, A. A., & Vogel, D. (2016). The effects of mandatory and voluntary regulatory pressures on firms' environmental strategies: Evidence from China. *Journal of Business Ethics, 138*(4), 731–748.

Cailou, Y., Zhao, J., & He, Z. (2021). Environmental regulation and corporate green innovation: Evidence from China. *Journal of Cleaner Production, 282*, 124–214.

China Securities Regulatory Commission. (2021). *Standards for the contents and formats of information disclosure by companies offering securities to the public no. 3*. China Securities Regulatory Commission Website.

Clarkson, P. M., Li, Y., Richardson, G. D., & Vasvari, F. P. (2008). Revisiting the relation between environmental performance and environmental disclosure: An empirical analysis. *Accounting, Organizations and Society, 33*(4–5), 303–327.

Cormier, D., & Gomez-Gutierrez, L. (2018). On the search for mimetic patterns in environmental disclosure: an international perspective. *International Journal of Sustainable Development & World Ecology, 25*(7), 655–671.

Cormier, D., & Magnan, M. (2013). Environmental disclosure quality and corporate governance: The role of the board of directors and audit committee. *Journal of Business Ethics, 113*(3), 389–402.

Deswanto, R. B., & Siregar, S. V. (2018). The association between environmental disclosures with financial performance, environmental performance, and firm value. *Social Responsibility Journal, 14*(1), 180–193.

Duvall, T. S., & Stern, T. (2020). Environmental regulation and disclosure: Lessons from the clean air act and the securities and exchange commission. *Environmental Policy Review, 6*(1), 34–45.

Gray, R., Javad, M., Power, D. M., & Sinclair, C. D. (2001). Social and environmental disclosure and corporate characteristics: A research note and extension. *Journal of Business Finance & Accounting, 28*(3–4), 327–356.

Gray, R., Kouhy, R., & Lavers, S. (1995). Corporate social and environmental reporting: A review of the literature and a longitudinal study of UK disclosure. *Accounting, Auditing & Accountability Journal, 8*(2), 47–77.

Griffin, P. A., Lont, D. H., & Sun, E. Y. (2017). The relevance to investors of greenhouse gas emission disclosures. *Contemporary Accounting Research, 34*(2), 1265–1297.

Hughes, K. E., Andersen, M. L., & Law, A. T. (2001). Corporate environmental disclosures: Are they useful in assessing environmental performance? *Journal of Accounting and Public Policy, 20*(3), 217–240.

Khaireddine, H., Salhi, B., Aljabr, J., & Jarboui, A. (2020). The impact of board characteristics on environmental disclosure: Evidence from the Gulf Cooperation Council. *Environmental Science and Pollution Research, 27*(21), 26232–26244.

Li, S., Wong, S. H., & Miao, M. (2021). State ownership, rent-seeking, and corporate environmental disclosure in China. *Journal of Cleaner Production, 281*, 124571.

Li, Y., Gong, M., Zhang, X. Y., & Koh, L. (2018). The impact of environmental, social, and governance disclosure on firm value: The role of CEO power. *The British Accounting Review, 50*(1), 60–75.

Liu, Z., Tang, Q., & Zhou, Y. (2021). Mandatory corporate social responsibility disclosure and corporate environmental responsibility: Evidence from China. *Journal of Business Ethics, 170*(2), 1–22.

Maama, H., & Appiah, K. (2019). Stakeholder influence on sustainability disclosure: An empirical examination of listed firms in Ghana. *Business Strategy and the Environment, 28*(5), 763–775.

Meng, J., Wang, L., Liu, X., Wu, J., Brookes, P. C., & Xu, J. (2013). Physicochemical properties of biochar produced from aerobically composted swine manure and its potential use as an environmental amendment. *Bioresource Technology, 142*, 641–646.

Meng, X. H., Zeng, S. X., & Tam, C. M. (2012). From voluntarism to regulation: A study on ownership, economic performance, and corporate environmental disclosure in China. *Journal of Business Ethics, 109*(2), 1–16.

Patten, D. M. (2002). The relation between environmental performance and environmental disclosure: A research note. *Accounting, Organizations and Society, 27*(8), 763–773.

Patten, D. M. (2005). The accuracy of financial report projections of future environmental capital expenditures: Evidence from the disclosure of potential environmental liabilities. *Journal of Accounting and Public Policy, 24*(5), 333–354.

Rahman, M. J., & Wu, J. (2023). M&A activity and ESG performance: Evidence from China. *Managerial Finance, 50*(1), 179–197.

Rahman, M. J., Wu, Q., & Zhu, H. (2024). Corporate social responsibility in times of social distancing: Evidence from China. *Business Ethics, the Environment & Responsibility*.

Rahman, M. J., Zhu, H., & Chen, S. (2023). Does CSR reduce financial distress? Moderating effect of firm characteristics, auditor characteristics, and Covid-19. *International Journal of Accounting & Information Management, 31*(5), 756–784.

Ramdhony, D. (2015). Environmental reporting: A synthesis of the normative, stakeholder and legitimacy theories. *Journal of Business Management and Economic Research, 7*(1), 45–57.

Rezaee, Z., Alipour, M., Faraji, O., Ghanbari, A., & Jamshidinavid, B. (2020). Firm performance and voluntary sustainability disclosures: Evidence from Asia. *Journal of Cleaner Production, 268*, 122296.

Rodrigue, M., Magnan, M., & Boulianne, E. (2013). Stakeholders' influence on environmental strategy and performance indicators: Evidence from Canadian firms. *Journal of Business Ethics, 112*(2), 207–224.

Schneider, T., Michelon, G., & Paananen, M. (2017). Environmental disclosure and economic performance: A study of large U.S. firms. *Journal of Business Ethics, 143*(2), 533–551.

Situ, H., & Tilt, C. A. (2018). The influence of corporate social responsibility on corporate environmental disclosure in Chinese firms. *Journal of Business Ethics, 148*(2), 347–361.

Tadros, H., & Magnan, M. (2019). How environmental performance shapes the relationship between environmental disclosure and economic performance: Evidence from global firms. *Journal of Business Ethics, 155*(3), 759–778.

Wang, L., & Zhang, J. (2019). The role of ownership structure in the relationship between corporate environmental disclosure and performance: Evidence from China. *Corporate Social Responsibility and Environmental Management, 26*(1), 19–35.

Wang, Y., & Bernell, D. (2013). Environmental disclosure and political economy in China: A new law and its implementation challenges. *Journal of Environmental Policy & Planning, 15*(2), 147–160.

Yekini, K. C., Adelopo, I., Wang, Y., & Song, D. (2019). The impact of environmental information disclosure on firm performance: Evidence from China. *Asian Review of Accounting, 27*(4), 496–519.

Zeng, S. X., Meng, X. H., & Tam, C. M. (2010). The relationship between corporate environmental responsibility and firm performance: Evidence from China. *Journal of Cleaner Production, 18*(12), 1142–1148.

Zeng, Z., Piao, S., Lin, X., Yin, G., Peng, S., Ciais, P., & Myneni, R. B. (2012). Global evapotranspiration over the past three decades: estimation based on the water balance equation combined with empirical models. *Environmental Research Letters, 7*(1), 014026.

Zhang, Y., Zhao, R., & Qu, H. (2010). Environmental information disclosure in China: A review. *Environmental Science and Pollution Research, 17*(1), 173–182.

Zhu, H., & Rahman, M. J. (2024). Ex-ante expected changes in ESG and future stock returns based on machine learning. *The British Accounting Review*, 101457.

9 Factors Influencing Sustainability Reporting

Evidence from Medicine Manufacturing Industry in China

Md Jahidur Rahman, Tarek Rana, Hongtao Zhu, and Zhou Dongying

1. Introduction

Due to increasingly complex businesses and intensified global transformation, companies strive to show their responsible and positive image by paying attention to the sustainability agenda and the disclosure of sustainability reports. According to the United Nations (UN) Brundtland Report, the extremely important role of sustainable development was proposed in 1987, to consider the satisfaction of contemporary needs without infringing on the interests of the next generation (World Commission on Environment and Development, 1987). Based on this, the European Commission has also emphasized the inevitability of the implementation of sustainable development strategies (Evans et al., 2006). Some Italian local governments are also increasingly willing to include sustainable development-related elements in their strategic plans (Evans et al., 2006). In addition, in Italy, the number of companies willing to disclose sustainable development goals is increasing, and new business models have been proposed based on these goals (World Commission on Environment and Development, 1987). Not only Italy but also New Zealand, Turkey, Malaysia, and Portugal follow the trend of sustainable development for disclosure of sustainability reports (Aktas et al., 2013; Gomes et al., 2015; Hamad et al., 2020; Montecalvo et al., 2018). In Europe, the number of sustainability reports in Portugal was second to none (Castelo Branco et al., 2014).

Compared to studies on sustainability reports abroad, in China, there are fewer studies, and there are still some aspects of sustainable development. In 2012, ahead of the UN summit in Rio+20, Ma Zhaoxu, Chinese assistant foreign minister, had already demonstrated China's importance and support for the concept of sustainable development (Xinhua et al., 2012). According to Bai et al. (2015), the international supply chain and market can promote the sustainable development of companies in China for dissemination and application. Not only that, but research also focuses on the results. Some companies' sustainability reports still show no consistency between saying and doing (Huang & Wang, 2010). These companies choose to miss out on some bad news in order not to damage their image. Moreover, there have been studies to analyse the annual reports of 642 listed companies in Shanghai through content analysis and to achieve a more thorough analysis of the

DOI: 10.4324/9781003488965-12

classification of sustainable development information (Li & Xiang, 2007). What's more, through descriptive research, Shen and Jin (2006) delved into the sustainability information presented in the annual reports of listed manufacturers in the petroleum, chemical, plastics, and other industries from 1999 to 2003.

Research in the area of the sustainable development report is still in its infancy in China. At present, in both China and abroad, majority of the research generally conducted content analysis or descriptive analysis of the sustainability reports of some companies to find the content and extent of disclosure by local companies. The leading authority on sustainability reporting recognized around the world, Global Reporting Initiative (GRI) has developed a "common framework for sustainability reporting" that is currently widely accepted (Dilling, 2010). China is one of the countries that follow GRI as a criterion for sustainability development reporting. However, there still exists a gap in the impact of GRI on the sustainability reports of listed companies in China. Besides, there are shortcomings in exploring the relationship between the level of disclosure of sustainability reports and the characteristics of companies from all angles.

Due to research gaps, sustainability reporting has yet to be deeply explored. Therefore, when comparing the impact of GRI on the sustainability reports of listed companies in China, it is necessary to classify the companies relying on industry categories for evaluation, which will be based on the statistical data from 2011 to 2020. What's more, further research will focus on the relationship among company characteristics, audit firm, and board composition and sustainability report disclosure in the medicine manufacturing industry.

In China, the economy is facing transformation and development. Thus, according to the Ministry of Foreign Affairs (2016), the Chinese government is also pushing hard for the adjustment of its development strategy and the achievement of sustainable development goals based on Transforming Our World: The 2030 Agenda for Sustainable Development, which was adopted at the UN General Assembly. As a form of corporate non-financial reporting, sustainability development reports play a pivotal role and has an influence on enhancing a company's image, catering to government policies, and promoting innovation. Besides, sustainability reporting is the basis for companies to develop sustainable development organization plans (Lozano et al., 2016). Therefore, for companies to achieve sustainability, they must pay attention to sustainability reporting. Based on Ang et al. (2021), pharmaceutical industries expect to practise more effective recycling programmes to achieve the goal of reducing ecosystem pollution, which is why they emphasize sustainable manufacturing and focus on sustainable development. Thus, it is necessary to explore the extent to which sustainability reports in the pharmaceutical industry are disclosed and the relationship between the various factors.

This study will contribute to the existing relevant literature on understanding the factors that affect the degree of disclosure of sustainability reports. As one of companies' key governance mechanisms, it is necessary and meaningful to explore the relationship between board composition and the degree of disclosure. In addition, the study will expand the literature contribution between other factors, namely, company characteristics and audit firms, and the degree of disclosure. This will be an important attempt to focus on the relationship among company characteristics, board composition, audit firm, and degree of disclosure.

This chapter is structured as follows. Section 1 will present the introduction. Previous relevant literature and the hypotheses set by this research will be described in Section 2. Section 3 will discuss the data and methodology used, and the research results will be analysed in detail in Section 4. Finally, Section 5 will summarize of the research and discuss the limitations and future research directions.

2.　Literature Review and Hypotheses

Research has found that companies in environmentally sensitive industries are more willing to disclose their sustainable development goals, and in Australia, very few companies make voluntary disclosures (Deegan & Gordon, 1996). In the study by Hossain et al. (2006), based on 107 non-financial companies listed on the Dhaka Stock Exchange, using the multiple regression analysis, they found that in Bangladesh, the degree of corporate sustainability report disclosure is positively related to the nature of the company and the company's net profit margins. A study of 87 Malaysian companies has found that sustainability report disclosure is significantly impacted by company size and stock exchange listings and ISO14001 certification (Sarumpaet, 2005). Among the 100 UK companies selected by the Centre for Social and Environmental Accounting Research (CSEAR), Gray et al. (2001) found that companies that disclose more information tend to be larger and more profitable.

According to Ernstberger and Grüning (2013), due to the information asymmetry between managers and other stakeholders, corporate governance and disclosure of sustainability reports are both crucial links that can effectively reduce information asymmetry. Effective corporate governance can also promote the increase in information transparency while achieving a relative balance between corporate and social interests (Sharif & Rashid, 2014). However, very few scholars have conducted in-depth research on these two. Moreover, the corporate governance mechanism is designed to reduce information asymmetry (Ernstberger & Grüning, 2013). Therefore, corporate governance and sustainability report disclosure can be linked through the analysis of agency theory (Allegrini & Greco, 2013).

2.1　Firm Age

According to the research by Cormier et al. (2005), as Europe's largest economy, in Germany, the degree of disclosure of a company's sustainability report is positively correlated with the age of its fixed assets. Based on the study by Sandhu et al. (2012), one of the reasons for the disclosure of the sustainability report by an organization is to match the legitimacy and gain the recognition of stakeholders (such as existing or potential customers). Disclosure of sustainable development information has a strong and positive effect on a company's reputation to a certain extent (Sandhu et al., 2012). Older companies will have to deal with known or unknown threats brought by society (Prasad et al., 2016). In addition, the longer a company's operating years, the more social problems it must face from the environment. More-over, disclosure of sustainability reports requires the addition of experienced com-panies and professionals. Therefore, the first hypothesis of this study is as follows:

H1: Firm age and its disclosure of sustainability reports have a positive association.

2.2 Firm Size

In addition to firm age, firm size also needs to be considered. Compared with small-scale companies, large-scale companies will have to bear the threat of le-gitimacy because of their larger scale (Prasad et al., 2016). Therefore, the second hypothesis will be that

H2: There is a positive association between firm size and its disclosure of sustain-ability reports.

2.3 Board Size

When facing the supervision of stakeholders, companies have also begun to attach importance to the needs of stakeholders and emphasize their sustainable develop-ment (Hassan et al., 2017). From the perspective of corporate governance, the size of the board of directors may affect a company's degree of disclosure (Ntim et al., 2013). However, based on the existing theoretical literature and empirical literature, the conclusions reached are contradictory. Based on agency theory, a larger board of directors not only has stricter monitoring capabilities but can also display richer and more diverse expertise and advice (Larmou & Vafeas, 2010; Uwuigbe, 2011; Sun et al., 2010). Moreover, in Sun et al. (2010), Lim et al. (2007), Buniamin et al. (2011), and other studies, the positivity between board size and disclosure level of the sustainability report can be found. The exact opposite is that the CEO has a strong control ability in a larger board (Jensen, 1993), and there are also studies showing a negative relationship between board size and the degree of disclosure (Kathyayini & Nagaraju, 2012; Fathi, 2013; Amran et al., 2014). Therefore, when studying the re-lationship between the size of a company's board of directors and the degree of dis-closure in the medicine manufacturing industry in China, three situations, positive, negative, or irrelevant, may appear. Consequently, the third hypothesis is as follows:

H3: There is a relationship between the size of the board of directors and the de-gree of disclosure of the sustainability report.

2.4 Board Independence

Based on corporate governance, not only the size of the board but also the degree of independence of the board is extremely critical (Amran et al., 2014). The board of directors, which is composed of dependent directors and independent directors, is the company's business executive body. Unlike dependent directors, independent directors can only rely on their own director positions to achieve indirect participation (Rouf, 2011). In addition, because the conflicts of interest among stakeholders are inevitable, the existence of dependent directors and independent directors can achieve a balance to a certain extent (Sharif & Rashid, 2014; Khan, 2010). Applying agency theory, the relationship between the proportion of independent directors in the board of directors and the supervisory ability of the board of directors is positive; that is, the higher the proportion of independent directors, the stronger the supervisory ability of the board of directors (Jizi et al., 2014; Liao et al., 2014; Chau & Gray, 2010). Thus, it can be reasonably speculated that the higher the proportion of independent directors, the higher the degree of disclosure of the company. Therefore, the fourth hypothesis is as follows:

H4: There is a positive association between board independence and the degree of sustainability report disclosure.

2.5   Board Gender Diversity

When emphasizing board diversity, gender always stands out (Liao et al., 2014). In terms of participation in important decisions and leadership roles, females also have the ability to contribute to the board of directors (Huse & Solberg, 2006). According to Coffey and Wang (1998), compared with males, the decisions made by females are less likely to be guided by their own interests. Therefore, it is possible for female directors on the board of directors to hold views contrary to that of male directors on the issue of disclosure. Moreover, previous studies can prove that women's awareness of sustainable development is stronger than that of men (Diamantopoulos et al., 2003; Mainieri et al., 1997; Wehrmeyer & McNeil, 2000). Thus, the fifth hypothesis can also be drawn as follows:

H5: The proportion of female directors is positively related to the degree of disclosure of the sustainability report.

2.6   Audit Firm

Deloitte, PricewaterhouseCoopers, Ernst & Young, and KMPG are the world's Big Four accounting firms (Christodoulou, 2011). Excellent auditors ensure the quality of audit services, which can also enable companies to display a positive and qualified image (Ahmad et al., 2016). In addition, the Big Four accounting firms attach great importance to the disclosure of sustainability reports (Haladu, 2018). Thus, the Big Four are more likely to have connections with companies with a high degree of disclosure. Therefore, the sixth hypothesis can be derived from this, which is as follows:

H6: There is a positive association between an audit firm and a company's sustainability report disclosure degree.

3. Data and Methodology

3.1   Data and Sample

In order to explore the relationship between company characteristics, namely, age and size, the composition of the board of directors and audit firm, and the degree of disclosure of sustainability reports, for this study we selected the medicine manufacturing industry from Chinese listed companies on the CSMAR database as the sample. The industry category is classified according to the 2012 edition of the China Securities Regulatory Commission's industry classification standards. Therefore, through screening, research is conducted on 64 Chinese listed companies. In addition, CSMAR collects and processes the annual social responsibility reports, environmental reports, sustainability reports, corporate citizenship reports and other reports regularly disclosed by listed companies every year, and then presents complete and objective data. Thus, this database will be the study's secondary source. More importantly, all data in this study are based on the data collected during a total of 10 years, from 2011 to 2020. Then, the processing and analysis of data will rely on the powerful data analysis tool, Stata 16, so that the degree of report disclosure can be determined. Therefore, further research can indicate the report disclosure rate of listed companies in the medicine manufacturing industry and can even dig out the impact of company characteristics, board composition, and audit firms on the degree of report disclosure in a simpler and clearer way.

3.2   Variables

3.2.1   Dependent Variable: Degree of Sustainability Report Disclosure

The dependent variable of this study is the degree of disclosure of sustainability reports of listed companies in China. According to the latest sustainability criteria mentioned in the GRI launched in 2013, this dependent variable can be represented by a simple average disclosure index (Haladu, 2018). As long as one item is disclosed in the company's report, one point will be awarded, but if it is not disclosed, a zero point will be shown on this indicator. Then, the final score will be displayed as an average value, which will be the index used by the dependent variable to connect the independent variable in this study.

3.2.2   Independent Variable

As already discussed in the Literature Review and Hypothesis, company characteristics, namely, age and size; composition of the board of directors, namely, its size, independence, and gender diversity; and audit firm are listed as independent variables. Table 9.1 summarizes the independent variables involved in this study and their measurement methods. The board-related data used are from CSMAR's basic information database of listed companies, the governance structure of listed companies database, the financial statement database of listed companies, and the social responsibility research database of listed companies.

Table 9.1 Summary of Independent Variables

Independent Variable	Code	Measurement
Firm age	FAG	Length of time between a firm's establishment date and 2020 (Rahman, Zhu, & Yue, 2024)
Firm size	FSIZE	Natural logarithm of total assets (Rahman, Zhu, & Jiang, 2024)
Board size	BSIZE	Total number of directors in the company (Rahman & Chen, 2023; Rahman, Zhu, & Yue, 2024)
Board independence	BIND	Percentage of dividing the number of independent directors by the total number of directors (Rahman & Chen, 2023; Rahman, Zhu, & Yue, 2024)
Board gender diversity	BGDIV	Percentage of female directors of the total number of directors (Rahman & Chen, 2023; Rahman, Zhu, Zhang & Hossain, 2024)
Audit firm	AFM	Variable that is equal to 1 if the company is audited by the Big Four audit firms and 0 otherwise (Rahman et al., 2023)

3.2.2.1 COMPANY CHARACTERISTICS

The firm age (FAG) is based on a company's establishment date. Because the data collected in this study are based on the period from 2011 to 2020, it can be measured by the length of time between its establishment date and 2020. Moreover, for easy comparation, the firm size (FSIZE) will rely on the value of its total assets. Therefore, the value obtained by calculating the natural logarithm of total assets can be used to measure FSIZE.

3.2.2.2 BOARD COMPOSITION

The board size (BSIZE) is decided by the number of people; thus, it is measured by the total number of directors in the company. In addition, board independence (BIND) means the proportion of independent directors, which is measured by the percentage of dividing the number of independent directors by the total number of directors. Board gender diversity (BGDIV) is used to describe the female proportion in the company's board of directors. Therefore, the percentage of female directors will be used to measure BGDIV.

3.2.2.3 AUDIT FIRM

Finally, when measuring the audit firm (AFM), the average method consistent with the dependent variables will be used. Companies audited by the Big Four audit firms can get 1 point; otherwise, 0 points. Therefore, the final value will be calculated by calculating the average value.

3.3 *Model and Method*

For the purpose of obtaining the relationship among the company characteristics, the board composition, and the audit firm and the degree of disclosure of the sustainability report, and to verify the correctness of the above six hypotheses, the empirical model is used. Besides, based on the ordinary least squares (OLS) model, the following equation will be established:

$$SRDI = a + \beta_1 FAG + \beta_2 FSIZE + \beta_3 BSIZE + \beta_4 BIND + \beta_5 BGDIV \qquad (1)$$
$$+ \beta_6 AFM + \varepsilon$$

Here,

SRDI = sustainability report disclosure index;
a = intercept of the equation;
$\beta 1$–$\beta 6$ = regression coefficients;
FAG = firm age;
FSIZE = firm size;
BSIZE = board size;
BIND = board independence;
BGDIV = board gender diversity;
AFM = audit firm;
ε = error term.

4. Results

4.1 Descriptive Statistics

The descriptive statics are shown in Table 9.2.

In this study, the range of the dependent variable SRDI is between 0.000 and 0.846. The degree of disclosure of the sampled companies has obviously quite large variations. Moreover, the mean value of 0.486 represents that the SRDI of the listed companies in China's medicine manufacturing industry is generally lower than the average. The disclosure level of most companies' sustainability reports is low, so there is considerable room for improvement in sustainability reports in this industry.

From the standpoint of independent variables, FAG, one of the company's characteristics, ranges from a minimum of 13 years to a maximum of 32 years, with an average of about 22 years. Another feature, FSIZE, which is measured by the

Table 9.2 Descriptive Statistics

Variable	Obs	Mean	Std. Dev.	Min	Max
SRDI	640	0.486	0.246	0.000	0.846
AFM	640	0.061	0.239	0.000	1.000
BSIZE	640	8.734	1.422	5.000	15.000
BIND	640	0.371	0.052	0.286	0.625
BGDIV	640	0.163	0.122	0.000	0.571
FSIZE	640	9.683	0.429	8.622	10.864
FAG	640	22.422	4.390	13.000	32.000

natural logarithm of total assets, fluctuates between 8.622 and 10.864, with an average of 9.683. This means that the average total assets of the listed companies in China's medicine manufacturing industry are RMB 4.819 billion.

Regarding the board composition, BSIZE ranges from as high as 15 directors to as low as 5 directors. On average, each of the companies has about 9 board directors. Besides, the average proportion of independent directors is only 37.1%, and the maximum is only 62.5%. This means that most of these companies have far fewer independent directors than non-independent directors. Moreover, the average percentage of female directors in companies is low, only 16.3%, and there are even companies without female directors.

As for AFM, the average percent of 6.1% is very low, which means that among the companies studied, only about 6.1% are audited by the Big Four accounting firms. Besides, the standard deviation value of 0.239 proves that the data distribution is relatively good. But this also means that nearly 94% of the company's annual financial statements are not audited by the internationally renowned Big Four accounting firms, which indicates that the audit quality of the companies studied is low. Moreover, the maximum and minimum values of AFM are 1 and 0, respectively.

4.2   Correlation Matrix

Table 9.3 provides the correlation data of each variable in the study. The correlation coefficient of 0.339 ($p < 0.001$) proves that there is a positive correlation between SRDI and FSIZE, which is consistent with H2. In addition, SRDI is also positively correlated with BGDIV, with a correlation coefficient of 0.211 ($p < 0.001$). Therefore, H5 is also verified. Moreover, the correlation coefficient between SRDI and AFM is 0.209 ($p < 0.001$), which means that there is a positive correlation between these two. Therefore, H6 can be accepted. On the other hand, the research results indicate that other independent variables (FAG, BSIZE, and BIND) are not statistically correlated with SRDI, which is inconsistent with the aforementioned hypotheses.

4.3   Regression Results

Table 9.4 shows the results of the OLS regression analysis. From the data presented in Table 9.4, the F-statistic value of 9.54, the significant p-value of 0.000,

Table 9.3 Correlation Matrix

Variable	SRDI	AFM	BSIZE	BIND	BGDIV	FSIZE	FAG
SRDI	1						
AFM	0.209***	1					
BSIZE	0.028	0.144***	1				
BIND	−0.010	−0.023	−0.451***	1			
BGDIV	0.211***	0.049	−0.128***	0.074*	1		
FSIZE	0.339***	0.253***	0.145***	−0.023	0.058	1	
FAG	−0.013	−0.010	0.148***	−0.043	−0.242***	0.302***	1

$*p < 0.05$, $**p < 0.01$, $***p < 0.001$.

Table 9.4 Results of OLS Regression Analysis

Variable	Coeff.	Std. Err.	t-value	p-value	VIF
AFM	0.1239804***	0.0352512	3.52	0.001	1.10
BSIZE	−0.002332	0.0124388	−0.19	0.852	1.33
BIND	−0.0999968	0.3127559	−0.32	0.750	1.26
BGDIV	0.3429436***	0.1309172	2.62	0.011	1.10
FSIZE	0.1835409***	0.0580146	3.16	0.002	1.22
FAG	−0.0037014	0.0057677	−0.64	0.523	1.21
_cons	−1.213977	0.5294079	−2.29	0.025	/
R^2	0.1705				
Adj. R^2	0.1626				
F	9.54				
p-value of F-statistic	0.0000				

and the *R2* value of 0.1705 can prove the statistical significance of the estimated model. These data mean that the variance of 17.05% in SRDI can be explained by the variance of firm size, firm age, board size, board independence, board gender diversity, and audit firms. Simultaneously, the adjusted *R2* is 0.1626, which means that the independent variables in the study can explain the 16.26% variability of SRDI. Moreover, one of the criteria for similar disclosure studies is low *R2* or low adjusted *R2* (Amran et al., 2014).

The results also show that there is no statistically significant relation between BSIZE, BIND, and FAG and SRDI. Thus, the research results reject H1, H3, and H4. The hypothesis, a larger company is expected to express higher SRDI, is confirmed. Statistically, the relationship between these two is positively significant ($p = 0.002$). Therefore, H2 can be accepted. At the 99% confidence level ($p = 0.011$), the results of the study show that companies with a higher proportion of female directors will increase the company's SRDI. Therefore, H5 can be accepted. In addition, companies audited by the Big Four accounting firms are expected to have higher SRDI. The relationship between these two is statistically positively significant ($p = 0.001$). Therefore, H6 can be accepted.

5. Discussion

The role of sustainability is not only to transmit relevant information about sustainable development but also to monitor management behaviours. Business organizations are facing huge challenges from providing sustainability reports. However, according to the results, it can be found that the SRDI of listed companies in China's medicine manufacturing industry is below average, which obviously has huge room for improvement. There may be two reasons for this result. Firstly, the company's strategic deployment for sustainable development is lacking. Secondly, sustainable development has not received the sufficient attention it deserves (Krechowicz & Fernando, 2009). Due to the content, quality, and importance of the report, it is necessary to explore these issues in depth.

5.1   AFM and SRDI

The results of the study found a positive and significant relationship between AFM and SRDI. This result confirms the hypothesis proposed in this article. The Big Four accounting firms will choose companies that disclose sustainability issues. Accounting firms that provide high-quality services represent the company's self-confident and complete image. Therefore, companies audited by the Big Four pay more attention to sustainable development issues and disclose these issues to the public. Although there is a positive relationship between these two, research shows that only 6.1% of the listed companies in China's medicine manufacturing industry are audited by the Big Four. These data mean that the majority of the companies' audit firms are less recognized. The reason for this result may be that in the context of China, the quality of audit services provided by the non-Big Four is lower than that of the Big Four because of their independence (Chen & Zhou, 2006). Moreover, due to China's current immature legal environment, even for low-quality audit services, the legal risks it faces are relatively low. Therefore, due to certain factors, the sampled companies may not choose the Big Four but choose low-quality audit services.

5.2   FAG and SRDI

According to the hypothesis, FAG should be positively associated with SRDI. However, the research results cannot prove this hypothesis. Listed companies are required to perform information disclosure obligations. Imperfect disclosure of information will hinder investors' access to information and lead to decision-making errors. In addition, sustainability reports are closely related to the operation of listed companies; thus, this information should also be disclosed. However, the results of the study show that the older companies did not make full disclosures. This may be because these older listed companies refuse to disclose information that is unfavourable to the company in order to maintain a good corporate image and selectively avoid punishment information.

5.3   FSIZE and SRDI

In this study, FSIZE is measured by the natural logarithm of the company's total assets. The research results can prove the correctness of the hypothesis that there is a positive relationship between FSIZE and SRDI. Larger companies disclose more sustainability information and pay more attention to the corporate image. When a company wants to choose the Big Four for auditing to improve its image, in order to meet the requirements of the Big Four, the company's disclosure of sustainable development information will be more complete. In addition, although the research results show that FSIZE is positively associated with SRDI, there is no significant relationship between FAG and SRDI. According to the study by Tang et al. (2008), in China, the relationship between the age and size of a company and its growth rate is just the opposite; that is, an older company will have a slower growth rate, but a larger company will have a faster growth rate. This means that the older age of a company does not necessarily equate to the larger size of the company. Therefore, this may be the reason for the different relationship between FSIZE, FAG, and SRDI.

5.4   *BSIZE and SRDI*

In guiding the sustainable development agenda, the board of directors has demonstrated its important role and promoted a tacit environment for the organization (Amran et al., 2014). However, without proper management and supervision, the existence of the board of directors may not be able to fully realize the benefits of resources, and the size of the board of directors may hinder its effective operation. The results of the study show that the mean value of BSIZE is about 9 people. These data have also been proven by Yermack (1996) that it is an ideal board size for effective operation. However, the final research results cannot show that there is a significant association between BSIZE and SRDI. Other studies, such as Lakhal (2005), have also proved that there is no relationship between these two. Therefore, it is possible that a consistency between the board's interests and the issue of sustainable development has not been achieved in practice. According to Amran et al. (2014), this is probably because the company believes that the board's management capabilities are good enough, and when results are achieved, it will reduce investment in disclosure.

5.5   *BIND and SRDI*

Based on sampled companies, the results prove that there is also no significant association between BIND and SRDI. This result is the opposite of the positive relationship results in previous studies (Jizi et al., 2014; Liao et al., 2014; Chau & Gray, 2010). This result may raise the question of whether independent directors' actions will be restricted in the sustainability report due to their lack of participation in the company's daily operations. Similar studies have also indicated that the role of independent directors has been questioned (Hashim & Devi, 2008).

5.6   *BGDIV and SRDI*

Although there is no significant relationship among BSIZE, BIND, and SRDI, consistent with the hypothesis, there is a positive association between BGDIV and SRDI. Therefore, this finding can support the viewpoint stated above. Female directors are more sensitive to sustainability issues than men. Without being affected by their own interests, female directors can better realize the importance of disclosure of sustainability issues. Therefore, female directors play a certain role in promoting this issue.

6.   **Robustness Test**

For the purpose of enhancing the robustness of the conclusions of this study, it is necessary to conduct tests to check the robustness of the model. Firstly, by

Table 9.5 Results of the Relationship between SRDI (2) and Other Factors

	SRDI		SRDI (2)	
	Coeff.	p-*value*	*Coeff.*	p-*value*
AFM	0.1239804***	0.001	0.170802***	0.001
BSIZE	−0.002332	0.852	−0.0023046	0.828
BIND	−0.0999968	0.750	−0.0778354	0.758
BGDIV	0.3429436***	0.011	0.3387379***	0.002
FSIZE	0.1835409***	0.002	0.1979109***	0.000
FAG	−0.0037014	0.523	−0.002096	0.658
_cons	−1.213977	0.025	−1.428353	0.000

calculating the variance inflation factor (VIF) value, multicollinearity can be checked. The results presented in Table 9.4 show that the VIF values of all variables are less than 2, which indicates that the regression results have no multicollinearity. In addition, in this study, a robustness test is conducted by changing the measurement method of the dependent variable.

In addition, robustness testing is further conducted through alternative measures for variables. The SRDI value can be obtained not only by the method employed in this study but also by other methods. When listed companies' annual reports, social responsibility reports, and environmental reports disclose sustainable development-related information, it is given 1 point, and otherwise 0. Obtaining the average value based on the scores of the three reports as the SRDI (2) value is also a feasible and meaningful measurement method (Amran et al., 2014). Therefore, under this method, Stata 16 is also used for further analysis, and results (Table 9.5) consistent with this research are obtained. From these analyses, it can be found that the robustness of this research conclusion is obvious.

7.   Limitations and Future Research

As with other studies, this study also has limitations. Firstly, in exploring the relationship between the corporate governance mechanism and the degree of disclosure of sustainability reports, this study only emphasizes the influence of board composition. Therefore, the future research direction can be to explore the impact of the audit committee on report disclosure to improve the research conclusions. Secondly, except for the six variables (FAG, FSIZE, BSIZE, BIND, BGDIV, and AFM) being explored, other factors are assumed to remain constant. Moreover, according to the results of the study, nearly 80% of the variance in SRDI cannot be explained. Therefore, there is still huge room for future research. The new dimension can be corporate profitability or financial leverage. Thus, these other factors can make the research of disclosure more comprehensive.

8. Conclusion

China's medicine manufacturing industry has developed vigorously in recent years. Simultaneously, sustainable development of enterprises is still the concern of investors and society. The sample of this study comes from the listed companies in the medicine manufacturing industry on CSMAR. Based on 64 listed companies, using Stata 16, this study explored the relationship between six independent variables (FAG, FSIZE, BSIZE, BIND, BGDIV, and AFM) and one dependent variable (SRDI).

By studying the data of the sample companies from 2011 to 2020, we can have a deeper understanding and explanation of the degree of disclosure of sustainability reports in the listed companies in China's medicine manufacturing industry. In reducing the information asymmetry of a company's stakeholders, the sustainability report plays a catalytic role. Research shows that there is still huge room for improvement in the degree of report disclosure of listed companies in this industry. In addition, the degree of disclosure of sustainability reports is affected by the size of the company and the audit firm. In order to enhance the corporate image, large enterprises are more willing to choose the Big Four to conduct audit work. Therefore, enterprises will also pay more attention to the sustainability report emphasized by the Big Four. Moreover, about the sustainability reports, there are still doubts about the conflict between the interests of the board of directors and sustainable development, and the restrictive issues of independent directors' behaviour. However, research can prove that the sensitivity of female directors to sustainable development has also facilitated the disclosure of the report. Generally speaking, in China, due to the lack of strict supervision and enforcement, there are still loopholes in the issue of sustainability report disclosure. In order to realize the sustainable development of enterprises and society, this issue still needs to be taken seriously.

References

Ahmed, M. I., & Che-Ahmad, A. (2016). Effects of corporate governance characteristics on audit report lags. *International Journal of Economics and Financial Issues, 6*(7), 159–164.

Aktas, N., De Bodt, E., & Roll, R. (2013). Learning from repetitive acquisitions: Evidence from the time between deals. *Journal of Financial Economics, 108*(1), 99–117.

Allegrini, M., & Greco, G. (2013). Corporate boards, audit committees and voluntary disclosure: Evidence from Italian listed companies. *Journal of Management and Governance, 17*(1), 187–216.

Amran, A., Lee, S. P., & Devi, S. S. (2014). The influence of governance structure and strategic corporate social responsibility toward sustainability reporting quality. *Business Strategy and the Environment, 23*(4), 217–235.

Ang, S. Y., Palaniappan, S., & Dugan, P. (2021). Sustainable pharmaceutical manufacturing: An overview of recycling and waste management practices. *Sustainability, 13*(9), 4712.

Bai, Y., Li, J., & Zhang, L. (2015). The impact of international supply chains on corporate sustainability practices in China. *Journal of Cleaner Production, 87*, 66–75.

Buniamin, S., Alrazi, B., Johari, N. H., & Rahman, N. R. A. (2011). Corporate governance practices and environmental reporting of companies in Malaysia: Finding possibilities of double thumbs up. *Jurnal Pengurusan, 32*(2011), 55–71.

Castelo Branco, M., Delgado, C., Ferreira Gomes, S., & Cristina Pereira Eugénio, T. (2014). Factors influencing the assurance of sustainability reports in the context of the economic crisis in Portugal. *Managerial Auditing Journal, 29*(3), 237–252.

Chau, G., & Gray, S. J. (2010). Family ownership, board independence and voluntary disclosure: Evidence from Hong Kong. *Journal of International Accounting, Auditing and Taxation, 19*(2), 93–109.

Chen, K. C., & Zhou, J. (2006). Audit committee, board characteristics and auditor switch decisions by Andersen's clients. *Contemporary Accounting Research, 23*(4), 1085–1117.

Christodoulou, D. (2011). The influence of the Big Four accounting firms on corporate reporting quality. *Journal of Accounting and Public Policy, 30*(4), 343–365.

Coffey, B. S., & Wang, J. (1998). Board diversity and managerial control as predictors of corporate social performance. *Journal of Business Ethics, 17*(14), 1595–1603.

Cormier, D., Magnan, M., & Van Velthoven, B. (2005). Environmental disclosure quality in large German companies: Economic incentives, public pressures or institutional conditions? *European Accounting Review, 14*(1), 3–39.

Deegan, C., & Gordon, B. (1996). A study of the environmental disclosure practices of Australian corporations. *Accounting and Business Research, 26*(3), 187–199.

Diamantopoulos, A., Schlegelmilch, B. B., Sinkovics, R. R., & Bohlen, G. M. (2003). Can socio-demographics still play a role in profiling green consumers? A review of the evidence and an empirical investigation. *Journal of Business Research, 56*(6), 465–480.

Dilling, P. F. A. (2010). Sustainability reporting in a global context: What are the characteristics of corporations that provide high quality sustainability reports – An empirical analysis. *International Business & Economics Research Journal, 9*(1), 19–30.

Ernstberger, J., & Grüning, M. (2013). How do firms adjust their board composition to changing external environments? An empirical analysis of corporate governance and disclosure strategies. *Journal of Business Finance & Accounting, 40*(5–6), 715–745.

Evans, S., Bertolini, M., & Zini, C. (2006). Corporate sustainability reporting: Current trends and future directions. *Business Strategy and the Environment, 15*(4), 293–305.

Fathi, J. (2013). The relationship between board size and level of sustainability disclosure: A study of large UK companies. *Corporate Governance: The International Journal of Business in Society, 13*(2), 200–213.

Gomes, P. S., Rodrigues, J. M., & Craig, R. (2015). Sustainability reporting in Portuguese public sector entities. *Sustainability Accounting, Management and Policy Journal, 6*(2), 129–156.

Gray, R., Owen, D., & Adams, C. (2001). Social and environmental reporting: Reporting standard or road less travelled? *Business Strategy and the Environment, 10*(2), 121–130.

Haladu, S. A. (2018). The impact of Big Four audit firms on corporate disclosure practices: Evidence from Nigeria. *African Journal of Business Management, 12*(5), 108–118.

Hamad, M., Dabic, M., & Stojanovic, M. (2020). The role of gender diversity in the sustainability of corporate boards: Evidence from Turkey. *Journal of Cleaner Production, 255*, 120217.

Hashim, H. A., & Devi, S. (2008). Board characteristics, ownership structure and earnings quality: Malaysian evidence. *Research in Accounting in Emerging Economies, 8*(97), 97–123.

Hassan, M. A., Khalil, A., Kaseb, S., & Kassem, M. A. (2017). Exploring the potential of tree-based ensemble methods in solar radiation modeling. *Applied Energy, 203*, 897–916.

Hossain, M., Islam, K., & Andrew, J. (2006). Corporate social and environmental disclosure in developing countries: Evidence from Bangladesh. *Academy of Taiwan Business Management Review, 2*(1), 257–268.

Huang, T., & Wang, A. (2010, October). Sustainability reports in China: Content analysis. In *2010 international conference on future information technology and management engineering* (Vol. 2, pp. 154–158). IEEE.

Huse, M., & Solberg, A. G. (2006). Gender-related boardroom dynamics: How Scandinavian women make and can make contributions on corporate boards. *Women in Management Review, 21*(2), 113–130.

Jensen, M. C. (1993). The modern industrial revolution, exit, and the failure of internal control systems. *The Journal of Finance, 48*(3), 831–880.

Jizi, M. I., Salama, A., Dixon, R., & Stratling, R. (2014). Corporate governance and corporate social responsibility disclosure: Evidence from the US banking sector. *Journal of Business Ethics, 125*(4), 601–615.

Kathyayini, H., & Nagaraju, N. (2012). Structure and catalytic activity relationships of Fe/Co supported on resistance different sources of Al (OH) 3 in the production of multiwall carbon nanotubes by catalytic chemical vapour deposition method. *Advanced Materials, 3*(4), 309–314.

Khan, A. (2010). Corporate social responsibility from an emerging market perspective: Evidences from the Indian pharmaceutical industry. *Corporate Social Responsibility and Environmental Management, 17*(2), 111–124.

Krechowicz, D., & Fernando, H. (2009). Sustainable development and corporate reporting in the European Union: From policy to practice. *Journal of Cleaner Production, 17*(11), 932–942.

Lakhal, F. (2005). Voluntary earnings disclosures and corporate governance: Evidence from France. *Review of Accounting and Finance, 4*(3), 64–85.

Larmou, S., & Vafeas, N. (2010). The relation between board size and firm performance in firms with a history of poor operating performance. *Journal of Management & Governance, 14*(1), 61–85.

Li, Y., & Xiang, E. (2007). The impact of corporate social responsibility on firm financial performance: A comparison of different industry sectors in China. *Journal of Applied Business Research, 23*(2), 61–72.

Liao, L., Luo, L., & Tang, Q. (2014). Gender diversity, board independence, environmental committee and greenhouse gas disclosure. *The British Accounting Review, 46*(1), 91–104.

Lim, S., Matolcsy, Z., & Chow, D. (2007). The association between board composition and different types of voluntary disclosure. *European Accounting Review, 16*(3), 555–583.

Lozano, R., Nummert, B., & Ceulemans, K. (2016). Elucidating the relationship between sustainability reporting and organisational change management for sustainability. *Journal of Cleaner Production, 125*, 168–188.

Mainieri, T., Barnett, E. G., Valdero, T. R., Unipan, J. B., & Oskamp, S. (1997). Green buying: The influence of environmental concern on consumer behavior. *The Journal of Social Psychology, 137*(2), 189–204.

Montecalvo, M., Farneti, F., & de Villiers, C. (2018). The influence of integrated reporting on corporate disclosure practices: A case study. *Journal of Business Ethics, 147*(3), 551–567.

Ntim, C. G., Lindop, S., & Thomas, D. A. (2013). Corporate governance and risk reporting in South Africa: A study of corporate risk disclosures in the pre-and post-King II periods. *International Review of Financial Analysis, 30*, 246–256.

Prasad, S., Tiwari, M., Pandey, A. N., Shrivastav, T. G., & Chaube, S. K. (2016). Impact of stress on oocyte quality and reproductive outcome. *Journal of Biomedical Science, 23*, 1–5.

Rahman, M. J., & Chen, X. (2023). CEO characteristics and firm performance: Evidence from private listed firms in China. *Corporate Governance: The International Journal of Business in Society, 23*(3), 458–477.

Rahman, M. J., Zhu, H., & Chen, S. (2023). Does CSR reduce financial distress? Moderating effect of firm characteristics, auditor characteristics, and Covid-19. *International Journal of Accounting & Information Management, 31*(5), 756–784.

Rahman, M. J., Zhu, H., & Jiang, X. (2024). Family firms, client importance, and auditor reporting behavior: Evidence from China. *Meditari Accountancy Research, 32*(2), 543–578.

Rahman, M. J., Zhu, H., & Yue, L. (2024). Does the adoption of artificial intelligence by audit firms and their clients affect audit quality and efficiency? Evidence from China. *Managerial Auditing Journal, 39*(6), 668–699.

Rahman, M. J., Zhu, H., Zhang, Y., & Hossain, M. M. (2024). Effect of female representation in audit committees on non-audit fees: Evidence from China. *Meditari Accountancy Research*.

Rouf, M. A. (2011). The relationship between corporate governance and value of the firm in developing countries: Evidence from Bangladesh. *Corporate Governance, 11*(1), 62–69.

Sandhu, S., Smallman, C., Ozanne, L. K., & Cullen, R. (2012). Corporate environmental responsiveness in India: lessons from a developing country. *Journal of Cleaner Production, 35*, 203–213.

Sarumpaet, S. (2005). The relationship between environmental performance and financial performance of Indonesian companies. *Journal of Accounting and Economics, 24*(1), 27–35.

Sharif, M., & Rashid, K. (2014). Corporate governance and corporate social responsibility (CSR) disclosure: An empirical study of the listed companies in Bangladesh. *Journal of Business Studies Quarterly, 6*(1), 1–14.

Shen, L., & Jin, J. (2006). Corporate governance and environmental reporting in China: An empirical analysis. *Journal of Cleaner Production, 14*(3), 341–350.

Sun, L. D., Cheng, H., Wang, Z. X., Zhang, A. P., Wang, P. G., Xu, J. H., . . . & Zhang, X. J. (2010). Association analyses identify six new psoriasis susceptibility loci in the Chinese population. *Nature Genetics, 42*(11), 1005–1009.

Tang, N., Song, W. X., Luo, J., Haydon, R. C., & He, T. C. (2008). Osteosarcoma development and stem cell differentiation. *Clinical Orthopaedics and Related Research, 466*(9), 2114–2130.

Uwuigbe, U. (2011). An examination of the relationship between management ownership and corporate social responsibility disclosure: A study of selected firms in Nigeria. *Research Journal of Finance and Accounting, 2*(6), 23–29.

Wehrmeyer, W., & McNeil, M. (2000). Gender issues in environmental management: Women as agents of change. *Women in Management Review, 15*(2), 61–69.

World Commission on Environment and Development. (1987). *Our common future*. Oxford University Press.

Xinhua, M., Ailin, J., Jian, T., & Dongbo, H. (2012). Tight sand gas development technology and practices in China. *Petroleum Exploration and Development, 39*(5), 611–618.

Yermack, D. (1996). Higher market valuation of companies with a small board of directors. *Journal of Financial Economics, 40*(2), 185–211.

10 Does Size Matter? Evaluating Company Environmental Information Disclosure in China's Food Industry

A Combined Approach of Quantity and Quality Measurement

Md Jahidur Rahman, Tarek Rana, Hongtao Zhu, and Ying Yilu

1. Introduction

As environmental pollution issues become increasingly severe, green economy has emerged as the mainstream approach, prompting China to urgently shift from resource- and pollution-intensive economic practices (Dong et al., 2021). The complexity of corporate environmental activities has created a mismatch between environmental information and traditional accounting principles, preventing its inclusion in financial statements (Ortas et al., 2014). Nevertheless, corporate environmental information disclosure (EID) helps address information asymmetry, serving as a key channel for external stakeholders to understand corporate environmental behaviour and responsibility. Since the publication of the Measures for the Disclosure of Environmental Information (MDEI) in 2008, the Chinese government has actively promoted corporate environmental governance, leading to an increase in sustainability disclosures through annual reports, standalone reports, and online platforms (KPMG, 2008, 2011). Despite the rise in EID among listed companies in China, 70% still fail to provide valid information (Lu et al., 2023).

Most studies focus on general motivations for EID across various industries, with limited attention given to the food sector, where environmental issues are equally pressing. Recent food safety crises have led many companies to bankruptcy and hindered the development of China's food industry. For food companies, corporate information disclosure is one of the most effective and direct ways for stakeholders to understand environmental protection measures, such as energy conservation and pollution control. Stakeholders can only make informed decisions based on a comprehensive understanding of this information. Furthermore, only by obtaining a summary of environmental information across the entire food industry can regulatory bodies fully grasp the current environmental status and implement effective measures. However, public interest tends to focus more on product quality reports than on EIDs.

DOI: 10.4324/9781003488965-13

Research has shown that the quality of EID is related to factors such as debt costs, media attention, and managerial preferences (Ji et al., 2020). Yet, few studies in China investigate the relationship between disclosure quality and quantity specifically in the food sector. This study aims to explore these relationships, questioning whether an increase in quantity equates to an improvement in quality.

This study contributes to the literature in several ways. Firstly, it utilizes content analysis to examine EID across all listed food companies in China, providing valuable insights that have been missing in broader industry studies. Secondly, it expands the research focus on environmental information within the food industry, investigating the relationship between disclosure quantity and quality, noting that a higher word count does not necessarily indicate improved quality. In China, where information dissemination channels are increasingly diverse, the public is becoming more adept at discerning useful information. Thirdly, this study utilizes the Environmental Research Database of Listed Companies in China [China Stock Market & Accounting Research (CSMAR)], which contains diverse sources of information to evaluate the quality of EID among food companies. Fourthly, sector-specific GRI guidelines are employed to measure and assess environmental information, addressing the limitations of general GRI frameworks that may not apply to specific industries, thereby promoting sustainable development reporting (Lee, 2017).

The study will analyse EID quality and quantity through content analysis, leveraging extensive data from various corporate reports. This approach will develop a quality scoring index based on the GRI framework (Lee, 2017; Skouloudis et al., 2012; Roca & Searcy, 2012). It is anticipated that a clear relationship exists between EID quantity and quality. Larger companies may disclose more comprehensive information across multiple channels, potentially correlating market value with EID quantity (Lee, 2017). Additionally, an increase in disclosed information may indicate a greater level of useful environmental details. To enhance practical relevance, the study will compare the best-performing ($n = 15$) and worst-performing ($n = 15$) companies to identify subtle differences in disclosure quality. In the following sections, we will review the literature on EID in China, outline the measurement methods, present the findings, and provide the conclusion.

2.   Literature Review

2.1   Evaluation of Environmental Information Disclosure

More companies are increasingly using various methods to disclose environmental information. In the 1980s, American companies faced growing public demand for ecological performance data. This pressure, along with environmental legislation, compelled companies to actively engage in environmental management plans. However, systematic methods for measuring environmental disclosure have been lacking (Beck et al., 2010; Unerman, 2000; Wiseman, 1982). One of the earliest attempts to evaluate EID involved developing a rating system for environmentally related items. Additionally, the Sustainability – United Nations Environment

Programme (UNEP) created guidelines in 1996 to index key factors related to disclosure quality. As corporate systems and economic conditions evolve, new evaluation methods have emerged. The Global Reporting Initiative (GRI) has played a crucial role in enhancing sustainability reporting, with its general guidelines serving as valuable tools for evaluating environmental information. However, due to their limited applicability across specific industry sectors, sector-based GRI guidelines are increasingly being utilized in academic research.

In China, influenced by government requirements for listed companies, more firms are adopting EID to showcase their environmental performance. Annual reports remain the predominant format, although corporate social responsibility (CSR) and environmental reports are also common. Existing studies employ various methods to measure environmental indicators. Research in China identifies two primary approaches: one evaluates the extent of environmental disclosure through content analysis and a scoring technique (Huang & Kung, 2010), while the other relies on external assessments from professional evaluation institutions and databases like CSMAR or Thomson Datastream Index Service. Due to the lack of an authoritative evaluation agency for EID, the content analysis method has gained popularity in this field (Ji et al., 2020). Several studies in China also investigate the relationship between environmental accounting and firm performance (Rahman & Wu, 2023; Rahman et al., 2023, 2024; Zhu & Rahman, 2024). Therefore, there is a positive relationship between environmental cost and firm performance.

To develop the scoring technique, the content of annual, CSR, or environmental reports is categorized and analysed based on different themes. These themes help define the quality of environmental information, considering factors like usefulness and relevance. Points are then assigned according to the level of each theme. While EID is influenced by industry-specific factors, environmental information can generally be classified into three categories: financial information, non-financial information, and descriptive information. In China, qualitative data are predominant, with non-monetary information being more common than financial data (Ji et al., 2020).

2.2 *Measurement of Environmental Information Disclosure*

In the past decade, the Chinese government has attached great importance to the issues of corporate EID, and system construction has been steadily advancing. The business community also recognizes the necessity of acting sustainably and the urgency of taking this responsibility. The significance of EID has aroused broad academic interests. As a technique to collect statistics, content analysis involves coding quantitative and qualitative information into different categories that have been pre-determined to get the pattern inherent in the reporting (Ji et al., 2020). When the disclosure-scoring technique is applied, one of the simple methods is to see whether the pre-defined categories exist in the environmental information. However, this method is too simple that it will cause the performances of some companies to be underestimated. Therefore, another assumption was brought up that "quantity of disclosures within a category signifies the importance of that category" (Unerman, 2000, p. 674). Many previous studies have scored the quantity of EID by accounting for the number of words, sentences, or pages (Deegan & Gordon, 1996; Unerman, 2000; Van Staden & Hooks, 2007). Different researchers have different opinions about the instruments used to measure the disclosure information. Some academics argued that counting sentences may be more meaningful than counting words, while others think that counting sentences will ignore the possibility of writing or expression differences (Beck et al., 2010; Milne & Adler, 1999; Unerman, 2000; Van Staden & Hooks, 2007). Therefore, there is no single or standard measurement method. Each method has pros and cons, and thus, the focus on both quality and quantity of environmental information can increase the accuracy and reliability of results.

3. Research Methods

3.1 Sample

The sample firms for this study are selected from all listed food processing companies that cooperated in China in 2019. Besides, all listed food companies are ranked by market capitalization to analyse the relationship between corporation size and the quantity (number of words) and quality of EID. Among 71 companies, we conducted initial information checking to finalize our sample. According to the CSMAR database of Environmental Carrier Information, we found there are 49 companies that disclosed some environmental information in 2019. After further examining all 2019 annual reports, CSR reports, and environmental reports, I found that six companies did not provide useful and meaningful environmental information in 2019. I excluded them to keep the results more reasonable. In total, 43 companies were finally selected for analysis (see Table 10.1). I used three data sources for environmental information: 2019 annual reports, CSR reports, and environmental reports. Besides, I used the CSMAR database and TianYanCha.com, and Eastmoney Securities to find companies' market capitalization and 2019 yearly operation revenue.

3.2 Environmental Disclosure Measurement

3.2.1 Quantity (Extent) of Environmental Disclosure

This study calculates the quantity of EID by counting the number of words of detailed and specified environmental information. Many previous studies measure the extent of environmental disclosure by measuring the presence or absence of disclosure items (e.g. Beck et al., 2010; Beck et al., 2010; Brammer & Pavelin, 2006, 2008; Cormier et al., 2005; Frost et al., 2005; Haniffa & Cooke, 2005). This measurement is simplistic and sometimes overlooks some qualitative differences, such as innovative pollution treatment projects. For instance, some companies' creative projects are of superior performance and can deal with more pollution than their peers, creating innovative projects with poor performance. At the same time, the

Table 10.1 Information of Sample Companies

Company	Symbol	Market Capital (CNY, M)	Revenue (CNY, M)	Sub-industry	Report Name and Year
Foshan Haitian Flavouring and Food Company Ltd.	603,288	494,100	19,797	Flavoured fermented product	2019 Annual Report
Inner Mongolia Yili Industrial Group Co., Ltd.	600,887	230,700	90,223	Drink milk	2019 Annual Report, CSR Report, Environmental Report
Henan Shuanghui Investment & Development Co., Ltd.	000,895	97,426	60,348	Flavoured fermented product	2019 Annual Report
Fujian Anjoy Foods Co., Ltd.	603,345	41,919	5,267	Food processing	2019 Annual Report
Byhealth Co., Ltd.	300,146	40,977	5,262	Food processing	2019 CSR Report
Angel Yeast Co., Ltd.	600,298	40,377	7,653	Flavoured fermented product	2019 Annual Report, CSR Report
Toly Bread Co., Ltd.	603,866	26,681	5,644	Leisure food	2019 Annual Report
Chongqing Fuling Zhacai Group Co., Ltd.	002,507	26,585	1,990	Flavoured fermented product	2019 Annual Report
Jonjee Hi-Tech Industrial and Commercial Holding Co., Ltd.	600,872	26,416	4,675	Flavoured fermented product	2019 Annual Report, CSR Report
Shanghai Milkground Food Tech Co., Ltd	600882	25191	1744	Drink Milk	2019 Annual Report, CSR Report
Chacha Food Company Ltd.	002,557	22,916	4,837	Leisure food	2019 Annual Report, CSR Report
Xiamen Kingdomway Group Company	002,626	19,456	3,192	Food processing	2019 Annual Report

(*Continued*)

Table 10.1 (Continued)

Company	Symbol	Market Capital (CNY, M)	Revenue (CNY, M)	Sub-industry	Report Name and Year
Qianhe Condiment and Food Co., Ltd.	603,027	16,359	1,355	Flavoured fermented product	2019 Annual Report
Bright Dairy & Food Co., Ltd.	600,597	16,078	22,563	Drink milk	2019 Annual Report, CSR Report
Jiangsu Hengshun Vinegar Industry Co., Ltd.	600,305	15,255	1,832	Flavoured fermented product	2019 Annual Report
Sanquan Food Co., Ltd.	002,216	14,287	5,986	Food processing	2019 Annual Report
Three Squirrels Inc.	300,783	14,115	10,173	Leisure food	2019 Annual Report
New Hope Dairy Co., Ltd.	002,946	13,408	5,675	Drink milk	2019 Annual Report
Guangzhou Restaurant Group Company Limited	603,043	11,244	3,029	Leisure food	2019 Annual Report
Shandong Longda Meat Foodstuff Co., Ltd.	002,726	10,331	16,822	Food processing	2019 Annual Report
Beijing Sanyuan Foods Co., Ltd.	600,429	8,881	8,151	Drink milk	2019 Annual Report, CSR Report
Shanghai Maling Aquarius Co., Ltd.	600,073	7,586	23,404	Food processing	2019 Annual Report
Jiangxi Huangshanghuang Group Food Co., Ltd.	002,695	7,154	2,117	Leisure food	2019 Annual Report
Sirio Pharma Co., Ltd.	300,791	7,041	1,580	Food processing	2019 Annual Report
Yanker Shop Food Co., Ltd.	002,847	6,874	1,399	Leisure food	2019 Annual Report
Xiwang Foodstuffs Co., Ltd.	000,639	5,419	5,227	Food processing	2019 Annual Report
Jiajia Food Group Co., Ltd.	002,650	5,242	2,040	Flavoured fermented product	2019 Annual Report

(Continued)

Table 10.1 (Continued)

Company	Symbol	Market Capital (CNY, M)	Revenue (CNY, M)	Sub-industry	Report Name and Year
Youyou Foods Co., Ltd.	603,697	5,180	1,008	Leisure food	2019 Annual Report
Zhejiang Huatong Meat Products Co., Ltd.	002,840	4,852	7,706	Food processing	2019 Annual Report
Guangdong Yantang Dairy Co., Ltd.	002,732	4,650	1,471	Drink milk	2019 Annual Report
Ganso Co., Ltd.	603,886	4,277	2,223	Leisure food	2019 Annual Report
Qingdao Richen Food Co., Ltd.	603,755	4,142	286	Flavoured fermented product	2019 Annual Report
Haoxiangni Health Food Co., Ltd.	002,582	4,081	5,961	Leisure food	2019 Annual Report, CSR Report
Qinghai Spring Medicinal Resources Technology Co., Ltd.	600,381	4,039	234	Leisure food	2019 Annual Report
Royal Group Co., Ltd.	002,329	3,996	2,253	Drink milk	2019 Annual Report
Xinjiang Tianrun Dairy Co., Ltd.	600,419	3,417	1,627	Drink milk	2019 Annual Report
Shandong Delisi Food Co., Ltd.	002,330	2,996	2,346	Food processing	2019 Annual Report
Xinjiang Western Animal Husbandry Co., Ltd.	300,106	2,659	649	Drink milk	2019 Annual Report
Shanghai Jiaoda Onlly Co., Ltd.	600,530	2,644	342	Food processing	2019 Annual Report
Haixin Foods Co., Ltd.	002,702	2,630	1,385	Food processing	2019 Annual Report,CSR Report
Lanzhou Zhuangyuan Pasture Co., Ltd.	002910	2517	814	Drink Milk	2019 Annual Report
Tianjin Guifaxiang 18th Street Mahua Food Co., Ltd.	002,820	2,107	507	Leisure food	2019 Annual Report
Shandong Huifa Foodstuff Co., Ltd.	603,536	1,710	1,210	Food processing	2019 Annual Report

quantity of EID of the former is also larger than the latter. Therefore, measuring the amount of information disclosed can more comprehensively show the company's environmental performance.

3.2.2   Unit of Analysis

Identification of the unit of analysis is vital when using content analysis (Neuendorf, 2002) because it is related to judgement (Krippendorff, 2004). In this study, a word is regarded as a measurement unit, and the number of words requires less judgement, resulting in less measurement error and more accurate results.

3.2.3   Quality of Environmental Disclosure

The quantity of EID does not show the actual performance of companies. Thus, measurement of the quantity is not enough and even misinterpret corporate environmental disclosure. Shortcomings can be mitigated by combining quantity and quality analysis of EID (Beck et al., 2010; Guthrie et al., 2004).

The content analysis contains two procedures in this study: (1) identification of the content of specified and detailed environmental information and (2) measurement of the quantity (number of words) of EID and calculation of the quality scores.

Previous studies mainly used the disclosure quality index to measure the quality of EID by monetary, non-monetary, and narrative information (e.g. Beck et al., 2010; Cormier et al., 2005; Guthrie & Farneti, 2008; Jones & Alabaster, 1999; Raar, 2002; Wiseman, 1982; Zeghal & Ahmed, 1990). However, sector-based GRI guidelines are more specific and avoid vague information than the general quality index.

Following GRI's food sector disclosure guidelines, we selected 12 environmental aspects, including 34 items (see Table 10.3). A maximum point of 4 was assigned for each item according to the rules adapted from Ki-Hoon Lee (2017; see Table 10.2).

Table 10.2 Five-Point Scale

Scale	Description
0	Not disclosed, no discussion on the environmental issue
1	Minimum coverage, little detail – general terms. Anecdotal or briefly mentioned
2	Descriptive: the impact of the company or its policies was clearly evident
3	Quantitative: the environmental impact was clearly defined in monetary terms or actual physical quantities. Clearly defined measurement methodology of environmental performance provided
4	Truly extraordinary. Positive and negative environmental disclosure included. Full website, full printed sustainability report, and annual reports provide same information. Benchmarking against best practice

Source: Adapted from Ki-Hoon Lee (2017)

4. Findings

4.1   Quantity and Quality of Environmental Disclosure

This study uses three types of reporting formats (annual reports, sustainability reports, and environmental information reports) to count the number of EID words (see Table 10.4). In total, companies reported on average 2,969 words in different reports in 2019, ranging between 49 and 18,193. Among 43 companies, the greatest volume of words of EID was made via the annual report. Nine companies utilized both annual reports and sustainability reports to disclose their environmental information. The most popular among the three methods is the use of annual reports to disclose environmental information. Around 80% of companies only used annual reports to disclose environmental information (on average, the companies reported 2,808 words, ranging between 48 and 18,193 words).

Table 10.3 Environmental Disclosure Quality Index and the Scale

Aspect	GRI Index (Enn)*	GRI Sub-index* (ENn-1–Enn-4)	Scale
Material	EN1 (materials used by weight or volume)	EN1-1–EN1-4	0–4
	EN2 (percentage of materials used that are recycled input materials)	EN2-1–EN2-4	0–4
Energy	EN3 (energy consumption within the organization)	EN3-1–EN3-4	0–4
	EN4 (energy consumption outside of the organization)	EN4-1–EN4-4	0–4
	EN5 (energy intensity)	EN5-1–EN5-4	0–4
	EN6 (reduction of energy consumption)	EN6-1–EN6-4	0–4
	EN7 (reductions in energy requirements of products and services)	EN7-1–EN7-4	0–4
Water	EN8 (total water withdrawal by source)	EN8-1–EN8-4	0–4
	EN9 (water sources significantly affected by withdrawal of water)	EN9-1–EN9-4	0–4
	EN10 (percentage and total volume of water recycled and reused)	EN10-1–EN10-4	0–4
Biodiversity	EN11 (operational sites owned, leased, managed in, or adjacent to, protected areas and areas of high biodiversity value outside protected areas)	EN11-1–EN11-4	0–4
	EN12 (description of significant impacts of activities, products, and services on biodiversity in protected areas and areas of high biodiversity value outside protected areas)	EN12-1–EN12-4	0–4
	EN13 (habitats protected or restored)	EN13-1–EN13-4	0–4
	EN14 (total number of IUCN red list species and national conservation list species with habitats in areas affected by operations, by level of extinction risk)	EN14-1–EN14-4	0–4

(Continued)

Table 10.3 (Continued)

Aspect	GRI Index (Enn)*	GRI Sub-index* (ENn-1–Enn-4)	Scale
Emissions	EN15 [direct greenhouse gas (GHG) emissions, Scope 1]	EN15-1–EN15-4	0–4
	EN16 (energy indirect GHG emissions, Scope 2)	EN16-1–EN16-4	0–4
	EN17 (other indirect GHG emissions, Scope 3)	EN17-1–EN17-4	0–4
	EN18 (GHG emission intensity)	EN18-1–EN18-4	0–4
	EN19 (reduction of GHG emissions)	EN19-1–EN19-4	0–4
	EN20 [emissions of ozone-depleting substances (ODS)]	EN20-1–EN20-4	0–4
	EN21 (nitrogen oxides, sulfur oxides, and other significant air emissions)	EN21-1–EN21-4	0–4
Effluent and Waste	EN22 (total water discharge by quality and destination)	EN22-1–EN22-4	0–4
	EN23 (total weight of waste by type and disposal method)	EN23-1–EN23-4	0–4
	EN24 (total number and volume of significant spills)	EN24-1–EN24-4	0–4
	EN25 (weight of transported, imported, exported, or treated waste deemed hazardous under the terms of the Basel Convention 2 Annexes I, II, III, and VIII, and percentage of transported waste shipped internationally)	EN25-1–EN25-4	0–4
	EN26 (identity, size, protected status, and biodiversity value of water bodies and related habitats significantly affected by the organization's discharges of water and run-off)	EN26-1–EN26-4	0–4
Product and Services	EN27 (extent of impact mitigation of environmental impacts of products and services)	EN27-1–EN27-4	0–4
	EN28 (percentage of products sold and their packaging materials that are reclaimed by category)	EN28-1–EN28-4	0–4
Compliance	EN29 (monetary value of significant fines and total number of non-monetary sanctions for non-compliance with environmental laws and regulations)	EN29-1–EN29-4	0–4
Transport	EN30 (significant environmental impacts of transporting products and other goods and materials for the Organization's operations, and transporting members of the workforce)	EN30-1–EN30-4	0–4
Overall	EN31 (total environmental protection expenditures and investments by type)	EN31-1–EN31-4	0–4
Supplier Environmental Assessment	EN32 (percentage of new suppliers that were screened using environmental criteria)	EN32-1–EN32-4	0–4

(Continued)

Table 10.3 (Continued)

Aspect	GRI Index (Enn)*	GRI Sub-index* Scale (ENn-1–Enn-4)
	EN33 (significant actual and potential negative environmental impacts on the supply chain and actions taken)	EN33-1–EN33-4 0–4
Environmental Grievance Mechanisms	EN34 (number of grievances about environmental impacts filed, addressed, and resolved through formal grievance mechanisms)	EN34-1–EN34-4 0–4

*The contents of ENn are subtracted from *G4 Sustainability Reporting Guidelines Implementation Manual GRI* (2013).
*Four categories in the sub-industry of all food companies.

Table 10.4 Market Cap, Quantity, and Quality of Environmental Disclosure by Report Type

		Annual Report & Sustainability Reports (n = 9)	*Annual, Sustainability, & Environmental Reports (n = 1)*	*Annual Report Only (n = 34)*	*Overall (n = 43)*
Market cap (million)	Mean	27,111	97,426	31,292	30,417
	Median	22,916	97,426	6,958	7,586
	Range	2,630–97,426	97,426	1,710–494,100	1,710–494,100
Quantity (number of words)	Mean	2,339	11,812	2,808	2,969
	Median	3,716	11,812	1,391	1,758
	Range	1,097–11,812	11,812	49–18,193	49–18,193
Quality (scores)	Mean	40.7	58	30.3	32.6
	Median	39	58	31.5	35
	Range	31–58	58	4–68	4–68

In comparison, only one corporation used all annual reports, sustainability reports, and environmental information reports. Besides, in Table 10.4, following GRI's G4 sustainability reports guidelines and the score scale that adopted for the food industry, the quality of EID in the quality index was scored from 12 aspects. We reported the aggregate scores for each element and each company. In terms of the quantity and the quality of EID, the corporation utilizing all three types of reports achieved the highest mean scores of 11,812 and 58, respectively.

In Table 10.5, materials (mean = 4.33, ranging between 4 and 7), water (mean = 5.12, ranging between 0 and 12), emissions (mean = 10.93, ranging between 0 and 20), effluent, and waste (mean = 3.6, ranging between 0 and 15) are the environmental aspects of EID with top four scores, and also ones that have larger than 3 points. The remaining scores did not exceed 2 points, at a low level. These values

Table 10.5 Quality Scores by Index

Max. Score	Mean	Median	Range
7	4.33	4	4–7
5	0.7	0	0–5
12	5.12	4	0–12
4	0.35	0	0–4
20	10.93	14	0–20
15	3.6	3	0–15
6	1.6	2	0–6
4	1.86	2	0–4
2	0.07	0	0–2
3	1.86	2	0–3
6	1.19	0	0–6
3	0.98	1	0–3
87			

reflect that the quality of EID of Chinese food companies has not reached the level required by international standards.

4.2   *Comparison between Quantity and Quality of Environmental Disclosure*

In this study, factors like number of words, quality scores, and market capitalization are utilized to evaluate the EID of companies. Besides, I correlate them against one another to find specific relationships.

Considering Panel A in Table 10.6, the correlation between market capitalization and number of words is significant ($R = 0.467$, $p < 0.01$). The correlation between market capitalization and quality scores is also significant at the 10% significance level ($R = 0.339$, $p < 0.1$). It is expected that the growth of companies' sizes will promote the development of the quantity and increase the quality of EID. Companies of large sizes tend to disclose more information than smaller companies. Those results are consistent with the findings of Ki-Hoon Lee (2017), a significant correlation between the market capitalization of companies and the quantity or quality of EID. Moreover, it is vital to notice a high and positive correlation between the number of words and EID quality ($R = 0.916$, $p < 0.01$). The increase in the number of words will improve the quality of the disclosed information because more valuable and different environmental information is publicized.

However, according to Hooks and van Staden (2011), the existence of "outliers" (the numerical distance between data points) may significantly influence the value of correlations. It is difficult to interpret the results without considering "outliers" in a small sample. Thus, we ranked 43 companies based on the number of words or quality scores to overcome this problem. We used Spearman's rank correlation for my analysis. According to Table 10.6 Panel B, the correlation between market cap rank and quantity rank ($R = 0.493$, $p < 0.01$) and between market cap rank and quality rank ($R = 0.485$, $p < 0.01$) are both higher than the correlation between actual items (Panel A). However, as for the correlation between quality rank and quantity

Table 10.6 Correlation: Market Cap, Extent, and Quality of Disclosure ($n = 43$ Companies)

Panel A. Market Cap, Quantity, and Quality

Correlation (p-Value)	*Market Cap*	*Quantity (Number of Words)*	*Quality (Scores)*
Market Cap	*1*		
Quantity (number of words)	0.467 (<0.01)***	1	
Quality (scores)	0.339 (<0.1)**	0.916 (<0.01) ***	1

Panel B. Market Cap Rank, Quantity Rank, and Quality Rank

Correlation (p-Value)	*Market Cap Rank*	*Quantity Rank*	*Quality Rank*
Market Cap Rank	*1*		
Quantity rank	0.493 (<0.01) ***	1	
Quality rank	0.485 (<0.01) ***	0.840 (<0.01) ***	1

*Because the variables are not normally distributed, the non-parametric Spearman correlation test is used in this study.

rank ($R = 0.840$, $p < 0.01$), it is lower than the actual one (Panel A). Overall, based on the results of Panels A and B, there is a significant correlation between sizes of companies and number of words or quality, and a positive and high correlation between the number of words and quality of EID.

Since there is a significant relationship between market capitalization and quality or quantity, we expect that companies with good performances and poor performances by rank may show different levels in the quality and quantity of EID. To verify this conjecture, we chose 15 top-ranked companies and 15 bottom-ranked companies. By conducting a Mann–Whitney U-test and comparing the means and standard deviations, we can check whether there are differences in quantity and quality measurements of the two groups. In Table 10.7, companies are ranked by three different elements. In Panel A, companies are ranked by market capitalization; the top-performer group is composed of 15 companies with highest market capitalization, and the bottom-performer group is the opposite. The Mann–Whitney U-test shows significant differences at the 1% level and 5% level between the top and bottom groups for these indexes. In Panel B, the test is repeated with the companies ranked by the quantity of EID, and the results are similar to Panel A. In Panel C, we repeat the test by using the companies ranked by the quality of EID, obtaining similar results to those in Panels A and B. All results indicate considerable differences between the two groups in the quality and quantity of EID.

5.   Conclusion

China, being the most populous country in the world, has an enormous demand for food, which drives the growth of food companies. With the ongoing enhancement

Table 10.7 Comparison of Disclosure between Top- and Bottom-Ranked Companies (*n* = 43 Companies)

Panel A. Companies Ranked by Market Cap

	Market Cap		Quantity (Number of Words)		Quality (Scores)	
	Top	Bottom	Top	Bottom	Top	Bottom
N	15	15	15	15	15	15
Mean	76,029	3,381	4,414	1,029	40	25
Std. deviation	128,069	978	3,435	1,078	16	14
Z-stat.	−4.666***		−3.422***		−2.491**	

Panel B. Companies Ranked by Quantity

	Market Cap		Quantity (Number of Words)		Quality (Scores)	
	Top	Bottom	Top	Bottom	Top	Bottom
N	15	15	15	15	15	15
Mean	36,613	9,465	6,699	332	47	14
Std. deviation	58,409	10,910	4,050	346	10	11
Z-stat.	−2.8***		−4.666***		−4.462***	

Panel C. Companies Ranked by Quality

	Market Cap		Quantity (Number of Words)		Quality (Scores)	
	Top	Bottom	Top	Bottom	Top	Bottom
N	15	15	15	15	15	15
Mean	35,043.20	9,604.13	5,361.80	529.47	49.40	13.33
Std. deviation	59,165.04	10,987.64	2,978.07	865.27	7.75	8.58
Z-stat.	−2.219**		−4.459***		−4.672***	

Std. dev. is the standard deviation.
Z-stat (TvB) is the Z-statistics from comparing the medians of the top and bottom companies using the Mann–Whitney *U*-test.
***Significant at the 0.01 level.
**Significant at the 0.05 level

of EID standards, companies must invest significant effort to improve the quality of their environmental disclosures to meet the requirements and establish themselves as trusted and responsible organizations in a sustainable world. Consequently, enhancing the quality of EID has become a crucial focus for these enterprises. The GRI guidelines help companies recognize their impacts on the environment and society, thereby increasing accountability and transparency in their environmental protection efforts. However, debates persist regarding the relationship between the quantity and quality of EID. Specifically, does an increase in the quantity of disclosures lead to higher quality? This research aims to address this question based on findings from the Chinese food industry.

Firstly, company size significantly influences both the quantity and quality of EID. This study examines all food companies in China, excluding those that do not disclose any useful environmental information. By analysing the market capitalization of 43 companies, we investigated the relationship between the quantity (measured by word count) and quality of EID. The results indicate a significant relationship between company size and both the quantity and quality of EID, with larger companies expected to disclose more high-quality environmental information. Generally, larger firms are likely to invest more effort and adhere to stricter sustainability standards compared to smaller ones. Additionally, a smaller sample of 43 companies was analysed to further explore differences between top and bottom performers in terms of EID quantity and quality. The findings reveal a strong correlation between the quantity and quality of EID, suggesting that increasing the amount of disclosed information can enhance its quality.

Secondly, this study contributes a valuable research direction to the existing Chinese literature. Few studies in China have utilized GRI guidelines to assess corporate environmental information quality. Currently, China lacks comprehensive and systematic standards for EID, which regulate corporate behaviour. The GRI guidelines, as international and independent principles, provide a robust framework for corporations worldwide, encompassing 12 aspects and 34 items to govern EID requirements. These guidelines could serve as an effective reference for the Chinese government in formulating standards. Due to the absence of stringent government mandates, the quality of information disclosed by Chinese food companies varies widely. Some companies may even withhold environmental information to conceal damage and protect their reputations. Furthermore, the GRI food sector guidelines and five-point scale offer a more precise and complete evaluation of EID quality.

Thirdly, utilizing multiple sources of environmental information enriches the analysis. This study gathers EID from annual reports, CSR reports, and environmental reports, providing a clearer and more comprehensive view of the current disclosure practices among Chinese food companies. Given that the environmental information disclosed by Chinese listed food companies is often limited, collecting data from multiple sources is essential.

However, this research has certain limitations. Firstly, it focuses exclusively on Chinese listed food companies, which are part of an emerging market and face different pressures regarding EID compared to firms in developed countries. Additionally, the scoring of EID quality and the methods employed are somewhat subjective, meaning that results may vary if other researchers replicate the study. The small sample size may also mean that the relationship between EID quantity and quality is influenced by other implicit factors, which warrants further exploration in future research. Moreover, besides the three report types examined here, many companies also use their websites to disclose environmental information, which could be evaluated in subsequent studies.

References

Beck, C., Campbell, D., & Shrives, P. (2010). Content analysis in environmental reporting research: Enrichment and rehearsal of the method in a British–German context. *British Accounting Review, 42*(3), 207–222.

Brammer, S., & Pavelin, S. (2006). Voluntary environmental disclosures by large UK companies. *Journal of Business Finance & Accounting, 33*(7–8), 1168–1188.

Brammer, S., & Pavelin, S. (2008). Factors influencing the quality of corporate environmental disclosure. *Business Strategy and the Environment, 17*(2),

Cormier, D., Magnan, M., & Van Velthoven, B. (2005). Environmental disclosure quality in large German companies: Economic incentives, public pressures or institutional conditions? *European Accounting Review, 14*(1), 3–39.

Deegan, C., & Gordon, B. (1996). A study of the environmental disclosure practices of Australian corporations. *Accounting and Business Research, 26*(3), 187–199.

Dong, S., Burritt, R., & Qian, W. (2021). Salient stakeholder focus and GRI G4-based sustainability reporting. *Accounting & Finance, 61*(2), 317–343.

Frost, G. R., Jones, S., Loftus, J., & Van Der Laan, S. (2005). A survey of sustainability reporting practices of Australian reporting entities. *Australian Accounting Review, 15*(35), 89–96.

Guthrie, J., & Farneti, F. (2008). GRI sustainability reporting by Australian public sector organizations. *Public Money and Management, 28*(6), 361–366.

Guthrie, J., Petty, R., & Ricceri, F. (2004). External intellectual capital reporting: Evidence from Hong Kong and Australia. *Journal of Intellectual Capital, 5*(2), 259–275.

Haniffa, R. M., & Cooke, T. E. (2005). The impact of culture and governance on corporate social reporting. *Journal of Accounting and Public Policy, 24*(5), 391–430.

Hooks, J., & van Staden, C. J. (2011). Evaluating environmental disclosures: The relationship between quality and extent measures. *The British Accounting Review, 43*(3), 200–213.

Huang, C. L., & Kung, F. H. (2010). Drivers of environmental disclosure and stakeholder expectation: Evidence from Taiwan. *Journal of Business Ethics, 96*(3), 435–451.

Ji, Z., Yu, X., & Yang, J. (2020). Analysis on the environmental information disclosure and stock price reactions of Chinese listed companies. *Environmental Science and Pollution Research, 27*(4), 3670–3681.

Jones, M. J., & Alabaster, T. (1999). Critical analysis of corporate environmental reporting: A case study of the water and energy industries in the UK. *Journal of Environmental Planning and Management, 42*(1), 23–45.

KPMG, L. (2008). For the fiscal year June 30, 2008.

KPMG, L. (2011). For the fiscal year June 30, 2011.

Krippendorff, K. (2004). *Content analysis: An introduction to its methodology* (2nd ed.). Sage.

Lee, K.-H. (2017). Does size matter? Evaluating corporate environmental disclosure in China's food industry. *Sustainability Accounting, Management and Policy Journal, 8*(4), 455–480.

Lu, L., Wang, D., Pu, C., Cao, Y., Li, Y., Xu, P., . . . & Liu, Y. (2023). High-performance flexible organic field effect transistors with print-based nanowires. *Microsystems & Nanoengineering, 9*(1), 80.

Milne, M. J., & Adler, R. W. (1999). Exploring the reliability of social and environmental disclosures content analysis. *Accounting, Auditing & Accountability Journal, 12*(2), 237–256.

Neuendorf, K. A. (2002). Defining content analysis. In *Content analysis guidebook* (pp. 10–27).

Ortas, E., Moneva, J. M., & Salvador, M. (2014). Do firms' environmental and social performances influence the market price of their stocks? *Business Strategy and the Environment, 23*(8), 486–502.

Raar, J. (2002). Environmental initiatives: Towards triple-bottom line reporting. *Corporate Communications: An International Journal, 7*(3), 169–183.

Rahman, M. J., & Wu, J. (2023). M&A activity and ESG performance: evidence from China. *Managerial Finance, 50*(1), 179–197.

Rahman, M. J., Wu, Q., & Zhu, H. (2024). Corporate social responsibility in times of social distancing: Evidence from China. *Business Ethics, the Environment & Responsibility.*

Rahman, M. J., Zhu, H., & Chen, S. (2023). Does CSR reduce financial distress? Moderating effect of firm characteristics, auditor characteristics, and Covid-19. *International Journal of Accounting & Information Management, 31*(5), 756–784.

Roca, L. C., & Searcy, C. (2012). An analysis of indicators disclosed in corporate sustainability reports. *Journal of Cleaner Production, 20*(1), 103–118.

Skouloudis, A., Evangelinos, K., & Kourmousis, F. (2012). Assessing non-financial reports according to the Global Reporting Initiative guidelines: Evidence from Greece. *Journal of Cleaner Production, 24*, 3–23.

Unerman, J. (2000). Methodological issues: Reflections on quantification in corporate social reporting content analysis. *Accounting, Auditing & Accountability Journal, 13*(5), 667–681.

Van Staden, C. J., & Hooks, J. (2007). A comprehensive comparison of corporate environmental reporting and response. *Accounting Forum, 31*(2), 169–185.

Wiseman, J. (1982). An evaluation of environmental disclosures made in corporate annual reports. *Accounting, Organizations and Society, 7*(1), 53–63.

Zeghal, D., & Ahmed, S. A. (1990). Comparison of social responsibility information disclosure media used by Canadian firms. *Accounting, Auditing & Accountability Journal, 3*(1), 38–53.

Zhu, H., & Rahman, M. J. (2024). Ex-ante expected changes in ESG and future stock returns based on machine learning. *The British Accounting Review*, 101457.

Part 4

Environmental Accounting and Disclosure Quality

11 Determinants of Environmental Accounting Disclosure Quality of Listed Chemical Manufacturing Companies in China

Md Jahidur Rahman, Tarek Rana, Hongtao Zhu, and Zhou Yanyan

1. Introduction

Over the past decade, worldwide attention regarding factors that influence environmental accounting disclosure has been booming in the field of accounting. Several categories of internal factors in different national contexts have been proven to have direct and significant relationships with the company's environmental disclosure, such as organizational stakeholders, which bolster the usage of the eco-control system (Abdel-Maksoud et al., 2021), and regulators with innovation strategies to mitigate carbon emissions (Yunus et al., 2020). Other internal factors also have different correlations with environmental disclosure: environmental accounting disclosure was negatively related to board size but positively related to company size (Tarus, 2020). Simultaneously, external factors have also been identified as critical sources to pressure companies' environmental accounting disclosure. For instance, external stakeholders like customers were revealed to play a leading role in promoting the economic and social development of small and medium-sized enterprises (Tahajuddin et al., 2020). On the contrary, many studies have explained that external forces are negatively or not directly related to environmental disclosure. It was stated that companies in countries with a culture of individualism, permissiveness, and long-term orientation would suppress environmental disclosure (Pucheta-Martínez & Gallego-Álvarez, 2020). Yunus (2020) also illustrated that non-regulatory stakeholders like media, creditors, and institutional shareholders do not correlate with carbon-emission disclosure.

Moreover, researchers have tried to find more relationships between environmental disclosure and corporate performance. Most of the existing studies were concentrated in developed countries (Dragomir, 2010) and even on the global scale (Doan & Sassen, 2020), while only a few studies focused on developing countries, such as Nigeria (Iliemena, 2020), Vietnam (Nguyen, 2021), and China (He & Loftus, 2014). Although a few studies were conducted in the context of China, the research objects were listed companies in the energy industry (Chiu et al., 2020) and the mining industry (Agyemang et al., 2021). In the context of Europe, Dragomir (2010) analysed a significant correlation between environmental performance and disclosure, but no association was found between financial performance and

DOI: 10.4324/9781003488965-15

environmental disclosure. Later, Chiu et al. (2020) and Agyemang et al. (2021) concluded that environmental disclosure has a significant and positive correlation with both environmental performance and financial performance with companies listed in China. Thus, the relationship between environmental disclosure and corporate financial performance can differ depending on times, national context, and industries (Dragomir, 2010; He & Loftus, 2014; Acar & Temiz, 2020; Doan & Sassen, 2020; Agyemang et al., 2021).

Currently, Chinese listed companies in the chemistry industry have never been considered before when promoting environmental disclosure awareness. Given that factors such as culture (Pucheta-Martínez & Gallego-Álvarez, 2020), media (Fan et al., 2020), stakeholders (e.g. Huang & Kung, 2010; Bellamy et al., 2020; Yunus et al., 2020), and government regulations (Nguyen, 2021) have been proven to have varying degrees of impact on companies' environmental accounting disclosure, more evidence is needed to evaluate how other more internal factors related to the board of directors and managers affect corporate environmental disclosure, which is still little understood. Finally, the inconsistency in the conclusions reached by research scholars regarding the association between environmental accounting disclosures and financial performance in different country contexts or industries has also revealed a literature gap in this study area. Several studies in China also investigate the relationship between environmental accounting and firm performance (Rahman & Wu, 2023; Rahman et al., 2023, 2024; Zhu & Rahman, 2024).

Therefore, in the specific context of China, the internal factors of listed companies in China's chemical manufacturing industry, such as stakeholder factors including ownership concentration, will be used to explore the association with environmental disclosure. Meanwhile, exploring the potential relationship between environmental disclosure and financial performance, taking firm size (Size), leverage (Lev), earnings per share (EPS), and return on assets (ROA) as indicators, is also one of the purposes of this study, under the circumstance that China has begun to pay attention to and attach importance to environmental disclosure while pursuing low-carbon goals.

Firstly, the chemical industry, a key energy-consuming sector in China, has the largest carbon dioxide emissions in the country and is one of the world's top five largest producers of chemical products (Jiahuey et al., 2019), so it is of great practical significance to take listed companies in China's chemical manufacturing industry as samples in this study to fill the gap. Secondly, this study can provide inspiration for the self-governance of companies within the industry and offer investors, shareholders, and other relevant groups an understanding of the prospects of the industry in China's sustainable development environment. Lastly, when the association between the three is established, enterprises are more likely to improve financial performance and meet environmental standards at the same time by changing the corresponding internal allocation of ownership.

This research makes contributions to the current literature in several ways. The context of most of the existing studies in the literature on environmental accounting disclosures was mainly in a broad industry or in countries with different levels of development or in some specific states and provinces (Huang & Kung, 2010;

Tahajuddin et al., 2020; Agyemang et al., 2021; Tahajuddin & Sulaiman, 2021), and few studies are focusing on one specific industry like chemical manufacturing industry. Therefore, this study takes the manufacture of raw chemical materials and chemical products in China as the research object and empirically analyses the factors affecting environmental accounting information disclosure, which has a particular realistic and innovative significance. Secondly, the board of directors and shareholders play an essential role in corporate strategy and performance. Therefore, considering ownership as an influencing factor can give insights to achieve economic and environmental benefits. Lastly, rather than focusing on environmental disclosure alone, that is, whether companies disclose environmental information, environmental disclosure index (EDI) is scored on the basis of each disclosure item in GRI in this study, so environmental disclosure can be revealed more comprehensively by analysing its quality.

The rest of the chapter is structured as follows: Section 2 includes a literature review and hypothesis development to provide theoretical support; Section 3 introduces the sample selection process and the methods used, including data collection, variables, and regression models; Section 4 describes and analyses the results obtained. Finally, a discussion and implication of the results, robustness tests, and limitations are presented in Sections 5, 6, and 7, respectively.

2. Literature Review and Hypothesis Development

According to environmental law, environmental accounting combines the fundamental theories of accounting and environmental economics and records costs and benefits from the perspective of social benefits. Therefore, by studying how different factors and financial performance indicators influence corporate environmental disclosure, environmental accounting can improve the overall social benefits by protecting the environment while developing the enterprise economy (Tahajuddin et al., 2020). Empirical studies have been conducted to investigate factors affecting environmental accounting disclosure on a global scale, ranging from developing and developed countries. In China, there have been many recent studies on environmental accounting considering stakeholder pressure (Huang & Kung, 2010; Bellamy et al., 2020; Yunus et al., 2020; Abdel-Maksoud et al., 2021; Tahajuddin & Sulaiman. 2021) and financial performance indicators (Chiu et al., 2020; Tarus, 2020; Agyemang et al., 2021) like ROA, firm size, EPS, and leverage.

2.1 Financial Performance Indicators

In general, the relationship between financial performance and environmental disclosure varies according to regional developments. On a global scale, Doan and Sassen (2020) integrated 62 previous studies and concluded that there is no association between financial performance and environmental disclosure. However, this research does not categorize a company's location and type, which could be applied to later research, yielding more meaningful insights. Similarly, Dragomir (2010) noticed that no association was found between environmental disclosure and company performance over the same period in the European context, while he also claimed that there are research finding a positive relationship between return of firms and disclosure of social and environmental information, and a negative relation between firm value and its disclosures. For developing countries like China, financial performance indicators, such as firm size and leverage, have been proven to have a significantly positive relationship with a company's environmental disclosure (Chiu et al., 2020; Agyemang et al., 2021).

2.1.1 Firm Size

Since total assets are the largest determinant of corporate financial performance, it is crucial to consider firm size as one of the variables. Some unique methods were used during the previous research. Nguyen (2021) conducted research on enterprises in the food and beverage industry in the context of Vietnam. Unlike most other studies, Nguyen (2021) used questionnaires to collect information about the target population, while face-to-face and email interviews with employees, managers, and accountants were also used in the investigation process. Besides, Ismail et al. (2018) examined the factors in a more rigorous and innovative way. The source of this research information showing the quality of environmental disclosure was not only an annual report like in most previous research (Chiu et al., 2020), but also three main media (annual reports, stand-alone environmental reports, and corporate homepages) were analysed using political economy theory, stakeholder theory, and legitimacy theory.

From the review of previous literature, there are different conclusions on whether firm size as an indicator of economic performance is associated with environmental disclosure. Some results confirmed the relationship, but others did not. Even in those conclusions with a relationship, they could be positive (Cormier & Magnan, 1999; Dragomir, 2010; He & Loftus, 2014; Yang et al., 2017; Deswanto & Siregar, 2018; Ismail et al., 2018; Sri & Arief, 2018; Tarilaye, 2018; Tarus, 2020; Abdel-Maksoud et al., 2021; Agyemang et al., 2021; Gerged et al., 2021; Nguyen, 2021) or negative (Baalouch et al., 2019). Therefore, we propose the following hypothesis:

H1a: Firm size has a positive association with EDI for the listed companies manufacturing raw chemical materials and chemical products in China.

2.1.2 Leverage

Based on the summary of previous studies, in most cases, the positive relationship between leverage of a company and quality of environmental disclosure has been proven. Companies with high leverage ratios tended to disclose more voluntary environmental information to gain legitimacy from creditors and reduce costs (Huang & Kung, 2010), which revealed that leverage is positively and significantly related to the level of quality of Corporate Environmental Disclosure (CED) (He & Loftus, 2014; Ismail et al., 2018; Tarilaye, 2018; Baalouch et al., 2019; Chiu et al., 2020; Fan et al., 2020; Yunus et al., 2020; Gerged et al., 2021).

However, a few studies have shown that the relationship is unrelated or negative. Agyemang et al. (2021) conducted an empirical study on mining companies listed in China and concluded a negative and insignificant correlation between debt levels and environmental information disclosure (EID). Therefore, we propose the following hypothesis:

H1b: Leverage has a positive association with EDI for the listed companies manufacturing raw chemical materials and chemical products in China.

2.1.3 Earnings per Share

Due to differences in research methods or research objects, there is still no conclusion as to whether the relationship between EPS and environmental disclosure is negative, positive, or neutral. Previous studies have shown a positive correlation between the two, in the context of Mongolia (Bayaraa, 2017), the United Kingdom (Samy et al., 2010), Bangladesh (Islam & Rahman, 2016), and Nigeria (Tarilaye, 2018), while an insignificant correlation has also been found in Pakistan (M. Ahmad et al., 2018). Therefore, we propose the following hypothesis:

H1c: EPS has a positive association with EDI for the listed companies manufacturing raw chemical materials and chemical products in China.

2.1.4 Return on Assets

The conclusions of research on the relationship between ROA and environmental disclosure are inconsistent. Companies with a positive correlation want to let the public know that they are willing to give back to society through environmental protection measures (Ahmad et al., 2017; Chiu et al., 2020), while those with a negative correlation shed light on the fact that companies with lower profits tend to disclose more corporate social responsibility (CSR) information in order to improve their image (Huang & Kung, 2010). At the same time, some studies have shown that there is an insignificant relationship or even no relationship between the two (Islam & Rahman, 2016; Feng et al., 2018; Sri & Arief, 2018). Therefore, we propose the following hypothesis:

H1d: ROA has a positive association with EDI for the listed companies manufacturing raw chemical materials and chemical products in China.

2.2 Stakeholder Factors

In most previous studies, stakeholder-related characteristics were common factors. Relevant literature revealed that some factors, such as external stakeholders, have been extensively studied, while other factors, such as internal stakeholders, have received little attention. Therefore, this study includes some factors that have not been paid attention in previous studies in the context of Chinese companies, such as ownership concentration.

Researchers mainly diverged thinking from two theories as a foundation: stakeholder theory and legitimacy theory. As the leading theory of social and political explanation, legitimacy theory examines the social value reflected in the process of disclosing environmental and social information. It is stated that environmental disclosure is the result of social value so that a good legitimacy theoretical model will consider the value of stakeholders in any case (Saini & Singhania, 2019).

Based on legitimacy theory, Fan et al. (2020) explored how the new media environment shapes the disclosure of environmental practices of highly polluting enterprises in China. Under the supervision of social media, environmental information is pushed to be disclosed because of the public pressure imposed on companies to take actions to deal with environmental problems. Through regression analysis and robustness test, the result turned out to be consistent with other research that the new media environment has a positive association with environmental disclosure quality (Fan et al., 2020). On the other hand, as supported by stakeholder theory, many people hold the view that a company's performance is largely influenced by its stakeholders. By focusing on various internal and external stakeholders, analysing how this group could affect environmental disclosure as supervisors can provide insights into managing enterprises in society. The findings generated by Nguyen (2021) indicated that stakeholders and leadership perceptions are two factors that have the most significant impact on environmental disclosure compared with others. Tahajuddin and Sulaiman (2021) conducted research on 90 Chinese small and medium enterprises and concluded through partial least squares-structural equation modelling (PLS-SEM) that operators with the idea of gaining more profit would be more active in promoting environmental disclosure, showing a significant and positive relationship between the two. For external stakeholders, customers were proven to have a positive relationship with environmental disclosure, while suppliers do not have such a linkage (Tahajuddin et al., 2020). In addition to analysing these common stakeholder factors that are easy to overlap, research on the ownership structure of shareholders can further meet the research needs.

2.2.1 Ownership Concentration

Existing studies have shown that a multi-shareholder ownership structure can promote social responsibility (X. Wang et al., 2021), and companies with dispersed ownership were more likely to voluntarily disclose environmental information (Huang & Kung, 2010; Feng et al., 2018). Research were also conducted in China's energy industry, which found that enterprises should optimize ownership concentration so as to strengthen the positive impact of social responsibility on financial performance (Ismail et al., 2018). Ismail et al. (2018) also examined an insignificant association between concentrated ownership and corporate social performance, including environmental disclosure. Therefore, we propose the following hypothesis:

H2a: Ownership concentration has a negative association with EDI for the listed companies manufacturing raw chemical materials and chemical products in China.

3. Methodology

3.1 Data Collection

The aim of this study was to investigate how stakeholder factors and financial performance indicators affect environmental accounting disclosure by sampling 307 Chinese listed firms manufacturing raw chemical materials and chemical products categorized in the Industry Classification of Listed Companies (2012 Revision).

When selecting the study population, it is considered that the industrial solid waste of four major industries, including the manufacture of raw chemical materials and chemical products, accounted for 95.1% of the total, according to the data revealed in the *China Environmental Statistics Annual Report 2015* (Yunus et al., 2020). According to Yunus et al. (2020), industrial emissions and solid waste production reached a plateau rather than exponential growth until 2011. Therefore, this study covers 10 fiscal years from 1 January 2011 to 31 December 2020, a period chosen considering the data availability.

In this study, information about each company's environmental disclosure was collected from the annual report, CSR report, and environmental report; data about the financial performance of a company were collected from financial statements, such as balance sheets and income statements. When collecting and combing financial data, a consolidated statement rather than a parent statement is used in the research to reflect the overall business situation and avoid the possibility of inflated performance. This study conducted an empirical study and adopted a quantitative research approach, and all data were mainly extracted from the China Stock Market & Accounting Research (CSMAR) database.

3.2 Variables

3.2.1 Environmental Disclosure Index

In regulating companies' environmental practices, several environmental reporting standards have been developed to provide consistent guidelines for standardizing environmental information, such as the Global Industry Classification Standard (GICS) (Sri & Arief, 2018), Public Environmental Reporting Initiative (PERI), Coalition for Environmentally Responsible Economies (CERES) (Jose & Lee, 2007), and Global Reporting Initiative (GRI) (Fan et al., 2020). Environmental disclosure quality can be measured by using content analysis on the specific item in different guidelines so that the authors may use annual reports only (Chiu et al., 2020) or a company's annual and sustainability reports (Sri & Arief, 2018), or combine the annual report, social responsibility report, and environmental report together (Fan et al., 2020). The high-quality, rigorous, and sustainable nature of the GRI has established a globally recognized standard for EID (Tarus, 2020). Therefore, in this study, the GRI guidelines, which consist of 33 items, were adopted as the basis, and each company is scored in accordance with the evaluation system. In the process of evaluating disclosure items in the guidelines, 0 represents no description, 1 represents qualitative description, and 2 represents quantitative description. Then all the scores of each item are added up to give an aggregated score as a reference for EDI score.

3.2.2   Earnings per Share

EPS, an important indicator showing corporate value and helping investors to make investment decisions, has been proven to have a positive impact on the financial performance of an organization (Bayaraa, 2017). EPS equals current net profit divided by paid-in capital at the end of the current period. It is reasonable to claim that the better the trend or the higher the number of EPS, the stronger the profitability of the company. Based on what was discussed, a positive correlation was found between EPS growth and CSR investment, including environmental protection actions (Islam & Rahman, 2016; Zakari, 2017). Therefore, EPS is a variable in this study.

3.2.3   Return on Asset

ROA is an important metric that reflects the level of a company's assets and the effectiveness of the company in generating profits from its assets, so Chiu et al. (2020) considered it as a proxy that demonstrates the profitability of the firm. ROA can be calculated by tiscal year-end net profit divided by year-end total assets. From scholars' research, a company's financial performance is a key determinant of environmental disclosure, and ROA can be the most dominant factor (Hansen & Wernerfelt, 1989; Cormier & Magnan, 1999). Therefore, ROA is a variable in this study.

3.2.4   Leverage

Leverage is an indicator of measuring risk, and higher leveraged companies have more debt than equity, which will have a higher risk of bankruptcy. Leverage is total debt divided by year-end total assets. According to related research, the increase in corporate debt would lead to increased requirements for information by stakeholders (Acar & Temiz, 2020). Therefore, leverage is a variable in this study.

3.2.5   Firm Size

Numerous studies have shown that firm size (Size) is one of the critical variables in environmental disclosure research (Brammer & Pavelin, 2006; Cormier & Magnan, 2007; Huang & Kung, 2010; Tarilaye, 2018; Chiu et al., 2020; Nguyen, 2021). Firm size can be calculated by the natural logarithm of the year-end total assets of a firm. They claimed that large companies that disclose environmental performance are more likely to clarify their determination and commitment to be more environmentally sensitive. Therefore, firm size is a variable in this study.

3.2.6   Ownership Concentration

According to Huang and Kung (2010), less information would be disclosed for enterprises with a more concentrated ownership structure because the cost of

Table 11.1 Variable Definition

Variable	Name	Definition
EDI	Environmental disclosure index	Corporate EID quality obtained by content analysis, built using the following formula: $$\text{Disclosure Index} = \frac{\text{Score of items disclosed}}{\text{Maximum Number of Items Disclosed}}$$
EPS	Earnings per share	Current net profit/paid-in capital at the end of the current period
ROA	Return on assets	Fiscal year-end net income/year-end total assets
LEV	Leverage	Total debt/year-end total assets
Size	Firm size	Natural log of total assets
LAR	Ownership concentration	Shareholding proportion of the largest shareholders

information disclosure will be lower. Therefore, it is reasonable to infer that these enterprises tend to produce less environmental disclosure. Ownership concentration (LAR) represents the percentage of shares held by the largest shareholder. The higher the ownership concentration, the higher the independent directors' proportion, also the lower the rights of non-independent directors. Because the independent director's control is enhanced, the possibility of environmental disclosure will be lower. However, the association between independent directors and CSR reporting, including environmental disclosure, had industry particularity (Ahmad et al., 2017). Therefore, ownership concentration is a variable in this study.

3.3   Model Design

This research uses empirical analysis to test the effect of influencing factors (independent variables) on environmental accounting information disclosure (dependent variable) and construct a multiple regression model as follows:

$$EDI_i = \beta_0 + \beta_1(SIZE) + \beta_2(LEV) + \beta_3(EPS) + \beta_4(ROA) + \beta_5(LAR) + \varepsilon$$

Here, β_0 is the constant term; β_1–β_5 are the regression coefficients; and ε is the error term.

4.   Results

Table 11.2 presents the summary descriptive statistics for the main variables used in the study. The ROA of the chemical manufacturing companies included in the sample is generally low, with a maximum value of 0.271. Besides, firm sizes of all the companies measured by the natural logarithm of total sales are greater than 20, suggesting that the sample companies consist of relatively larger firms. The results regarding director ownership illustrate that there is a huge difference between the minimum (11.98%) and maximum (72.15%) shareholding proportion of the

Table 11.2 Descriptive Statistics

Variable	Obs.	Mean	Std. Dev.	Min	Max
EPS	313	.434	1.338	−16.748	4.868
ROA	313	.039	.133	−2.071	.271
LEV	313	.446	.219	.014	2.290
Size	313	22.621	1.260	20.109	25.619
LAR	313	33.304	13.562	11.980	72.150
EDI	313	.392	.151	.026	.789

Table 11.3 Correlation Coefficients among Selective Variables

	EPS	ROA	LEV	Size	LAR	EDI
EPS	1					
ROA	0.874***	1				
LEV	−0.465***	−0.600***	1			
Size	0.027	−0.09	0.561***	1		
LAR	0.033	0.024	−0.044	0.126**	1	
EDI	0.085	0.120**	−0.188***	0.079	−0.038	1

$* p < 0.05$, $** p < 0.01$, $*** p < 0.001$.

largest shareholders, and on average, the largest shareholder accounts for 33.304% of total company shareholding. Moreover, the results also show that the minimum and maximum values of EDI are 0.026 and 0.789, respectively, with a standard deviation of 0.151, revealing that there is a subtle difference in the quality of environmental information disclosed by different companies. The average of 0.392 indicates that the sample firms report information on nearly 40% of the 33 environmental items considered in our research to construct the EDI, implying that the average quality of corporate environmental accounting disclosure is generally low.

As shown in Table 11.3, for the period between 2011 and 2020, the positive correlation coefficient between EDI (environmental disclosure score) and ROA is 0.120 and is statistically significant at 5%. This result is highly consistent with previous studies (Samy et al., 2010; Ahmad et al., 2017; Chiu et al., 2020). Also, LEV shows a highly negative relationship with EDI and is statistically significant at 1%, consistent with the previous studies (Gao & Connors, 2011; Fontana et al., 2012; Setyorini & Ishak, 2012). However, the correlation results indicate that there is no significant association between EDI score and EPS ($R = 0.085$), firm size ($R = 0.079$), and LAR ($R = −0.038$).

The panel data of the sample companies from 2011 to 2020 were analysed using regression analysis. The results of the regression analysis with four financial performance variables (EPS, ROA, LEV, and Size) and one stakeholder variable (LAR) as dependent variables are given in Table 11.4. The results imply that there is a statistically significant and positive relationship between EDI and Size ($\beta = 0.0411758$, $t = 4.73$, $p = 0.000 < 0.01$). The results support H1a, which states that when a chemical manufacturing company has a larger size, meaning a larger scale or volume of operation, it will likely have a higher environmental disclosure information quality in its annual report, CSR report, and environmental report. Contrary

Table 11.4 Regression Analysis

EDI	Coeff.	Std. Dev.	t-value	p-value
EPS	−0.0134745	0.0128617	−1.05	0.296
ROA	−0.0125029	0.1420845	−0.09	0.93
LEV	−0.3080043	0.0610713	−5.04	0
Size	0.0411758	0.0086996	4.73	0
LAR	−0.0010738	0.0006142	−1.75	0.081
_cons	−0.3599218	0.1770272	−2.03	0.043
N	313			
R^2	0.1034			
Adj. R^2	0.0888			

to H1b, the relationship between leverage and environmental disclosure was proven to be negative ($\beta = -0.3080043$, $t = -5.04$, $p = 0.000 < 0.01$). However, a p-value greater than 0.5 indicates that the distribution of the sample is not significantly different from the normal distribution (Dragomir, 2010). Thus, EDI is not statistically significant with EPS, ROA, and LAR ($p > 0.05$), representing that hypotheses H1c, H1d, and H2a are not supported.

5. Discussions

EDI can be one of the indications that while these high-polluting companies have developed an awareness of environmental disclosure, the quality of their reporting is somewhat random and unstandardized, revealing that most companies do not have a systematic environmental accounting information disclosure procedure. Although environmental regulation in China has begun to increase in recent years with the development of a relatively comprehensive set of national environmental laws, it still has a long way to go. From the perspective of environmental information disclosure quality, chemical manufacturing companies are more inclined to qualitative disclosure. Some companies reported their environmental activities based on disclosure items like environmental management systems, environmental performance indicators, and environmental profiles. Compared with the development of a corporate economy, environmental regulation remains a low political priority. As Wang (2021) points out, "local protectionism" makes central leaders powerless to achieve their goals at the local level. Therefore, Chinese chemical manufacturing companies should make greater efforts to improve their CSR and environmental performance to develop a sustainable competitive advantage.

As shown in Table 11.4, some of the variables investigated are able to capture the relationship with the quality of EID of Chinese chemical manufacturing companies, while some variables are not. Based on the regression test, the results show that firm size has a significant positive effect on EID quality. It is reasonable to claim that firm size can be one of the effective mechanisms to boost environmental accounting information reporting. Therefore, the finding is consistent with the previous studies on firm size (measured by total assets) and EDI quality (N. Ahmad et al., 2017; Baalouch et al., 2019; Chiu et al., 2020; X. Wang et al.,

2021). This result provides some support for the legitimacy theory, which posits that corporate disclosure is a response to environmental factors that legitimize corporate behaviour and ensure acceptance in society (Guthrie & Parker, 1989). According to Ismail et al. (2018), in this theory, large companies are assumed to be highly scrutinized and need to accentuate their corporate image through EID to earn or maintain their social status and goodwill. Similarly, according to Ullmann's stakeholder theory, to be successful, the company must find ways to satisfy stakeholders' needs, for example, by publishing social responsibility reports such as environmental disclosures to obtain enough capital from the market for the company to function properly (Kent & Chan, 2009).

A higher leverage ratio (debt-to-equity) can indicate that a company is riskier because it may have a significant negative impact on the company's ability to operate in the future. Better environmental performance, including the quality of environmental disclosure, in some ways, increases debt capacity (Gao & Connors, 2011). This is attributable to the high cost of environmental disclosures and environmental improvements and the increasing interest payments on a firm's growing debt, which will reduce net income, leading to higher leverage. In this case, the level of corporate investment may be severely affected, due to the detriment of corporate financing and improving corporate performance. Because of the high cost of environmental performance, the decision to disclose information about a company's social and environmental activities can reduce the company's revenues, becoming a burden to the company. Thus, higher debt compared to equity can cause companies to reduce their ability to manage and handle environmental resources effectively, and highly leveraged companies are more likely to reduce their level of social and environmental disclosure.

However, the remaining independent variables, including EPS, ROA, and LAR, have been found have statically insignificant effects on EID quality. These findings are inconsistent with previous studies (Hansen & Wernerfelt, 1989; Cormier & Magnan, 1999; Samy et al., 2010; Islam & Rahman, 2016; Bayaraa, 2017; Tarilaye, 2018) on the relationship between EDI and these factors while supporting several other studies (Ismail et al., 2018; M. Ahmad et al., 2018). This is not an exceptionally surprising result, as mixed conclusions of these factors have been mentioned in a previous literature review. Both EPS and ROA can be indicators of financial performance, showing the ability to gain profitability. None of the literature on EPS and ROA gives a possible reason why these two factors have only a neutral effect on the quality of environmental disclosure. However, total assets (firm size) was found to have a significantly positive effect on EID quality in this study. This may indicate that firm size has become one of the most critical variables affecting the quality of environmental disclosure. Besides, a number of studies also did not indicate any significant relationship with ownership concentration (Halme & Huse, 1997; Huang & Kung, 2010). Indeed, according to stakeholder theory, stakeholders are an important factor in the achievement of a company's objectives because they are able to participate in and control the company's planning, management, and operational policies. However, it is precise: because shareholders within a firm seek good financial performance, they should make trade-offs with environmental performance.

Therefore, the impact of this group on a company's environmental disclosure cannot be accurately quantified. Huang and Kung (2010) claimed that other considerations, such as operating based on the principle of cost-effectiveness, have a more vital influence on shaping management strategies related to environmental disclosure. Therefore, it can be substantiated that the shareholder group can influence the quality of a company's environmental disclosure.

6.   Robustness Tests

This study repeats the above analysis by replacing the environmental disclosure quality score with a CSR index to assess the robustness of the results of the previous test study, following an earlier procedure to re-regress the model (Fan et al., 2020). CSR can be calculated by the number of types of items disclosed by enterprises divided by the total number of items. CSR evaluation includes items such as environment and sustainable development, public relations and social welfare, and employee and consumer rights protection. The results obtained using this alternative approach (presented in Table 11.5) were similar in nature to the main regression analysis results. Therefore, the consistent findings with the dependent variable alternative measures further corroborate the robustness of the study results.

Given the results of the calculated and investigated variance inflation factor (VIF) presented in Table 11.6, which is a test for the presence of multicollinearity among the variables in the sample, the values of this statistics in the regression models are all less than 10, which are acceptable values for multicollinearity,

Table 11.5 Robustness Test

Variable	EDI	CSR
EPS	−0.0135 (−0.0114)	−0.0253 (−0.0179)
ROA	−0.0125 (−0.106)	−0.0476 (−0.167)
LEV	−0.308*** (−0.0616)	−0.338*** (−0.0884)
Size	0.0412*** (−0.00882)	0.0469*** (−0.0127)
LAR	−0.00107 (−0.000648)	−0.000284 (−0.000896)
_cons	−0.360* (−0.175)	−0.557* (−0.252)
N	313	313
R^2	0.103	0.073

Standard errors in parentheses.* $p < 0.05$, ** $p < 0.01$, *** $p < 0.001$.

Table 11.6 Tests of Multicollinearity

Variable	VIF	1/VIF
ROA	5.38	0.186008
EPS	4.45	0.224568
LEV	2.69	0.371333
Size	1.80	0.554099
LAR	1.04	0.959148
Mean VIF	3.07	

indicating that there is no harmful multicollinearity problem. Besides, the average VIF value is less than 5, suggesting that multicollinearity is not a threat to the computational accuracy of the models.

7.   Limitations and Future Research

Several limitations can be revealed from this study. Firstly, there are many types of companies in the manufacturing industry in China, such as state-owned companies and private companies, and these categories are not discussed individually in this study. Besides, the sample in this study focuses on only one of the broad industry sectors in China, so the results of this sample are not applicable to other major industries in China. In addition, there are no specialized EDI score databases in China; thus, factors that can reflect environmental disclosure are collected manually, and the score is calculated by using content analysis based on the GRI guidelines. Lastly, in this study, only internal stakeholders are investigated, while external interest groups, such as customers, suppliers, and the media, may also play a vital role in influencing the environmental accounting disclosure quality.

Considering these limitations, future research could take new directions to enrich the currently limited research in this area. Future research could consider different research designs to further explore the environmental disclosures of corporations with different natures at various stages of time and graphically show the trends, so as Abdel-Maksoud et al. (2021) stated, longitudinal design and multi-informant design can play a vital role. Future research could address the limitation by incorporating internal and external factors into the investigation to explore how stakeholder behaviours or expectations and the implementation of some policies may affect environmental disclosure.

8.   Conclusion

Due to the exploitation of natural resources and inadequate conservation practices by Chinese companies for a long time, a polluted environment and an out-of-balance ecology have in turn constrained economic development. Under such circumstances, EID will help the management address some environmental issues and ensure the implementation of sustainable development strategies. At this stage, although some listed companies in heavily polluting industries have started to disclose environmental information voluntarily, the level of disclosure varies. Therefore, a systematic study of the factors affecting the quality issues of corporate EID can promote the standardization of EID in China.

Using corporations' financial and environmental data from 2011 to 2020, this study explores the relationship that exists between EID and financial performance for chemical manufacturing companies listed in China. The empirical results indicate several findings: firstly, based on the regression analysis, firm size and leverage have a positive and negative relationship with environmental disclosure quality, reflecting that it is more likely for companies with a lower level of financial risk and higher asset level to put more effort and capital in disclosing the environmental situation and improving related operation strategy. Secondly, other factors, including EPS, ROA, and LAR, do not have a noticeable impact on environmental disclosure quality. In other words, the level of corporate profits does not prompt companies to become more active in environmental disclosure or to take greater corporate responsibility. Besides, internal governance mechanisms, which can be seen from ownership concentration, do not have sufficient control to influence the management of enterprises in terms of environmental governance. In terms of hinting at reality, this study is able to reflect the shortcomings of environmental disclosure at this stage, such as non-standardization and lack of comparability. Therefore, this study can help policymakers, investors, and companies to some extent to realize the importance and reference of environmental disclosure for sustainable development and green economy development.

References

Abdel-Maksoud, A., Elbanna, S., Mahama, H., & Pollanen, R. (2021). The use of eco-control systems in firms' environmental innovation strategies: A stakeholder theory perspective. *Business Strategy and the Environment, 30*(3), 1249–1262.

Abdel-Maksoud, A., Jabbour, M., & Abdel-Kader, M. (2021, January). Stakeholder pressure, eco-control systems, and firms' performance: Empirical evidence from UK manufacturers. In *Accounting forum* (Vol. 45, No. 1, pp. 30–57). Routledge.

Acar, M., & Temiz, H. (2020). An empirical investigation on the relationship between environmental performance and financial performance: The case of Borsa Istanbul. *Asian Journal of Business and Management Sciences, 9*(2), 19–28.

Agyemang, M., Agyei-Mensah, B. K., & Assibey, E. O. (2021). Determinants of voluntary environmental disclosures: Evidence from listed mining companies in China. *Environmental Science and Pollution Research, 28*(5), 5648–5663.

Ahmad, M., Hassan, Y. M., & Ezat, A. N. (2018). Environmental disclosure and financial performance: Evidence from Islamic banks in Pakistan. *Journal of Islamic Accounting and Business Research, 9*(3), 379–394.

Ahmad, N., Lodhi, M. S., Zaman, K., & Naseem, I. (2017). Knowledge management: A gateway for organizational performance. *Journal of the Knowledge Economy, 8*, 859–876.

Baalouch, F., Ayadi, S. D., & Hussainey, K. (2019). A study of the determinants of environmental disclosure quality: evidence from French listed companies. *Journal of Management and Governance, 23*(4), 939–971.

Bayaraa, B. (2017). The effect of earnings per share on stock price in the context of Mongolia. *Mongolian Journal of Accounting, 1*(1), 52–61.

Bellamy, L., Gunningham, N., & Grabosky, P. (2020). Regulating the regulators: The regulatory impact of New Public Management reform on state environmental agencies. *Law & Policy, 42*(4), 401–423.

Brammer, S., & Pavelin, S. (2006). Voluntary environmental disclosures by large UK companies. *Journal of Business Finance & Accounting, 33*(7–8), 1168–1188.

Chiu, T. K., Cho, S. S., & Luo, M. (2020). Environmental information disclosure and firm financial performance. *Sustainability, 12*(15), 6014. https://doi.org/10.3390/su12156014

Cormier, D., & Magnan, M. (1999). Corporate environmental disclosure strategies: Determinants, costs and benefits. *Journal of Accounting, Auditing & Finance, 14*(3), 429–451.

Cormier, D., & Magnan, M. (2007). The revisited contribution of environmental reporting to investors' valuation of a firm's earnings: An international perspective. *Ecological Economics, 62*(3–4), 613–626.

Deswanto, R. B., & Siregar, S. V. (2018). The associations between environmental disclosures with financial performance, environmental performance, and firm value. *Social Responsibility Journal, 14*(1), 180–193.

Doan, T. P., & Sassen, R. (2020). Environmental disclosure, financial performance, and firm value: Evidence from Vietnam. *Journal of Sustainability Research, 2*(1), e200004.

Dragomir, V. D. (2010). Environmentally sensitive disclosures and financial performance in a European setting. *Journal of Accounting & Organizational Change, 6*(3), 359–388.

Fan, Y., Yang, L., & Li, S. (2020). The impact of media coverage on corporate environmental disclosure quality: Evidence from China. *Journal of Cleaner Production, 243*, 118–555.

Feng, T., Wang, Q., & Gong, Y. (2018). Stakeholder pressure, supply chain integration, and circular economy practices: The moderating role of governance mechanisms. *Journal of Cleaner Production, 206*, 160–169.

Fontana, V., Silva, P. S., Gerlach, R. F., & Tanus-Santos, J. E. (2012). Circulating matrix metalloproteinases and their inhibitors in hypertension. *Clinica Chimica Acta, 413*(7–8), 656–662.

Gao, L. S., & Connors, E. (2011). Corporate environmental performance, disclosure and leverage: An integrated approach. *International Review of Accounting, Banking and Finance, 1*.

Gerged, A. M., Matthews, L., & Elheddad, M. (2021). How does environmental information disclosure relate to firms' financial performance? Evidence from the Gulf Cooperation Council countries. *International Journal of Finance & Economics, 26*(1), 334–364.

Guthrie, J., & Parker, L. D. (1989). Corporate social reporting: A rebuttal of legitimacy theory. *Accounting and Business Research, 19*(76), 343–352.

Halme, M., & Huse, M. (1997). The influence of corporate governance, industry and country factors on environmental reporting. *Scandinavian Journal of Management, 13*(2), 137–157.

Hansen, G. S., & Wernerfelt, B. (1989). Determinants of firm performance: The relative importance of economic and organizational factors. *Strategic Management Journal, 10*(5), 399–411.

He, C., & Loftus, J. (2014). Does environmental disclosure affect environmental performance? Evidence from China. *Environmental Economics, 5*(4), 14–25.

Huang, X., & Kung, F. H. (2010). Drivers of environmental disclosure and stakeholder expectation: Evidence from China. *Journal of Business Ethics, 99*(3), 417–434.

Iliemena, R. O. (2020). Environmental accounting practices and corporate performance: Study of listed oil and gas companies in Nigeria. *European Journal of Business and Management, 12*(22), 58–70.

Islam, M. A., & Rahman, M. A. (2016). The relationship between corporate social responsibility and firm performance: Evidence from Bangladesh. *International Journal of Finance & Banking Studies, 5*(2), 25–34.

Ismail, M., Leventis, S., & Rizopoulos, Y. (2018). The impact of environmental disclosure quality on the cost of equity capital: Evidence from Canada. *Journal of Accounting and Public Policy, 37*(1), 52–71.

Jiahuey, Y., Liu, Y., & Yu, Y. (2019). Measuring green growth performance of China's chemical industry. *Resources, Conservation and Recycling, 149*, 160–167.

Jose, A., & Lee, S. M. (2007). Environmental reporting of global corporations: A content analysis based on website disclosures. *Journal of Business Ethics, 72*(4), 307–321.

Kent, P., & Chan, C. (2009). Application of stakeholder theory to corporate environmental disclosures. *Corporate Ownership & Control, 7*(1), 24–30.

Nguyen, T. D. (2021). Firm characteristic and fair value adoption in the measurement of biological assets. *European Journal of Business and Management Research, 6*(6), 201–204.

Pucheta-Martínez, M. C., & Gallego-Álvarez, I. (2020). An international approach of the relationship between board attributes and the disclosure of corporate social responsibility issues. *Corporate Social Responsibility and Environmental Management*, *27*(5), 2300–2316.

Rahman, M. J., & Wu, J. (2023). M&A activity and ESG performance: evidence from China. *Managerial Finance*, *50*(1), 179–197.

Rahman, M. J., Wu, Q., & Zhu, H. (2024). Corporate social responsibility in times of social distancing: Evidence from China. *Business Ethics, the Environment & Responsibility*.

Rahman, M. J., Zhu, H., & Chen, S. (2023). Does CSR reduce financial distress? Moderating effect of firm characteristics, auditor characteristics, and Covid-19. *International Journal of Accounting & Information Management*, *31*(5), 756–784.

Saini, A., & Singhania, M. (2019). Determinants of corporate environmental disclosure: Evidence from an emerging economy. *Environmental Management*, *63*(3), 344–365.

Samy, M., Odemilin, G., & Bampton, R. (2010). Corporate social responsibility: A strategy for sustainable business success. An analysis of 20 selected British companies. *Corporate Governance: The International Journal of Business in Society*, *10*(2), 203–217.

Setyorini, C. T., & Ishak, Z. (2012). Corporate social and environmental disclosure: A positive accounting theory view point. *International Journal of Business and Social Science*, *3*(9).

Sri, W. I. F., & Arief, B. M. (2018). Relationship between company financial performance, characteristic and environmental disclosure of ASX listed companies. In *E3S web of conferences* (Vol. 73, p. 10024). EDP Sciences.

Tahajuddin, S., Basuony, M. A., & Mohamed, E. K. (2020). The determinants of corporate environmental disclosure in Saudi Arabia. *International Journal of Business Governance and Ethics*, *14*(2), 156–179.

Tahajuddin, S., & Sulaiman, N. N. (2021). Malaysian government choice of fiscal and monetary policies during covid-19 pandemic: Preliminary insight. *International Journal of Advances in Engineering and Management*, *3*(1), 248–253.

Tarilaye, D. (2018). Environmental disclosure and firm financial performance: Evidence from Nigerian manufacturing companies. *Research Journal of Finance and Accounting*, *9*(10), 85–96.

Tarus, D. K. (2020). Environmental disclosure and financial performance: Evidence from Kenya. *International Journal of Economics, Commerce and Management*, *8*(2), 540–558.

Wang, C. (2021). Departmental protectionism and local protectionism in China's WTO disputes. *Hong Kong Law Journal*, *51*, 273.

Wang, X., Zhang, Y., & Hao, Z. (2021). The effect of ownership structure on corporate social responsibility in China: A new perspective. *Sustainability*, *13*(1), 400.

Yang, M., Wang, F., & Yang, M. (2017). Corporate environmental performance and disclosure: An empirical study in China. *Social Responsibility Journal*, *13*(2), 251–265.

Yunus, S., Elijido-Ten, E., & Abhayawansa, S. (2020). Determinants of carbon management strategy adoption: Evidence from Australia. *Australian Accounting Review*, *30*(1), 4–19.

Zakari, M. (2017). The impact of earnings per share on stock price in the context of Nigerian banking industry. *International Journal of Finance and Accounting*, *6*(4), 106–113.

Zhu, H., & Rahman, M. J. (2024). Ex-ante expected changes in ESG and future stock returns based on machine learning. *The British Accounting Review*, 101457.

12 Quality Evaluation of Corporate Environmental Accounting and Disclosure in China's Steel Industry

Md Jahidur Rahman, Tarek Rana, Hongtao Zhu, and Shao Zichun

1. Introduction

Environmental accounting information disclosure (EAID) is very important to promote corporate environmental initiatives. On the one hand, disclosure of enterprise environmental accounting information can promote the regulation of corporate environmental behaviours. EAID helps managers make sustainable decisions that take environmental and social considerations into account and formulate effective environmental protection and business strategies (Deegan, 2004; Kong & Wang, 2016). Besides, environmental disclosure is considered a suitable instrument to disseminate a good public image and demonstrate that the company is operating within the boundaries and norms of the society. On the other hand, stakeholders and government departments could make better decisions to promote sustainable development of industries. Better disclosure helps firms establish more stable relationships with stakeholders, thereby reducing their business risks and operation costs (Wu & Shen, 2013), and thus government departments could formulate better regulation policies and investors could invest in more sustainable companies.

As the largest emerging country, China is experiencing environmental deterioration at an alarming speed. To achieve sustainable development, the Chinese central government has growingly issued a series of guidelines and standards to encourage enterprise environmental information disclosure. For example, the *Guidelines on Environmental Information Disclosure for Listed Companies (Draft for Comment)* published by the Ministry of Ecology and Environment requires that all listed companies in the Shanghai and Shenzhen stock exchanges to disclose environmental information. Especially, the listed companies in 16 heavily polluted industries (e.g., thermal power, steel, and cement) should regularly disclose environmental accounting information. The disclosure system in China is gradually improving; nevertheless, the current lack of regulatory system results in relatively arbitrary disclosure and uneven disclosure quality.

China began the preliminary research on environmental accounting in the 1990s, while the theoretical and empirical research on environmental accounting and EAID began in the 1970s (Beams, 1971; Ernst & Ernst, 1978; Guthrie & Parker, 1989). Now, the development of the EAID quality evaluation system in China is still in its infancy. Compared with European and American researchers,

DOI: 10.4324/9781003488965-16

Chinese researchers focus more on qualitative information disclosure (Qi et al., 2021). At present, in China, there is little systematic research on EAID evaluation, and there is no unified conclusion on the establishment and selection of evaluation indicators or methods (Liu & Liu, 2021).

This research also fills the gaps in the application of the projection pursuit (PP) model in the EAID literature. Distinctively, the PP model can be applied on multidimensional, non-linear, and non-normally distributed data with an insufficient sample size (Liu & Liu, 2021). So far, the PP model has been widely used in many fields, such as agriculture, water conservancy, geology, and medicine, and the model is also suitable for EAID evaluation. However, the literature on accounting information evaluation is nearly blank. Zhibin Liu and Ming Liu (2021) innovatively introduced the PP model for EAID evaluation, considering only two criteria, relevance and reliability. However, according to prior studies and regulations on accounting information disclosure, quality evaluation should consider both the environment of information generation and compliance with relevant laws and regulations (Jones, 2010; Bahra & Cadez, 2018; Eiler, 2015). The availability of the application of the PP model on EAID needs further verification, and the subjectivity brought by the content analysis method has not been completely overcome yet.

In China, although there have been studies designing and applying a systematic evaluation system (Liu & Anbumozhi, 2009; Liu & Liu, 2021), there has not been any on the EAID of China's steel industry. The disclosure level of corporate environmental accounting information may vary greatly from industry to industry (Guthrie & Farneti, 2008). According to Guthrie and Farneti (2008), the main reason for this is the distinctly different levels of social and environmental impacts on environmental information disclosure. The environmental information of listed companies in heavily polluting industries has gradually become the reference for creditors' grants of loans (Luo et al., 2019). Considering the importance of EAID in heavily polluting industries, it is necessary to have a discussion on the steel industry and fill the gaps.

There is a lack of systematic research on EAID quality evaluation. Without evaluation, the weaknesses of disclosure reports cannot be found, the credibility of disclosures becomes questionable, and improvements cannot be made properly. The aim of this research was to evaluate the quality of EAID of listed enterprises in the steel industry of China and fill the gaps in the literature on the application of the PP model in the field of environmental accounting. For this research, data were collected from 20 listed companies in the steel industry of China, an index system consisting of 22 indicators was established to evaluate disclosure quality, and an accelerated genetic algorithm was used to process data.

This study complements and extends the prior work by focusing on a newly introduced evaluation method, broadening the evaluation indicators of EAID and filling the gaps in systematic quality evaluation of EAID in the literature of China's steel industry. At present, nearly all studies on theoretical development and empirical analyses were conducted in the developed economies, while very limited quantification is being done in developing countries like China, which have different industrial and socio-economic fabrics (Luo, 2019). The lack of a comprehensive and widely accepted evaluation system causes difficulty in measuring the operating risks created by environmental issues and in formulating regulations to promote sustainable development.

Besides, this study firstly applies the PP model on the evaluation of EAID quality in the steel industry. In China, common methods proposed on EAID quality evaluation, like the factor analysis method, the principal component analysis method, and the entropy weight method, have high requirements for the amount and structure of the sample data. If the sample size is not sufficient or the data distribution is not normal, the analysis results would not fully cover the internal relationships of the data. Liu and Liu (2021) innovatively used the PP model in the EAID quality evaluation in the thermal power industry, but the availability of the PP model is still in need of verification. Besides, the evaluation index system in their research was not comprehensive enough to cover environmental accounting information of the whole industry.

This research also considers the importance of discussing EAID quality by industry. Firms in environmentally sensitive industries, like steel, mining, and oil companies, tend to disclose more environmental information than firms in other industries, as doing so consolidates the reputation and legitimacy of the firms within the stakeholder communities (Cho & Patten, 2007; Manes-Rossi et al., 2018). The characteristics of the steel industry's environmental information disclosure are revealed.

In this research, we applied the PP model and accelerated genetic algorithm to process sample data. The PP model has its special advantages in processing high-dimensional data with non-normal distribution, and the accelerated genetic algorithm improves the operating quality of the model. The sample data collection is based on real number coding, and the best projection vector is essentially the weight of each evaluation indicator. Environmental accounting information is collected from the annual reports and social responsibility reports of 20 listed companies in the steel industry of China.

The overall EAID quality of China's listed steel enterprises is poor. The environmental protection system and EAID system are incomplete, and the main manifestation is the overall weak awareness of environmental protection. Government pressure is a very important factor in enterprises deciding whether or not to disclose environmental information.

This research contributes to the existing literature as follows. This research fills the gap in the application of the PP model in the environmental accounting area, broadens the evaluation indicators, and discusses the systematic evaluation of EAID quality in the steel industry of China. In line with the study by Liu and Liu (2021), this research uses the PP model to improve the comparability of EAID among companies. Besides, this research broadens the evaluation indicators in the literature of environmental accounting information. To increase the objectivity of evaluation results, this study seeks to make the evaluation index more comprehensive. Compliance is added into the index system to reflect whether the accounting activities and disclosure scope of enterprises complied with the laws and regulations. A total of 22 evaluation indicators are selected and designed to reflect the disclosure quality from these three aspects. Detailed descriptions and explanations of all indicators are provided.

The remainder of this chapter is organized as follows. Section 2 reviews the associated literature. Section 3 describes the research methodology and the selection and design of the indicator system. Section 4 presents the main results of data processing and discusses the results. Section 5 states the limitations of the research and the recommendations for further research. Section 6 reviews the whole research and provides conclusions.

2. Literature Review

2.1 *Background Knowledge of Environmental Accounting and EAID*

In the 1970s, the concept of environmental accounting and the scope of environmental accounting information were firstly discussed (Beams, 1971; Ernst & Ernst, 1978; Guthrie & Parker, 1989). Environmental accounting information includes financial, environmental performance, and policy aspects (Jasch, 2000). In terms of the financial aspect, environmental accounting information refers to environmental expenditures devoted to the preservation of the environment, including environmental expenses and investments, environmental liabilities in the balance sheet, income statements, and notes added to financial statements or disclosed in annual reports (Senn & Giordano-Spring, 2020).

Different from the traditional accounting information system, environmental accounting information system creates a new scope including both directors of accounting and sustainable development. Environmental accounting is a broader concept of accounting that supports enterprise-level decision-making on environmental issues, while traditional financial and cost accounting fails to manage environmental information, particularly the environmental cost information (Burritt et al., 2002; Christ & Burritt, 2013). Traditional income statements may simply limit the environmental costs to separately identified items such as fines and penalties and expenditure to remediate past environmental damage, and thus the environmental costs are underestimated (Abdel-Kader, 2011).

Considering the importance of environmental accounting information, many studies have been conducted to analyse EAID. However, the establishment of environmental accounting information norms is still considered to be in its infancy (Senn & Giordano-Spring, 2020). Non-financial information (like environmental accounting information) disclosure is changing from a sporadic phenomenon to a systematic activity (Milne & Gray, 2007; Cho & Patten, 2007; Patten 2021), and the driver is the social and political pressures from the external environment. In fact, even in a regulated context, the disclosure of environmental accounting information is unsatisfactory due to the lack of a complete and systematic EAID evaluation system. The weak definitions in regulation and the absence of accounting guidance negatively impact a company's decision to disclose environmental accounting information (Senn & Giordano-Spring, 2020). Voluntary disclosure of corporate non-financial information is still found to be more likely a reputation enhancement tool instead of a significant accountability tool (Patten & Zhao, 2014), and the voluntary nature of EAID has mitigated these benefits due to credibility, transparency, and irrelevance drawbacks (Matuszak & Różańska, 2017; Venturelli et al., 2017). Besides, in determining the scale and scope of a company's disclosure strategy, managers tend to consider the perceived economic interests and the needs of creditors and potential investors. However, compared with most other actions of information disclosure, assessing the effect of a company's environmental disclosure policy on stock markets is rather difficult (Aerts et al., 2007). Due to the lack of a reliable and comprehensive quality evaluation system of EAID, stakeholders, especially investors, have difficulty measuring the corporate operating risk caused by corporate environmental actions. The establishment and improvement of environmental accounting evaluation systems and evaluation index systems are still strongly demanded.

2.2   *Review of EAID Evaluation Systems*

In the developed countries, many efforts have been made to study EAID evaluation (Fekrat et al., 1996; Kreuze et al., 1996; Patten & Trompeter, 2003; Chaklader & Gulati, 2015). For instance, Kreuze et al. (1996) included 10 aspects for evaluation and identified 17 information demands of environmentally conscious investors. The identified information demands were organized into two categories: environmental friendliness of a corporation's output and internal environmental philosophy and actions. Besides, they thought that EAID should provide specific information concerning how a company formulates regulations, satisfies environmental obligations, and estimates the environmental expenditures. Fekrat et al. (1996) constructed a scoring procedure covering 18 items in four categories: accounting and financial factors, environmental litigation, environmental pollution abatement, and other environment-related accounting measurements. Patten and Trompeter (2003) believed that listed enterprises should disclose their compliance with environmental protection laws and regulations, their environmental risks, and their pollution control actions in annual environmental information. Aerts et al. (2007) emphasized the importance of fully disclosing environmental accounting information including corporate environmental policies, environmental risks, compliance with environmental regulations, sustainable development status, and environmental expenditures. Chaklader and Gulati (2015) concluded that the evaluation should be based on five aspects: the disclosure of the environmental system of the enterprise's location, environmental governance strategies, environmental costs, specific environmental protection programmes, and enterprise environment-related rewards or penalties. Several studies in China also investigated the relationship between environmental accounting and firm performance (Rahman & Wu, 2023; Rahman et al., 2023, 2024; Zhu & Rahman, 2024).

2.3   *Previous Research on the Influencing Factors and Disclosure Channels of EAID*

Apart from systematic evaluation, previous efforts have also been made into the research of factors influencing the quality of EAID (e.g., Bewley & Li, 2000; Arif &

Tuhin, 2013; Lewis et al., 2014; Liu & Anbumozhi, 2019). For example, the educational background and tenure of top managers are proved to have a significant influence on their willingness to make EAID (Lewis et al., 2014). Liu and Anbumozhi (2019) found that only a company's environmental sensitivity (a proxy of the pressure from the Chinese government) and its size are significantly related to the corporate environmental information disclosure effort, while shareholders and creditors play weak roles in the test. Bewley and Li (2000) found that public knowledge of the corporate pollution propensity and firm size are the two significant variables in explaining EAID level. Besides, environmental accounting information is found to be less predictable than general environmental information. One interpretation is that such disclosures are subject to accounting principles, and the management has less discretion in disclosing environmental accounting information. Arif and Tuhin (2013) considered three corporate attributes, size, age, and profitability, in measuring their effect on the disclosure level of listed banks. This research reveals that bank size has a significant positive relationship with the voluntary disclosure level, which is consistent with previous research (e.g., Bewley & Li, 2000).

Besides studies on which factors could influence EAID quality and how to systematically evaluate corporate EAID quality, studies on the disclosure channels are also worth noting. Environmental accounting information is scattered in many parts of annual reports, social responsibility reports, investment prospectuses, etc., but not all disclosures are always considered to be reliable. Annual reports are recognized as the principal means of corporate communication with shareholders and are also considered as the primary source of environmental reporting (Wiseman, 1982; Kreuze et al., 1996; Fekrat et al., 1996). In recent years, there is a trend that European companies use integrated reports, instead of separate ones, to consolidate financial and non-financial information (Manes-Rossi et al., 2018). Nevertheless, previous accounting research under the "social responsibility" paradigm found that corporate environmental disclosures do not reflect the actual environmental performance (Ingram & Frazier, 1980; Wiseman, 1982; Cho et al., 2012; Patten & Zhao, 2014). There has been empirical evidence on the use of stand-alone social responsibility reports as an image/reputation enhancement tool rather than as a significant accountability tool (Cho et al., 2012; Patten & Zhao, 2014). Cho et al. (2012) analysed sustainability reports in different countries and pointed out that the stand-alone social responsibility reports are "legitimacy tools". Patten and Zhao (2014) indicated that the stand-alone corporate social responsibility reports do not appear to be about increasing social accountability or providing relevant performance data. Instead, these reports tend to focus more on discussing programmes and initiatives. Besides, the studied retail firms that already engaged in stand-alone social responsibility reporting were assessed to have worse environmental performance, although these companies had higher environmental reputations than non-reporting companies. While collecting sample data to evaluate corporate EAID quality, researchers should be cautious about the environmental information disclosed in the social responsibility reports.

3. Methodology

3.1 Sample Selection

For this research, we collected sample data from 20 steel companies that have gone public in the Shenzhen and Shanghai stock exchanges, excluding special treatment (ST) enterprises. The original data were gathered from the selected enterprises' annual reports and social responsibility reports during 2010–2020. After being sorted and extracted, the data were scored based on the real number coding method. Part of the sample data was derived from the China Stock Market & Accounting Research (CSMAR) database.

The steel industry is among the list of 16 heavily polluted industries according to the *Guidelines on Environmental Information Disclosure for Listed Companies (Draft for Comment)* published by the Ministry of Ecology and Environment. The disclosure level of the selected companies could to some extent reflect the achievements of promoting environmental disclosure among Chinese enterprises.

3.2 Establishment of the Indicator System

The purpose of evaluating the quality of the disclosure of enterprise environmental accounting information is to measure the reliability, comprehension, and compliance of the information. The index system, shown in Table 12.1, was constructed according to the results in the existing research (Clarkson et al., 2008; Jones, 2010; He, 2014; Eiler et al., 2015; Bahra & Cadez, 2018; Li et al., 2020; Liu & Liu, 2021).

3.2.1 Design of "Relevance" Level Indicators

This relevance index is to measure whether the environmental information disclosure is in line with an enterprise's financial situation, business performance, and development targets (He, 2014). At this layer, environmental policies, environment performance, and environmental financial information are considered. Environmental policy information (C1) mainly refers to the relevant disclosures of enterprise information including the company regulations and goals that are related to the environmental protection issues. Environmental performance information (C2) mainly refers to the enterprise information disclosure, representing the results and costs of the governance of the production and operating activities in the field of the environment. Environmental financial information (C3) refers to the disclosure of sources and destination of funds used for the environmental protection work.

3.2.2 Design of "Reliability" Level Indicators

The reliability index is used to measure whether the environmental information has been disclosed according to the actual events (He, 2014). Environmental information disclosure process (C4) refers to the various internal control descriptions and other related reports. Environmental information disclosure integrity (C5) refers to the completeness of environmental information disclosure.

3.2.3 Design of "Compliance" Level Indicators

The compliance index is used to measure whether the EAID complies with related laws and regulations in China. The evaluated information includes the content and scope of enterprise disclosure, as well as the related financial information.

3.3 Projection Pursuit Model Build-Up

The PP model and accelerated genetic algorithm are used in this research to process data. In the 1970s, Kruskal and Shepard (1974) firstly used the PP model to process high-dimensional data with non-normal distribution. By converting high-dimensional data to low-dimensional space, the model can help find the best projection vector to reflect the characteristics of the data (Fu et al., 2003). To optimize the PP model and improve the operating quality, this research uses the genetic algorithm. The genetic algorithm, a kind of adaptive global optimization probability search algorithm, imitates the evolutionary processes and natural selection and is often used to solve for the fittest solutions. The genetic algorithm requires no extra data about the given problem, and thus this method could deal with the lack data on features like continuity, derivatives, and linearity (Kinnear, 1994).

In step 1, to eliminate the dimension of each indicator value and unify the variation range of each indicator value, formula (1) is used to normalize the extreme values. In step 2, the index is formalized by taking a product of the spread term S_z and the local term D_z. The spread term S_z partly normalizes the index against scale effects, while the local density term D_z serves to capture structure in the data (Jones & Sibson, 1987). In step 3, this research uses the real-coded accelerated genetic algorithm (RAGA) to optimize the PP model via the p-dimensional variable Z. During the course of RAGA, the parent generation scale (n) is 400. The cross-over probability (pc) is 0.80. The mutation probability (p_m) is 0.05. The number of excellent individual is 11 ($\alpha = 0.05$). By accelerating 20 times, we can obtain the best projection value, which can be used to sort and compare the EAID quality of the sample enterprises. The data processing process using the PP model is as follows (Liu & Liu, 2021).

Step 1: Normalize the sample evaluation indicator set. Let the sample set of each indicator value be

$$\{x^*(i,j) \mid i=1,2,\ldots,n; j=1,2,\ldots,p\}.$$

The formalization formula is

$$x(i,j) = \frac{x^*(i,j) - x_{min}(j)}{x_{max}(j) - x_{min}(j)}. \tag{1}$$

$x_{max}(j)$ and $x_{min}(j)$ are the maximum and minimum values of the j indicator values, respectively, and $x(i, j)$ is a normalized sequence for the indicator eigenvalues.

Step 2: Construct a projection indicator function. The PP method is $\{x^*(i,j) \mid i=1,2,\ldots,p\}$. This is synthesized into a one-dimensional projection value $z(i)$, with $a = \{a(1), a(2), a(3), \ldots, a(p)\}$ as the projection direction, that is:

$$z(i) = \Sigma_{j=1}^{p}\, a(j)x(i,j), i = 1,2,\ldots,n \tag{2}$$

Then, the values are classified according to the one-dimensional walking graph of $\{z(i) \mid i=1,2,\ldots,n\}$. In Equation (2), a is the unit length vector. Therefore, the projection indicator function can be expressed as

$$S_z = \sqrt{\frac{\left(\Sigma_{i=1}^{n}\left(z(i)-E(z)\right)^2\right)}{n-1}}, \tag{3}$$

$$D_z = \Sigma_{i=1}^{n} \Sigma_{i=1}^{p} \left(R - r_{ij}\right) u\left(R - r_{ij}\right), \tag{4}$$

$$Q(a) = S_z D_z. \tag{5}$$

In Equation (4), $E(z)$ is the average value of the sequence $\{z(i) \mid i=1,2,\ldots,n\}$; R is the window radius of the local density; $r(i, j)$ represents the distance between samples, $r(i,j) = |z(i) - z(j)|$; and $u(t)$ is the unit step function. When $t \geq 0$, $u(t)$ is 1, and when $t \leq 1$, $u(t)$ is 0.

Step 3: Optimize the projection indicator function.
Maximize the objective function:

$$Q(a) = S_z D_z. \tag{6}$$

The restriction is

$$\Sigma_{j=1}^{p}\, a_j^2 = 1. \tag{7}$$

Step 4: Order the results. The optimal projection direction $a^*(j)$ can be obtained by using the projection indicator function $Q(a)$ and running the eight steps of the RAGA. The projection indicator function $Q(a)$ in the PP model is used as the objective function, and the projection $a(j)$ of each indicator is used as the optimization variable. After entering the data into Equation (2), the projection value $z(i)$ of each sample can be obtained. Sorting $z(i)$ from large to small sorts the samples from good to bad, respectively.

4.   Results and Discussion

This research evaluates the quality of EAID of the Chinese listed steel enterprises. Based on an indicator system (see Table 12.1), the related enterprise disclosure

Table 12.1 Evaluation Index System of EAID Quality

Target Layer A	Criterion Layer B	Sub-criterion Layer C	Indicator Layer D
EAID quality evaluation (A)	Relevance (B1)	Environmental policy (C1)	Independent social responsibility report (D1)
			Environmental protection principles, goals, and systems (D2)
			Disclosure and implementation of environmental laws and regulations (D3)
			Environmental protection plans and environmental issues (D4)
			Independent environmental report (D5)
			Environmental management structure and status (D6)
			Environmental protection honour reward (D7)
			Propaganda and education on environmental protection concepts (D8)
		Environmental performance information (C2)	Energy consumption and efficiency (D9)
			Environmental petition letter case (D10)
			"Three wastes" emissions (D11)
			"Three simultaneities" implementation (D12)
		Environmental financial information (C3)	Environmental liabilities (D13)
			Environmental investment (D14)
			Internal control of environmental work (D15)
	Reliability (B2)	Environmental information disclosure process (C4)	Third party audits (D16)
			Enterprise internal control system and its implementation (D17)
			Major environmental accidents (D18)
		Environmental information disclosure integrity (C5)	Environmental litigation (D19)
			Negative media reports on enterprise environment (D20)
			Pollution emission standard (D21)
	Compliance (B3)	Timeliness (C6)	Integrity of accounting processes and contents of financial information according to laws and regulations (D22)

information is gathered and scored. The indicators are scored quantitatively (i.e., the indicator is scored as 1 if corresponding information is provided; otherwise, the score is 0). Then the PP model and the RAGA are used to process the data. Firstly, to unify the variation range of each indicator value, the data are normalized. Secondly, a mathematical model of accounting information quality evaluation

Table 12.2 Indicator Weights

Rank	Indicator Layer D	Indicator Weight
D13	Environmental liabilities	0.9152
D9	Energy consumption and efficiency	0.9133
D11	"Three wastes" emissions	0.9128
D8	Propaganda and education on environmental protection concepts	0.8277
D15	Internal control of environmental work	0.7648
D4	Environmental protection plans and environmental issues	0.7050
D7	Environmental protection honour reward	0.6732
D12	"Three simultaneities" implementation	0.6731
D22	Integrity of accounting processes and contents of financial information according to laws and regulations	0.6454
D20	Negative media reports on enterprise environment	0.6214
D3	Disclosure and implementation of environmental laws and regulations	0.5762
D19	Environmental litigation	0.5760
D1	Independent social responsibility report	0.5760
D16	Third-party audits	0.5728
D18	Major environmental accidents	0.5128
D10	Environmental petition letter case	0.5126
D2	Environmental protection principles, goals, and systems	0.5121
D5	Independent environmental report	0.5028
D17	Enterprise internal control system and its implementation	0.4797
D14	Environmental investment	0.4189
D21	Pollution emission standard	0.4172
D6	Environmental management structure and status	0.4107

standards is established for the data by constructing the projection indicator function $Q(a)$.

The component values of the best projection direction represent the weights of the corresponding indicators (see Table 12.2). The evaluation weights of the evaluation indicators are different for different case samples. In this case, there are three indicators that have a greater impact on the EAID quality. The three indicators are D13 ("Environmental liabilities"), D9 ("Energy consumption and efficiency"), and D11 ("'Three wastes' emissions"), whose projection directions are 0.9192, 0.9133, and 0.9128, respectively. In contrast, indicators D6 ("Environmental management structure and status"), D21 ("Pollution emission standard"), and D14 ("Environmental investment") have very low weights, which are 0.4107, 0.4172, and 0.4189, respectively. Usually, having a better disclosure status does not mean a better weight in deciding the evaluation results. For example, D21 and D6 have little impact on EAID quality, but they have very good disclosure status. Also, 83.18% of the enterprises disclosed D21 in time, and 54.55% of the enterprises disclosed D6 in time (as seen in Table 12.3).

The projection values can be used to sort and compare the EAID quality of the sample enterprises (see Table 12.5). The top two are "Baoshan Steel" and "Taiyuan Iron and Steel", and the projection values are 6.6162 and 6.6029, respectively.

Table 12.3 Difference between an Indicator's Standard Value and Mean Value

	Indicator Layer D	Value
D21	Pollution emission standard	0.1500
D11	"Three wastes" emissions	0.1636
D13	Environmental liabilities	0.4136
D8	Propaganda and education on environmental protection concepts	0.4182
D6	Environmental management structure and status	0.4455
D1	Independent social responsibility report	0.5227
D17	Enterprise internal control system and its implementation	0.5545
D2	Environmental protection principles, goals, and systems	0.6318
D12	"Three simultaneities" implementation	0.6909
D7	Environmental protection honour reward	0.9655
D4	Environmental protection plans and environmental issues	0.7136
D9	Energy consumption and efficiency	0.7864
D22	Integrity of accounting processes and contents of financial information according to laws and regulations	0.8364
D3	Disclosure and implementation of environmental laws and regulations	0.8455
D15	Internal control of environmental work	0.8591
D14	Environmental investment	0.9500
D20	Negative media reports on enterprise environment	0.9818
D16	Third-party audits	0.9909
D5	Independent environmental report	0.9909
D18	Major environmental accidents	0.9955

Both companies behave especially well in indicator D22 ("Integrity of accounting processes and contents of financial information according to laws and regulations"), indicating that the content and scope of enterprise EAID highly conform to the related laws and regulations. Besides, "Taiyuan Iron and Steel" does very well in organizing activities to promote environmental protection concepts and environmental education. Generally, companies like "Shandong Steel", "Nanjing Iron and Steel", and "Maanshan Steel" have a strong awareness of constructing an enterprise management system to prompt environmental protection work, but most companies still lack the disclosure of the indicators measuring reliability, such as D10, D16, D18, and D19.

"Fushun Special Steel" has the worst disclosure quality with a projection value of 0.3602. From 2010 to 2020, "Fushun Special Steel" disclosed few kinds of environmental accounting information, but it did very well in indicators D11 ("'Three wastes' emissions"), D13 ("Environmental liabilities"), and D21 ("Pollution emission standard"). "Shagang Group" has a projection value of only 0.6067, which is the second lowest value, and it has a very similar disclosure condition to that of "Fushun Special Steel". Differently, "Shougang Group", with the third lowest projection value of 0.6143, also behaves well in D22 ("Integrity of accounting processes and contents of financial information according to laws and regulations").

Regarding the overall situation of the projection values, the average evaluation value is 2.7652, and the perfect score is 14.0548. The extreme difference of the evaluation is 8.8387, indicating that the overall EAID quality of the listed steel enterprises in China is poor and uneven.

The following will analyse the causes of the low EAID quality of China's steel enterprises from different aspects and compare the results with the conclusions in previous research.

Firstly, enterprises are less likely to voluntarily disclose negative environmental information. As seen in Table 12.4, examples include D10 ("Environmental petition letter case"), D19 ("Environmental litigation"), and D18 ("Major environmental accidents"). This finding is in line with the previous research (Wiseman, 1982; Cho et al., 2012; Patten & Zhao, 2014; Matuszak & Różańska, 2017). At present, the voluntary nature of EAID means that the companies disclose related information only if the information is positive. Before disclosure, company managers may well consider the social influences and make choices in order to enhance company reputation. Furthermore, many of the indicators with the high weights (D13 and D8) are all about enterprise environmental policies. Disclosing the related information greatly benefits the companies for social responsibility reasons. Thus, the quality of EAID is highly related to the enterprise voluntariness.

Secondly, the government's mandatory requirement has a great impact on disclosure quality. Examples include D9, D11, and D12. D12 indicates the disclosure status of "three simultaneities" implemented. The "three simultaneities"

Table 12.4 Enterprise's Points to Total Points

Rank	Indicator Layer D	Percentage
D21	Pollution emission standard	83.18%
D6	Environmental management structure and status	54.55%
D1	Independent social responsibility report	47.73%
D17	Enterprise internal control system and its implementation	44.09%
D2	Environmental protection principles, goals, and systems	36.82%
D7	Environmental protection honour reward	30.00%
D4	Environmental protection plans and environmental issues	28.64%
D11	"Three wastes" emissions	20.00%
D9	Energy consumption and efficiency	20.00%
D22	Integrity of accounting processes and contents of financial information according to laws and regulations	17.27%
D13	Environmental liabilities	16.82%
D8	Propaganda and education on environmental protection concepts	16.82%
D3	Disclosure and implementation of environmental laws and regulations	15.54%
D15	Internal control of environmental work	13.64%
D12	"Three simultaneities" implementation	10.00%
D14	Environmental investment	4.09%
D20	Negative media reports on enterprise environment	1.36%
D6	Environmental management structure and status	0.91%
D5	Independent environmental report	0.91%
D18	Major environmental accidents	0.45%
D19	Environmental litigation	0.00%
D10	Environmental petition letter case	0.00%

Table 12.5 Projection Values

Company	Projection Value
Baoshan Steel	6.6102
Taiyuan Iron and Steel	6.6029
Magang Group	5.5940
Nanjing Iron and Steel	5.5940
Shandong Steel	5.0964
Maanshan Steel	5.0573
Hebei Steel	4.3747
Xining Special Steel	2.1541
Fangda Steel	2.1210
SGIS Songshan	1.5631
Baotou Steel	1.5542
Hualing Steel	1.5322
Xinyu Iron and Steel	1.5192
Hangzhou Steel	1.4640
Benxi Steel	1.0853
Lingyuan Steel	0.9586
CITIC Pacific Special Steel	0.8410
Shougang Group	0.6143
Shagang Group	0.6067
Fushun Special Steel	0.3602

system is China's earliest environmental protection stipulation for construction projects. To ensure the effective implementation of the "Three simultaneities" system, China set a series of administrative orders and regulations. Besides, information disclosure of "three wastes" emissions (D11) is required by *Trial Standards for Industrial "Three Wastes" Emissions*, and information disclosure of energy consumption and efficiency (D9) is required by *Emergency Management Measures for Emergent Environmental Incidents*. Although the information mentioned above may expose the weakness of an enterprise's performance in environmental protection work, the disclosure qualities of these indicators are above the average level. Thus, the government plays an important role in the disclosure of environmental performance information. This finding is in line with previous research (Liu & Anbumozhi, 2019). However, the Chinese government has not made enough laws and regulations to press enterprises to disclose environmental accounting information. Currently, there is no completed regulatory system for EAID.

Thirdly, EAID quality is not significantly related to company size. According to the result, "Baoshan Steel" has the greatest projection value, and it also has the biggest market value. However, "CITIC Pacific Special Steel", "Baotou Steel", and "Hualing Steel", the largest steel companies that are preceded only by "Baoshan Steel", all have projection values lower than the average. In contrast, "Shougang Group" has the lowest projection value, but its market value ranked seventh in China in 2020. "Shandong Steel" ranks fourteenth among all the 20 enterprises, but its projection value is second only to "Baoshan Steel". In the case of China's steel industry, larger company size does not mean higher disclosure quality. This finding is not in conformity with previous research (Bewley & Li, 2000; Arif & Tuhin, 2013; Liu & Anbumozhi, 2019), which indicates that company size is one of the most important variables of determining environmental information disclosure quality.

5. Limitations and Future Research

Although a case study was conducted, further verification of other enterprises is needed. The environmental accounting information disclosed by unlisted enterprises needs to be gathered and analysed. Moreover, the evaluation index system is mainly designed for enterprises in heavily polluted industries, and it might not fit another specific industry well. In future research, the evaluation index system can be refined to adapt to the characteristics of different industries.

6. Conclusions

In this research, an evaluation index system was constructed from the aspects of relevance, reliability, and compliance and the quality of EAID in China's steel industry evaluated. A case study was conducted to analyse the evaluation system and devise methods for evaluating EAID quality. The listed enterprises in China's steel industry are chosen as the sources of sample data. Considering the characteristics of primary data, the PP model and RAGA are used to aggregate the evaluation information. After thorough analysis, it is found that the overall EAID quality in China's steel industry is poor. In China, most companies do not have a strong awareness of disclosing environmental accounting information, and the high quality of EAID relies on the pressure from the government. Usually, to enhance reputation, companies disclose environmental information to show their enterprise culture and social responsibility. Few companies voluntarily disclose negative information. In view of the problems, this study suggests the government to introduce a series of policies to encourage enterprise EAID.

References

Abdel-Kader, R. F. (2011). Hybrid discrete PSO with GA operators for efficient QoS-multicast routing. *Ain Shams Engineering Journal, 2*(1), 21–31.

Aerts, W., Cormier, D., Gordon, I. M., & Magnan, M. (2007). Performance disclosure on the web: An exploration of the impact of managers' perceptions of stakeholder concerns. *International Journal of Accounting Information Systems, 8*(1), 1–24.

Arif, M., & Tuhin, M. H. (2013). Factors determining corporate environmental information disclosure practices: Evidence from Bangladesh. *Business Review, 5*(2), 106–112.

Bahra, H. K., & Cadez, S. (2018). Accounting for climate change: A literature review and research agenda. *Journal of Cleaner Production, 10*(4), 34–56.

Beams, J. D. (1971). Social accounting: Theory, issues, and cases. In M. F. Gynther (Ed.), *Social accounting* (pp. 16–30). McGraw-Hill.

Bewley, K., & Li, Y. (2000). Disclosure of environmental information by Canadian manufacturing companies: A voluntary disclosure perspective. *Advances in Environmental Accounting and Management, 1*, 201–226.

Burritt, R. L., Hahn, T., & Schaltegger, S. (2002). Towards a comprehensive framework for environmental management accounting – Links between business actors and environmental management accounting tools. *Australian Accounting Review, 12*(27), 39–50.

Chaklader, B., & Gulati, P. (2015). A study of corporate environmental disclosures and reporting practices of listed companies in India. *IIMB Management Review, 27*(3), 295–306.

Cho, C. H., Michelon, G., Patten, D. M., & Roberts, R. W. (2012). CSR disclosure: The more things change . . .? *Accounting, Auditing & Accountability Journal, 25*(3), 486–517.

Cho, C. H., & Patten, D. M. (2007). The role of environmental disclosures as tools of legitimacy: A research note. *Accounting, Organizations and Society, 32*(7–8), 639–647.

Christ, K. L., & Burritt, R. L. (2013). Critical environmental concerns in wine production: An integrative review. *Journal of Cleaner Production, 53*, 232–242.

Clarkson, P. M., Overell, M. B., & Chapple, L. (2008). Environmental reporting and its relation to corporate environmental performance. *Abacus, 44*(4), 423–440.

Deegan, C. (2004). Environmental disclosures and shareholder value: The cases of BHP and Esso. In *Financial accounting and management control* (pp. 21–39). Springer.

Eiler, L. A., Miranda-Lopez, J., & Tama-Sweet, I. (2015). The impact of accounting disclosures and the regulatory environment on the information content of earnings announcements. *The International Journal of Accounting, 50*(2), 142–169.

Eiler, R. G. (2015). *Environmental accounting: An introduction and practical guide*. John Wiley & Sons.

Ernst & Ernst. (1978). *Social responsibility disclosure: 1978 survey of fortune 500 annual reports*. Ernst & Ernst.

Fekrat, M. A., Inclan, C., & Petroni, D. (1996). Corporate environmental disclosures: Competitive disclosure hypothesis using 1991 annual report data. *International Journal of Accounting, 31*(2), 175–195.

Fu, G., Liu, Y., & Yuan, M. (2003). Application of projection pursuit model in comprehensive evaluation of environment quality. *Environmental Monitoring and Assessment, 83*(3), 151–164.

Guthrie, J., & Farneti, F. (2008). GRI sustainability reporting by Australian public sector organizations. *Public Money and Management, 28*(6), 361–366.

Guthrie, J., & Parker, L. D. (1989). Corporate social reporting: A rebuttal of legitimacy theory. *Accounting and Business Research, 19*(76), 343–352.

He, C. (2014). Environmental disclosure and financial performance: Evidence from China. *Journal of Business Ethics, 124*(4), 491–507.

Ingram, R. W., & Frazier, K. B. (1980). Environmental performance and corporate disclosure. *Journal of Accounting Research, 18*(2), 614–622.

Jasch, C. (2000). Environmental performance evaluation and indicators. *Journal of Cleaner Production, 8*(1), 79–88.

Jones, M. C., & Sibson, R. (1987). What is projection pursuit? *Journal of the Royal Statistical Society. Series A (General), 150*(1), 1–37.

Jones, M. J. (2010). Accounting for the environment: Towards a theoretical perspective for environmental accounting and reporting. In A. Hopwood, H. Sjögren & L. Henningsson (Eds.), *Accounting for the environment: More talk and little progress* (pp. 70–99). Springer.

Kinnear, K. E. (1994). *Advances in genetic programming*. MIT Press.

Kong, D., & Wang, J. (2016). Environmental accounting: A new component of the accounting system. *Sustainability Accounting, Management and Policy Journal, 7*(3), 351–365.

Kreuze, J. G., Newell, G. E., & Newell, S. J. (1996). Environmental disclosures: What companies are reporting. *Management Accounting, 77*(1), 37–40.

Kruskal, J. B., & Shepard, R. N. (1974). A nonmetric variety of linear factor analysis. *Psychometrika, 39*(2), 123–157.

Lewis, L. S., Parker, J., & Evans, J. (2014). The influence of top managers' environmental attitudes on environmental management in high-tech companies. *Business & Society, 53*(3), 395–420.

Li, T., Fang, Y., Zeng, D., Shi, Z., Sharma, M., Zeng, H., & Zhao, Y. (2020). Developing an indicator system for a healthy City: Taking an urban area as a pilot. *Risk Management and Healthcare Policy*, 83–92.

Liu, X., & Anbumozhi, V. (2009). Determinant factors of corporate environmental information disclosure: An empirical study of Chinese listed companies. *Journal of Cleaner Production, 17*(6), 593–600.

Liu, Z., & Anbumozhi, V. (2019). Determinants of corporate environmental information disclosure: An empirical study of Chinese listed companies. *Journal of Cleaner Production, 32*(1), 31–41.

Liu, Z., & Liu, M. (2021). The application of projection pursuit model in the evaluation of environmental accounting information disclosure quality: Evidence from China's thermal power industry. *Environmental Impact Assessment Review, 87*, 106512.

Luo, L. (2019). The influence of the regulatory environment on corporate environmental disclosure and performance. *Journal of Environmental Management, 249*, 109374.

Luo, X., Tong, S., Fang, Z., & Qu, Z. (2019). Frontiers: Machines vs. humans: The impact of artificial intelligence chatbot disclosure on customer purchases. *Marketing Science, 38*(6), 937–947.

Manes-Rossi, F., Tiron-Tudor, A., Nicolò, G., & Zanellato, G. (2018). Ensuring more sustainable reporting in Europe using non-financial disclosure—De facto and de jure evidence. *Sustainability, 10*(4), 1162.

Matuszak, L., & Różańska, E. (2017). Impact of environmental accounting information on financial performance of companies: Evidence from Poland. *Sustainability, 9*(2), 230–241.

Milne, M. J., & Gray, R. (2007). Future prospects for corporate sustainability reporting. In *Sustainability accounting and accountability* (pp. 184–207). Routledge.

Patten, D. (2021). Legal crime: An analytical framework for studying international criminogenic polices. *International Journal of Comparative and Applied Criminal Justice, 45*(4), 405–422.

Patten, D. M., & Trompeter, G. (2003). Corporate environmental disclosures in response to public policy pressure: The case of the 1990 Clean Air Act amendments. *Journal of Accounting and Public Policy, 22*(1), 83–94.

Patten, D. M., & Zhao, N. (2014). Standalone CSR reporting by US retail companies. *Accounting Forum, 38*(2), 132–144.

Qi, L., Yao, X., & Liu, Z. (2021). Environmental information disclosure quality evaluation: A study of Chinese listed companies. *Journal of Cleaner Production, 278*, 123456.

Rahman, M. J., & Wu, J. (2023). M&A activity and ESG performance: evidence from China. *Managerial Finance, 50*(1), 179–197.

Rahman, M. J., Wu, Q., & Zhu, H. (2024). Corporate social responsibility in times of social distancing: Evidence from China. *Business Ethics, the Environment & Responsibility*.

Rahman, M. J., Zhu, H., & Chen, S. (2023). Does CSR reduce financial distress? Moderating effect of firm characteristics, auditor characteristics, and Covid-19. *International Journal of Accounting & Information Management, 31*(5), 756–784.

Senn, J., & Giordano-Spring, S. (2020). *Environmental accounting: A managerial emphasis*. McGraw-Hill Education.

Venturelli, A., Caputo, F., & Leopizzi, R. (2017). The state of the art of corporate social disclosure prior to the introduction of non-financial reporting directive: A cross country analysis. *Social Responsibility Journal, 13*(1), 85–97.

Wiseman, J. (1982). An evaluation of environmental disclosures made in corporate annual reports. *Accounting, Organizations and Society, 7*(1), 53–63.

Wu, J., & Shen, M. (2013). The impact of corporate social responsibility on the cost of equity capital: Evidence from China. *Journal of Business Ethics, 114*(3), 487–493.

Zhu, H., & Rahman, M. J. (2024). Ex-ante expected changes in ESG and future stock returns based on machine learning. *The British Accounting Review*, 101457.

13 ESG Accounting and Firm Value during COVID-19

Evidence from China

Md Jahidur Rahman, Tarek Rana, Hongtao Zhu, and Cheng Yibo

1. Introduction

Based on the study by Khanchel et al. (2023), examining the impact of corporate social responsibility (CSR) on the French stock market during COVID-19, we explored similar issues in the Chinese market context. Our study addresses two key questions: (1) were firms that disclose ESG reports more resilient during the COVID-19 pandemic? (2) Do companies that actively disclose ESG perform better than those that do not?

This research is prompted by ongoing debates about whether companies should prioritize maximizing shareholder value or broader societal relevance (Friedman, 1970). The COVID-19 pandemic has disrupted global operations, compelling companies to seek innovative solutions (Diab-Bahman & Al-Enzi, 2020). During this period, public focus on corporate image intensified, making it crucial for firms to cultivate a positive image to enhance performance (Miller et al., 2022). Companies are increasingly pressured to disclose ESG performance amid evolving green development standards. ESG ratings, assigned by professional organizations, significantly influence investor decisions (Patel et al., 2021; Drempetic et al., 2020). ESG practices are now central to corporate strategies aimed at meeting sustainability policies (Li et al., 2021).

The pandemic heightened public concern for social responsibility, benefiting companies that actively engage in such practices. Theories suggest that firms engaging in social responsibility bolster their reputation and stakeholder relationships, leading to better performance (Gray et al., 1995; Lanis & Richardson, 2012). Studies indicate that firms attentive to social and environmental needs foster long-term investor relationships (Kim et al., 2019). The link between corporate social responsibility (CSR) and corporate value strengthens in contexts demanding higher CSR (Griffin et al., 2020). Companies perceived as socially responsible tend to maintain better stock performance during crises.

According to Morgan Stanley, 79% of financial industry participants expressed interest in CSR investments during COVID-19, underscoring the importance of socially responsible investments during emergencies. Research shows that US non-financial firms with high ESG scores outperformed their peers financially during the pandemic (Lins et al., 2017). Evidence suggests that ESG investments

DOI: 10.4324/9781003488965-17

helped companies mitigate the pandemic's impact on market capitalization (Garel & Petit-Romec, 2021). Furthermore, investor sensitivity to CSR can favourably influence company valuation during crises (Palma-Ruiz et al., 2020). Notably, companies that had previously disclosed ESG policies were better equipped to handle the pandemic's effects.

Our study analyses data from Chinese firms listed from 2011 to 2021, focusing on firm performance as the dependent variable and using ESG scores to assess performance. To address potential endogeneity, we employed instrumental variables, the Heckman method, propensity score matching (PSM), and fixed effects. The results indicate that firms adopting ESG policies were better able to withstand the pandemic's impact, showing greater resilience in stock prices compared to non-ESG firms. Our endogeneity and robustness tests support these findings.

This study contributes to the ESG literature by demonstrating its protective effect on firm value during COVID-19. While ESG has gained traction as an investment strategy, previous research had not compared firm performance during the pandemic with earlier ESG prevalence in China. Our findings reveal that companies with strong ESG reports build better investor relationships over the past decade, providing valuable insights for industry practitioners. Additionally, our results confirm that ESG continues to help firms resist crisis effects, expanding the literature on its value. This information is crucial for investors and managers considering ESG's role during COVID-19. Lastly, we assessed the epidemic's impact using the natural logarithm of newly confirmed diagnoses in target cities. Our findings suggest that firms with high ESG scores perform better financially as pandemic severity increases, enhancing our understanding of ESG performance during varying crisis stages. This knowledge is beneficial for practitioners interested in ESG initiatives.

The chapter is outlined as follows: Section 2 introduces the background of the study and the environment of ESG policies in the Chinese market. Section 3 introduces the theoretical framework. Section 4 describes the literature review and hypothesis development. Section 5 explains the research design. Section 6 discusses the empirical results. Section 7 presents the endogeneity test. Section 8 discusses the robustness check, and this section is divided into two parts to make it more convincing. Section 9 presents the discussion, wherein the relationship with previous literature, theoretical implications, and practical implications are analysed. Section 10 concludes this study.

2. Background: ESG and COVID-19 Crisis in China

ESG is a development concept that has emerged in recent years to address the ongoing deterioration of the global environment. While ESG gained traction in the Western world during the 20th century, it wasn't until 2010 that Chinese companies began to recognize its importance (Gao et al., 2021). As awareness of climate change's detrimental effects on ecosystems has grown, ESG frameworks have been introduced, requiring listed companies to submit ESG reports as mandated by various governments. Notably, 83 countries have set ambitious goals to reduce global carbon emissions by 74% (Tian et al., 2022), yet China, as a developing country, contributes 27% of annual carbon emissions.

According to Global Sustainable Investment Alliance statistics, ESG investment in Asian countries stands at a mere 0.8%, significantly lower than that of developed nations, reflecting the ongoing development stage of most Asian economies. Although ESG investment is becoming mainstream, China's ESG policies are still evolving, having started relatively late without a unified evaluation standard. Furthermore, the Chinese A-share market does not mandate compulsory ESG report publication, limiting the feasibility of ESG-based investments.

In 2016, the Chinese government introduced the "Guidance on Establishing a Green Financial System" to promote sustainable development. This guidance initiated the development of ESG policies in China; however, the lack of uniform standards has hindered effective ESG information disclosure in the domestic securities market. Consequently, the ESG data provided by individual companies often fail to accurately reflect their actual ESG performance.

Differences in understanding ESG's connotations have led to variations in rating systems. Internationally, ESG indicators emphasize green development and biodiversity, while China prioritizes energy conservation and emission reduction (Berg et al., 2022). China's development strategy has shifted from focusing on speed to prioritizing quality, and the market is increasingly adopting new development concepts to support sustainable growth (Zhao et al., 2021). The trend towards emphasizing ESG policies is evident, with a growing number of Chinese companies publishing ESG reports since 2011. By 2020, a quarter of listed companies had disclosed such reports.

As the initial epicentre of the COVID-19 outbreak, China faced devastating losses. The pandemic not only threatened public health but also severely impacted the socio-economic landscape, resulting in a historic GDP decline of 6.8% (Wang & Su, 2020). The hotel and restaurant industry suffered particularly, experiencing a one-third drop in revenue, while one in five new labour market entrants lost their jobs (Chetty et al., 2020). Travel restrictions decimated the tourism sector, with visitor numbers plummeting and revenues decreasing by 60% (Hoque et al., 2020). China's A-class tourist attractions reported losses nearing 65% in 2020 (Wang et al., 2022). The pandemic's spread and subsequent control measures caused significant turmoil in China's financial markets, exemplified by the CSI 300 index's dramatic drop from 5,200 to 4,600 when the stock market reopened on February 3.

3. Theoretical Framework

Stakeholder theory and shareholder theory are central to discussions about how ESG influences a company's value (Freeman, 1994). Stakeholder theory emphasizes the importance of various groups affected by the firm, including investors, suppliers, customers, and employees (Phillips et al., 2003). These stakeholders can impose specific demands based on their interests (Al Amosh et al., 2021). According to stakeholder theory, firms that prioritize the interests of their social environment can better meet stakeholder needs, ultimately enhancing their performance in a virtuous cycle (Friedman & Miles, 2002).

In contrast, shareholder theory prioritizes the rights and interests of shareholders (Freeman, 2010). Since ESG returns often manifest over the long term, ESG initiatives might be viewed as risky investments that fail to deliver immediate value growth (Cornell, 2021). During crises, a company's ability to maintain its relationships within the social environment can come under scrutiny (Coombs & Holladay, 2015). Stakeholders may become sceptical of a firm's actions, viewing them as self-serving. Consequently, firms often concentrate on stakeholder relations during crises to mitigate pressure and protect their interests. If a company actively engages in ESG practices, it is more likely to garner stakeholder support rather than face backlash (Campbell, 2007).

Firms that adopt this strategy can effectively build trust with stakeholders, especially during disasters (Gromis di Trana et al., 2022). By investing in ESG, companies can secure stakeholder backing during crises. For instance, many firms embraced their social responsibilities during the COVID-19 pandemic, demonstrating a proactive stance towards stakeholder welfare (Asante Antwi et al., 2021). This approach serves as a vital strategy for protecting companies in challenging times.

4. Literature Review and Hypothesis Development

4.1 ESG, Firm Value, and COVID-19

COVID-19 represents an unprecedented event, characterized as an unanticipated shock with significant economic repercussions (Yarovaya et al., 2021). Such events, often termed Black Swan events, are rare, extreme, and traceable (Taleb, 2007). These characteristics offer companies a unique opportunity to make strategic investments that can help them navigate the crisis and mitigate severe negative market reactions (Akhtaruzzaman, 2022). Given the description of ESG policies, investments targeting ESG can serve as a viable solution.

To develop a theoretical framework, we draw insights from the crisis characteristics, ESG policies, and firms' market performance. The rapid spread of COVID-19 led to widespread, severe lockdown measures globally (Lassoued & Khanchel, 2021). The sudden emergence of the virus had immediate and far-reaching consequences, with approximately $6 trillion lost from global markets within just four days (Ozili & Arun, 2023). The crisis necessitated urgent national responses to control its impact, complicating accurate assessments of its consequences. Factors such as regional environmental conditions made comprehensive evaluations resource-intensive. Additionally, stakeholders became increasingly sensitive to market reactions, particularly those in vulnerable situations. Several studies in China also investigated the relationship between environmental accounting and firm performance (Rahman & Wu, 2023; Rahman et al., 2023, 2024; Zhu & Rahman, 2024).

In this context, ESG policies can enhance the stability of the global economy amid crisis. While COVID-19 has caused widespread disruption, it also presented opportunities for companies to respond proactively to threats against their operational environment. Research indicates that stocks with strong ESG performance exhibit superior price elasticity during the pandemic (Broadstock et al., 2021). Thus, a more aggressive adoption of ESG measures correlates with greater resilience in firms (Boubaker et al., 2020).

Furthermore, the role of ESG may have fundamentally shifted due to the pandemic. During this period, firms focusing on ESG prioritized stakeholder interests, implementing measures such as avoiding layoffs and providing telecommuting options and paid sick leave (Boubaker et al., 2022). Companies that had been actively engaged in ESG practices prior to the outbreak demonstrated notable resilience during the crisis (Huang et al., 2020). However, it has been observed that the resources allocated to socio-environmental governance tend to be limited during crises (Qiu et al., 2021). This leads us to hypothesize that firms investing less in ESG during the epidemic can still outperform those that do not disclose any ESG information. Then, we propose the following hypothesis.

H1: Difference exists in the impact of COVID-19 on companies that have adopted ESG policies and those that have not.

4.2   Impact Intensity of COVID-19

Some studies have considered COVID-19 as a health crisis that affected the world more than the economic crisis (Roubini, 2020). There are also previous studies that argue that ESG and financial performance do not positively affect each other (Bénabou & Tirole, 2010). Empirical studies have shown that for firms with more outstanding performance in corporate responsibility activities during the financial crisis, its financial performance was better during the crisis (Lins et al., 2017). This study analysed the study question, taking into account the gap areas in previous studies. We argued that a firm receives a different shock during an epidemic than it would have received during an economic crisis. During the outbreak, a firm's value had great uncertainty; it was trapped by the unknown nature of the epidemic. Specifically, the population did not have access to accurate information at the time of a crisis similar to the epidemic, and as a result, the national embargo caused most people to suffer varying degrees of economic loss, which was caused by the different intensities of the epidemic in different regions. Then, the following hypothesis is proposed:

H2: When the impact of the epidemic increased, companies with high ESG scores were less affected.

5.   Research Design

5.1   Sample and Data

This study focuses on examining the impact of corporate ESG, epidemic shocks, and firm performance. Drawing on the ideas of Broadstock et al. (2021) and Zhou and Zhou (2021), this study analyses ESG, epidemic shock, and firm performance

Table 13.1 Variable Screening Procedure

Selection Criterion	Number of Companies	Rejection Sample
Excluding financial sector	94	825
Excluding ST companies	529	1,185
Excluding samples with missing key variables	290	4,730

Note: This table shows the variable filtering process for the final sample.

Source: Table by authors

under the same framework in order to carry out the identification of the relationship among ESG, firm value, and COVID-19. Specifically, this study uses the data of A-share listed companies in China from 2011–2021 for empirical analysis. The financial data of the firms were obtained from the China Stock Market & Accounting Research (CSMAR) database, while the ESG data were based on the relevant ratings of China Securities, Hexun, and Bloomberg databases. For the raw data, the following data processing was carried out according to the usual research practice: (1) excluding special treatment (ST) enterprises; (2) excluding financial listed companies; (3) excluding samples with asset–iability ratios greater than 1; (4) excluding samples with missing key variables; and (5) to avoid the influence of outliers, all continuous variables were subjected to bilateral 1% tail-shrinking treatment. After the above processing, this study finally obtained a total of 4,024 enterprises containing 30,112 observations of unbalanced panel data. The corresponding variable screening process is shown in Table 13.1.

5.2　Variable Measurements

5.2.1　Dependent Variable

The dependent variable is corporate performance. This study used research ideas of Shen et al. (2020) and Rahman et al. (2022) for reference to analyse corporate performance from both the financial and capital market performance aspects. In this study, we analysed both financial performance and capital market performance. Financial performance is measured by ROA, and capital market performance is measured by the annual individual share return of individual shares without considering cash dividends (return). To avoid the interference of indicators on the model estimation results, robustness tests were also conducted by replacing the variable measures.

5.2.2　Independent Variables

The independent variables are ESG and COVID-19. By combing with the ESG-related studies, it can be seen that academics currently mainly use ESG ratings of enterprises by external organizations to reflect the ESG level of enterprises, but in general, there is no consensus on the definition of ESG selection ratings. In general, most scholars use ESG ratings from China Securities, Hexun, or Bloomberg database for their analysis (Deng, 2019; He et al., 2022; Zhang et al., 2021). Considering the relatively large number of samples included in the ratings, this study mainly adopted ESG ratings for analysis, and further robustness tests were conducted by replacing the ESG metrics in the robustness analysis section.

Following the idea of Jin et al. (2022), this study used the natural logarithm of the number of new diagnoses in the city where the company is located to reflect the strength of the impact of the epidemic. In general, the higher the number of confirmed new infections in a firm's city, the more pronounced the impact of new infections on the firm. In this study, robustness tests were also conducted by replacing the measures.

5.2.3   Control Variables

To accurately identify the relationship between ESG, COVID-19, and firm performance, this work draws on the study by Jin et al. (2022) to further control for the following control variables. These include firm size (Size), measured by the natural logarithm of total firm assets; firm gearing (Lev), calculated by total firm liabilities divided by total assets; cashflow (CFO), calculated by net operating cash flow divided by total assets; and growth capacity (Growth), measured by the growth rate of firms' operating income. Board size (Board) is calculated by the natural logarithm; board independence (Indep) is measured by the ratio of independent directors to the board members; duality (Dual) is measured by the value of 1 in the situation where the chairman and CEO are both appointed by one person, or it is measured by the value of 0; concentration of equity (Top10) is measured by the shareholdings of the ten largest shareholders; and nature of ownership (SOE) is measured by the shares of the ten largest shareholders divided by total shares. SOE is taken as 1 if the enterprise is state-owned; otherwise, it is taken as 0. Age is the natural logarithm of the year of establishment.

5.3   Model Specification

To test the relationship among ESG, COVID-19, and firm performance, we set the following econometric model, drawing on the research idea of Zhang et al. (2023):

$$ROA_{i,t} = \alpha_0 + \alpha_1 ESG_{i,t} + \alpha_2 Covid_{i,t} + \alpha_3 ESG_{i,t} * Covid_{i,t} + \sum Control$$
$$+ \sum IND + \sum YEAR + \varepsilon_{i,t} \tag{1}$$

$$Return_{i,t} = \beta_0 + \beta_1 ESG_{i,t} + \beta_2 Covid_{i,t} + \beta_3 ESG_{i,t} * Covid_{i,t}$$
$$+ \sum Control + \sum IND + \sum YEAR + \varepsilon_{i,t} \tag{2}$$

Here, i and t in the above equations mean firms and years. $ROA_{i,t}$ denotes firm i's operating performance in year t. $Return_{i,t}$ denotes firm i's individual stock return in year t. $\sum Control$ is the set of control variables, including the relevant control variables mentioned above. $\sum IND$ and $\sum YEAR$ are industry- and time-fixed effects, respectively. $\varepsilon_{i,c,t}$ is a random disturbance term indicating there are factors that disturb empirical testing process firm performance. This study focused on coefficients α_3 and β_3 of the interaction term between ESG and epidemic. If the coefficients are positive, it indicates that companies with better ESG reports also perform better in financial performance when an impact crisis rises.

6. Empirical Results

6.1   Descriptive Statistics

Table 13.2 presents the descriptive statistics of the data. The two extreme values of corporate financial performance are 0.223 and −0.253, which indicates that the gap in the financial performance of Chinese listed companies is relatively large. The mean value of 0.0375 indicates that most companies are able to obtain positive financial performance reports, and the standard deviation of 0.0641 indicates that the financial performance of these companies is more balanced. The maximum of individual stock returns is 2.334, and the minimum is −0.572, which means that these corporations' performance in the capital market is also characterized by large extreme differences, while the mean value of 0.158 indicates that most firms have positive annual individual returns and the standard deviation of 0.491 indicates that the differences in individual stock returns among firms are higher than the differences in financial performance. The epidemic also shows a large extreme difference and a large standard deviation. In addition, the standard deviation of only firm size among the control variables is relatively large. Overall, the descriptive

Table 13.2 Descriptive Statistics

Variable	Mean	p50	Std. Dev.	Min	Max	N
ROA	0.0375	0.0378	0.0641	−0.253	0.223	30,112
Return	0.158	0.0545	0.491	−0.572	2.334	30,112
ESG	4.093	4	1.034	1.250	6.250	30,112
COVID	0.966	0	2.120	0	10.84	30,112
Size	22.24	22.05	1.295	19.83	26.24	30,112
Lev	0.429	0.421	0.206	0.0559	0.903	30,112
Cashflow	0.0455	0.0448	0.0689	−0.161	0.241	30,112
Growth	0.174	0.108	0.416	−0.574	2.625	30,112
Board	2.124	2.197	0.199	1.609	2.708	30,112
Indep	0.377	0.364	0.0539	0.333	0.571	30,112
Dual	0.279	0	0.449	0	1	30,112
Top10	0.579	0.587	0.152	0.228	0.902	30,112
SOE	0.345	0	0.475	0	1	30,112
Age	2.907	2.944	0.329	1.792	3.526	30,112

statistics of this study are closer to those of existing studies, which indicates that the results of this study are reliable.

Table 13.3 presents the correlation coefficients between the variables in this study, where the results are significantly positive, while the correlation coefficient with capital market performance is negative but insignificant. The correlation coefficient between the epidemic and financial performance is negative and insignificant, while the correlation coefficient with individual stock returns is significantly positive.

Table 13.4 lists the VIF values of each variable in the empirical analysis of this study, and the VIF test helps avoid the interference of multicollinearity in the empirical analysis on the estimation results. The results show that the VIF values of each variable are around 1.4, the maximum VIF value does not exceed 2, and the mean value is 1.2. Results suggest the problem of multicollinearity does not exist among the variables in the empirical analysis of this study and helps improve the accuracy of the results of this study.

6.2 Baseline Regression Results

The first and second columns of Table 13.5 show the estimation results with financial performance and individual stock returns as dependent variables, respectively, when the model controls only for industry- and year-fixed effects. The coefficient of COVID-19 in the first column is −0.0042 and is significant at the 1% level, indicating that epidemic shocks reduce the financial performance of firms. The results in the second column indicate that the coefficient of the epidemic shock remains significantly negative when individual stock returns are the dependent variable, which in turn indicates that epidemic shocks reduce firms' financial performance and capital market performance. The regression coefficients of the interaction term ESG*COVID in the first and second columns are both significantly positive, thus indicating that the financial performance and individual stock returns of firms with better ESG ratings are relatively higher when subjected to epidemic shocks; that is, ESG can help firms withstand epidemic shocks. The third and fourth columns include firm-level control variables, where the regression coefficient of the epidemic shock is still significantly negative, while the interaction coefficient ESG*COVID is still significantly positive, thus indicating that ESG can still help firms withstand the negative effects of the epidemic shock when controlling for other factors that may affect firm performance.

This is consistent with the previous analysis that ESG helps firms counteract the negative effects of epidemic shocks. In this regard, this study argues that it can be explained by both customer relationship and investor reflection. On the one hand, ESG helps firms form stable customer relationships. Those firms with better ESG performance face fewer production shocks when they are hit by an epidemic because their supply chains and customer relationships are more stable, and thus their financial performance declines relatively less. On the other hand, ESG helps send positive signals to investors that the company's operations and internal governance are relatively good, making it easier for companies to be trusted by investors in the event of an epidemic, which in turn leads to less risk of stock price declines and more stable individual stock returns.

Table 13.3 Correlation Coefficient Analysis of Variables

Variables	Roa	Return	ESG	Covid	Size	Lev	Cashflow	Growth	Board	Indep	Dual	Top10	Soe	Age
Roa	1.000													
Return	0.150***	1.000												
ESG	0.253***	−0.002	1.000											
Covid	−0.001	0.097***	0.027***	1.000										
Size	0.031***	−0.044***	0.233***	0.076***	1.000									
Lev	−0.358***	−0.030***	−0.066***	−0.000	0.494***	1.000								
Cashflow	0.389***	0.099***	0.093***	0.040***	0.077***	−0.163***	1.000							
Growth	0.249***	0.114***	0.010*	−0.032***	0.041***	0.015***	0.021***	1.000						
Board	0.021***	−0.044***	0.043***	−0.058***	0.263***	0.144***	0.041***	−0.016***	1.000					
Indep	−0.021***	0.018***	0.064***	0.035***	−0.005	−0.009*	−0.009*	−0.001	−0.548***	1.000				
Dual	0.033***	0.027***	−0.031***	0.039***	−0.168***	−0.128***	−0.016***	0.029***	−0.180***	0.112***	1.000			
Top10	0.232***	−0.007	0.159***	−0.009	0.162***	−0.082***	0.130***	0.095***	0.027***	0.030***	0.024***	1.000		
Soe	−0.074***	−0.063***	0.098***	−0.026***	0.351***	0.274***	0.001	−0.072***	0.260***	−0.059***	−0.294***	0.005	1.000	
Age	−0.087***	0.009	−0.044***	0.201***	0.172***	0.172***	0.020***	−0.049***	0.034***	−0.011*	−0.100***	−0.183***	0.166***	1.000

***p<0.01,**p<0.05,*p<0.1

Note: This table shows the correlation coefficients between the variables in this paper; *, ** and *** are significant at the level of 10%, 5% and 1%, respectively.

Source: Table by authors

Table 13.4 Multicollinearity

Variable	VIF	1/VIF
Size	1.75	0.5704
Board	1.65	0.6053
Lev	1.52	0.6583
Indep	1.49	0.6714
SOE	1.30	0.7705
Age	1.15	0.8672
Top10	1.13	0.8816
ESG	1.13	0.8850
Dual	1.12	0.8924
Cashflow	1.07	0.9304
COVID	1.06	0.9409
Growth	1.02	0.9795
Mean VIF	1.2	

Note: The table shows the VIF values of each variable for the empirical analysis in this study and the corresponding estimation results.

Source: Table by author

7.   Mitigating Endogeneity Concerns

7.1   Instrumental Variables Method

The estimation results in this study may have endogeneity issues, as epidemic shocks are relatively exogenous. Thus, the epidemic shocks themselves do not have endogeneity issues, but there may be a reverse causality relationship problem in this study. For example, those firms with better corporate performance may be more likely to be noticed by ESG rating agencies and receive higher ESG scores; that is, there is reverse causality in firm performance in this study, which in turn affects ESG. To reduce the possibility of bias in the estimation results of this study due to endogeneity issues, the idea of Breuer et al. (2018) is borrowed here, and the mean ESG values of other firms in the same region in the same year are used as instrumental variables. Since the business characteristics and market environment of firms in the same region are closer, we can see the instrumental variables have a relationship with dependent variables, but firms usually affect firm performance only by affecting the ESG of firms; that is, the instrumental variables cannot directly affect firm performance. This study focuses on the interaction term between ESG and previous shocks, and thus introduces instrumental variables in the form of interaction terms, and corresponding estimation results are in the first and second columns of Table 13.6. These numbers indicate the results remain significantly positive when the mean ESG of the same industry is used as the instrumental variable, which in turn indicates that ESG still helps firms resist the negative effects of epidemic shocks after accounting for endogeneity issues.

Table 13.5 Baseline Regression Results

	(1)	(2)	(3)	(4)
	ROA	*Return*	*ROA*	*Return*
	Coeff. (t-value)	*Coeff. (t-value)*	*Coeff. (t-value)*	*Coeff. (t-value)*
ESG	0.0157*** (24.1446)	−0.0075*** (−3.2978)	0.0082*** (16.4595)	−0.0061*** (−2.6094)
COVID	−0.0042*** (−4.3841)	−0.0314*** (−6.1680)	−0.0031*** (−3.9377)	−0.0269*** (−5.4949)
ESG*COVID	0.0009*** (4.6803)	0.0065*** (6.0216)	0.0007*** (4.3246)	0.0059*** (5.6546)
Size			0.0081*** (16.0811)	−0.0159*** (−7.1971)
Lev			−0.1171*** (−37.2568)	0.0200 (1.5616)
Cashflow			0.2798*** (35.6627)	0.5719*** (15.4910)
Growth			0.0342*** (32.1469)	0.1749*** (22.6501)
Board			0.0007 (0.2406)	−0.0181 (−1.3933)
Indep			−0.0316*** (−3.3231)	0.0299 (0.6573)
Dual			0.0019* (1.9544)	0.0033 (0.6565)
Top10			0.0430*** (13.5608)	0.0193 (1.3778)
SOE			−0.0032*** (−2.8597)	−0.0254*** (−5.5483)
Age			0.0036** (2.2169)	0.0054 (0.7940)
Ind	No	No	Yes	Yes
Year	No	No	Yes	Yes
Cons	−0.0278** (−2.4863)	−0.2875*** (−11.7018)	−0.1581*** (−10.4398)	0.0206 (0.3667)
N	30,112	30,112	30,112	30,112
Adj. R^2	0.111	0.352	0.392	0.382

Source: Table by author

Table 13.6 Regression Results for Instrumental Variables

	(1)	(2)
	ROA	*Return*
	Coeff. (t-value)	*Coeff. (t-value)*
ESG	0.0082*** (16.4960)	−0.0061*** (−2.6098)
COVID	−0.0031*** (−3.9338)	−0.0264*** (−5.4904)
ESG*COVID	0.0007*** (4.3328)	0.0057*** (5.6625)
Size	0.0081*** (16.1051)	−0.0159*** (−7.2207)
Lev	−0.1171*** (−37.3242)	0.0200 (1.5661)
Cashflow	0.2798*** (35.7281)	0.5719*** (15.5202)
Growth	0.0342*** (32.2028)	0.1749*** (22.6869)
Board	0.0007 (0.2402)	−0.0181 (−1.3975)
Indep	−0.0316*** (−3.3311)	0.0298 (0.6547)
Dual	0.0019* (1.9569)	0.0032 (0.6560)
Top10	0.0430*** (13.5888)	0.0194 (1.3846)
SOE	−0.0032*** (−2.8642)	−0.0254*** (−5.5577)
Age	0.0036** (2.2204)	0.0054 (0.7945)
Ind	Yes	Yes
Year	Yes	Yes
_cons	−0.1581*** (−10.4553)	0.0211 (0.3763)
N	30,112	30,112
Adj. R^2	0.392	0.382

Source: Table by author

7.2 Heckman Method

To further strengthen the credibility of the estimation results, the Heckman method is used to solve the possible endogeneity problem in model estimation, drawing on the research idea of Tang et al. (2023). Specifically, firstly, according to the ESG rating, we divided enterprises into three groups and treatment variables (Treat) were set, where the group with the highest ESG is the experimental group and the group with the lowest ESG is the control group, and the logit model is estimated with this variable as the dependent variable, and then the inverse Mills ratio (IMR) was estimated, where the estimation results of the first step are shown in the first column of Table 13.7. Secondly, the IMR is in the baseline regression, and the second and third columns of Table 13.7 are the results. The results show ESG*COVID is still significantly positive after the estimation by the Heckman method, which in turn indicates that the estimation results have a strong robustness.

7.3 *Propensity Score Matching Test and Replacement of Fixed Effects*

In addition, those firms with better ESG ratings and those with worse ratings may themselves have large systematic differences and thus lead to endogeneity problems caused by sample bias. To strengthen the credibility of the estimation results,

Table 13.7 Heckman Regression Results

	(1)	(2)	(3)
	Treat *Coeff. (t-value)*	*ROA* *Coeff. (t-value)*	*Return* *Coeff. (t-value)*
ESG		0.0085*** (16.3043)	−0.0046* (−1.7932)
COVID		−0.0032*** (−3.8159)	−0.0248*** (−4.7145)
ESG*COVID		0.0007*** (4.1970)	0.0053*** (5.0436)
Size	0.4288*** (20.6659)	0.0043** (2.0664)	−0.0018 (−0.2007)
Lev	−1.9364*** (−16.6643)	−0.1007*** (−9.6750)	−0.0310 (−0.7114)
Cashflow	1.3462*** (6.2130)	0.2586*** (22.1091)	0.6196*** (11.3930)
Growth	−0.0951*** (−3.8672)	0.0352*** (24.9949)	0.1592***(17.0104)
Board	0.2306* (1.8798)	−0.0002 (−0.0529)	−0.0152 (−0.8879)
Indep	2.5475*** (6.2757)	−0.0559*** (−3.3659)	0.0789 (1.0632)
Dual	−0.0250 (−0.6082)	0.0022* (1.7881)	0.0067 (0.9907)
Top10	0.9310*** (6.7990)	0.0343*** (5.9386)	0.0514** (2.0391)
SOE	0.2581*** (4.8545)	−0.0056*** (−2.9541)	−0.0188** (−2.3453)
Age	−0.3018*** (−3.9594)	0.0076*** (3.1854)	−0.0054 (−0.4872)
IMR		−0.0131* (−1.7102)	0.0483 (1.6280)
Ind	Yes	Yes	Yes
Year	Yes	Yes	Yes
Cons	−11.0587*** (−16.8193)	−0.0487 (−0.7925)	−0.3060 (−1.2417)
N	18,353	18,353	18,353
Adj. R^2		0.394	0.359

the PSM approach is further used for the analysis. Specifically, the treatment variable (Treat) set as the dependent variable according to the above approach was estimated by the logit model, and the principle of 1:1 matching of nearest neighbours was used, thus screening out samples with more similar characteristics of various categories in the experimental and control groups. Based on the screened samples, the baseline regression was estimated again, and Table 13.8 presents the corresponding results, which shows that ESG*COVID is still significantly positive after the PSM regression, thus indicating that the systematic bias between samples does not affect the findings of this study.

Finally, in this study, we conducted an endogeneity analysis by replacing the fixed effects of the model. In the previous estimation results, we used industry- and year-fixed effects for the analysis, but there may be a problem of individual unobservable factors, which leads to endogeneity problems caused by omitted variables. Based on this, the individual and time-fixed effect models were estimated, and Table 13.9 presents the corresponding estimation results. The results show ESG*COVID is still significantly positive after using individual and time-fixed effects, which in turn indicates that the estimation results of this study are robust.

Table 13.8 PSM Regression Results

	(1)	(2)
	ROA	*Return*
	Coeff. (t-value)	*Coeff. (t-value)*
ESG	0.0078*** (13.2879)	−0.0014 (−0.4216)
COVID	−0.0029*** (−2.8569)	−0.0151** (−1.9958)
ESG*COVID	0.0006*** (3.0782)	0.0039***(2.6898)
Size	0.0062***(10.0132)	−0.0098*** (−2.6819)
Lev	−0.1043*** (−24.0961)	0.0316 (1.3490)
Cashflow	0.2956*** (28.8226)	0.6964*** (11.5726)
Growth	0.0334*** (20.9087)	0.1946*** (15.5199)
Board	0.0009 (0.2869)	−0.0316 (−1.5887)
Indep	−0.0312*** (−2.7399)	0.0050 (0.0735)
Dual	0.0008 (0.6007)	0.0043 (0.5133)
Top10	0.0385*** (9.8834)	0.0020 (0.0874)
SOE	−0.0051*** (−3.5147)	−0.0244*** (−3.1039)
Age	0.0025 (1.2496)	−0.0172 (−1.5029)
Ind	Yes	Yes
Year	Yes	Yes
Cons	−0.1122*** (−5.8796)	−0.0045 (−0.0493)

8.   Robustness Check

In order to prevent errors in the results of this study, further robustness tests were conducted. Firstly, replacing the measures of the dependent variables in the previous estimation results, ROA and individual stock returns without considering dividends were used as dependent variables in this study for analysis. To reduce the likelihood that the choice of variable measures will affect the model estimation results, ROE was chosen to measure the financial performance of firms, and the return on individual shares with dividends (Rtu1) was chosen to measure capital market performance. We found the results remained significantly positive after replacing the measure of the dependent variable.

Secondly, replacing the measure of COVID-19 shock, the number of new deaths (Covid1) was selected to measure the epidemic shock, and the third and fourth columns of Table 13.10 present the results of the corresponding estimation. We found the results remained significantly positive after replacing the measuring method for epidemic shock, thus indicating that the estimation results of this study are robust.

Finally, the ESG measure was replaced. In the benchmark regressions, in this study we used ESG ratings of Huawei for analysis, in order to avoid systematic differences caused by ESG ratings. The ESG measures were further replaced by ESG ratings of Hexun (ESG_HX) and Bloomberg (ESG_PB) databases, and Table 13.11

Table 13.9 Estimated Results of Replacement Fixed Effects

	(1)	(2)
	ROA	*Return*
	Coeff. (t-value)	*Coeff. (t-value)*
ESG	0.0047*** (7.7866)	0.0005 (0.1240)
COVID	−0.0023*** (−2.7907)	−0.0326*** (−6.0676)
ESG*COVID	0.0004** (2.3965)	0.0056*** (5.0037)
Size	0.0163*** (13.0593)	−0.0737*** (−10.2936)
Lev	−0.1664*** (−32.0681)	0.0533** (1.9984)
Cashflow	0.1710*** (21.7459)	0.5815*** (12.0930)
Growth	0.0312*** (30.7342)	0.1551*** (19.1869)
Board	−0.0010 (−0.2241)	−0.0406 (−1.4733)
Indep	−0.0137 (−1.0891)	0.0110 (0.1341)
Dual	0.0016 (1.2911)	−0.0119 (−1.3984)
Top10	0.0354*** (5.9876)	0.2264*** (6.5230)
SOE	−0.0090*** (−2.8871)	−0.0384** (−2.3946)
Age	0.0207** (2.4785)	0.2728*** (6.0802)
Firm	Yes	Yes
Year	Yes	Yes
Cons	−0.3252*** (−9.5473)	0.5016** (2.5084)
N	30,112	30,112
Adj. R^2	0.273	0.398

presents the corresponding results. The results show that the coefficients of the interaction term between ESG and COVID-19 are significantly positive in both the ESG form with Hexun ESG and the ESG form with Bloomberg ESG measures, which in turn indicates that ESG helps reduce the impact of the coronavirus pandemic on corporate value.

9.   Discussion

9.1   Relationship with Previous Literature

Currently no study combines the time from 2010 to 2019 in investigating the impact of ESG on firm value during the pandemic from 2019 to 2021. Our study compared the financial performance of companies prior to the outbreak and found that companies that implemented ESG policies performed better under the impact of COVID-19. Previous studies did not focus on the period prior to the outbreak, but there are some studies on the role played by ESG in the COVID-19 pandemic. Some studies argued that firms need to take some new measures in response to the crisis, and ESG is shown in COVID-19 to have a significant increase in firm profitability (Armani et al., 2020). Other studies considered ESG as a value-added behaviour that helps companies achieve better profits and create a better social environment (Ferrell et al., 2016). Since 2011, Chinese companies have started to

Table 13.10 Robustness Tests (COVID-19)

	(1)	(2)	(3)	(4)
	ROE	*Rtu1*	*ROA*	*Return*
	Coeff. (t-value)	*Coeff.* (t-value)	*Coeff.* (t-value)	*Coeff.* (t-value)
ESG	0.0160***	−0.0019 (−0.7885)	0.0079*** (15.2686)	−0.0045* (−1.9418)
COVID	−0.0064*** (−3.3498)	−0.0187*** (−3.7390)		
ESG*COVID	0.0011*** (2.8530)	0.0026** (2.4318)		
Covid-1			−0.0018*** (−4.2606)	−0.0085*** (−3.2292)
ESG*Covid-1			0.0004*** (4.9401)	0.0018*** (3.7833)
Size	0.0222*** (18.4755)	−0.0047** (−2.0886)	0.0081*** (16.1433)	−0.0159*** (−7.1772)
Lev	−0.1899*** (−22.5904)	0.0118 (0.8933)	−0.1171***(−37.2975)	0.0202 (1.5739)
Cashflow	0.4846*** (29.7525)	0.6134*** (15.9798)	0.2798*** (35.6776)	0.5716*** (15.4693)
Growth	0.0729*** (29.8354)	0.1784*** (22.3811)	0.0341*** (32.1054)	0.1751*** (22.6602)
Board	−0.0010 (−0.1530)	−0.0141 (−1.0601)	0.0007 (0.2471)	−0.0178 (−1.3713)
Indep	−0.0639*** (−3.0818)	0.0407 (0.8657)	−0.0315*** (−3.3148)	0.0291 (0.6395)
Dual	0.0036* (1.8071)	0.0029 (0.5665)	0.0019* (1.9362)	0.0031 (0.6152)
Top10	0.0779*** (11.6258)	0.0182 (1.2572)	0.0428*** (13.5321)	0.0199 (1.4187)
SOE	−0.0046* (−1.7893)	−0.0254*** (−5.4468)	−0.0032*** (−2.8469)	−0.0252*** (−5.5047)
Age	0.0113*** (3.3905)	0.0073 (1.0485)	0.0036** (2.1912)	0.0056 (0.8225)
Ind	Yes	Yes	Yes	Yes
Year	Yes	Yes	Yes	Yes
Cons	−0.4761*** (−15.0376)	−0.2804*** (−4.8581)	−0.1573*** (−10.3952)	0.0126 (0.2239)
N	30,079	30,112	30,112	30,112
Adj. R^2	0.285	0.374	0.392	0.381

Table 13.11 Robustness Tests (ESG_HX)

	ROA	Return	ROA	Return
	Coeff. (t-value)	Coeff. (t-value)	Coeff. (t-value)	Coeff. (t-value)
ESG_HX	0.0013*** (35.6755)	−0.0003* (−1.8346)		
COVID	−0.0102*** (−16.5365)	−0.0232*** (−5.1689)	−0.0025** (−2.1764)	−0.0242** (−2.4754)
ESG_HX*COVID	0.0005*** (23.0638)	0.0013*** (9.3524)		
ESG_PB			−0.0000 (−0.0984)	0.0018*** (2.9579)
ESG_PB*COVID			0.0001*** (2.7798)	0.0005** (2.0132)
Size	0.0034*** (6.9802)	−0.0171*** (−7.0199)	0.0090*** (10.5892)	−0.0268*** (−6.6291)
Lev	−0.0962*** (−31.7576)	0.0249* (1.8638)	−0.1437*** (−27.4688)	0.0237 (1.0045)
Cashflow	0.2150*** (29.4259)	0.5397*** (13.8472)	0.3149*** (24.4768)	0.6812*** (10.5161)
Growth	0.0278*** (29.2166)	0.1558*** (19.8821)	0.0311*** (18.4588)	0.2321*** (15.3418)
Board	−0.0011 (−0.4018)	−0.0208 (−1.5518)	−0.0021 (−0.5285)	−0.0117 (−0.5327)
Indep	−0.0242*** (−2.6090)	0.0170 (0.3627)	−0.0066 (−0.4848)	0.0779 (1.0584)
Dual	0.0016* (1.7581)	0.0071 (1.3676)	0.0028* (1.6848)	0.0246** (2.4833)
Top10	0.0312*** (10.3499)	0.0574*** (3.9415)	0.0354*** (7.1381)	0.0312 (1.2412)
SOE	−0.0032*** (−3.0258)	−0.0277*** (−5.7823)	−0.0063*** (−3.6868)	−0.0275*** (−3.5014)
Age	0.0016 (1.0254)	0.0070 (1.0152)	0.0035 (1.3723)	−0.0295** (−2.2285)
Ind	Yes	Yes	Yes	Yes
Year	Yes	Yes	Yes	Yes
Cons	−0.0549*** (−3.7748)	0.0427 (0.6891)	−0.1189*** (−5.2543)	0.2302** (2.3056)
N	26,643	26,643	11,155	11,155
Adj. R^2	0.480	0.418	0.472	0.342

embrace ESG and use it as an important investment strategy, which has a significant impact on the sustainability of the Chinese market. In turn, investors will be more inclined to trust companies that actively implement ESG.

9.2   *Theoretical Implications*

We support the stakeholder theory through our findings, for example, that investors prefer ESG firms because ESG firms tend to have better stability in crises to ensure that stakeholders' assets do not take a huge hit. Stakeholders' trust in ESG companies places the latter in a stable social environment, constituting a positive cycle (Friedman & Miles, 2002). This study also confirms that ESG does not harm shareholders' wealth, as the financial performance and capital market performance of companies that have adopted ESG tend to be better in crisis performance.

9.3   *Practical Implications*

This study offers a new possibility for the responses of users of financial information and corporate managers in a crisis. Users of financial information can use ESG as a strategic policy for their companies in times of high uncertainty, thus helping stakeholders and the companies themselves to maintain good growth trends. In addition, this study provides a new direction for regulators. Regulators can improve the market environment in China by standardizing the ESG scoring criteria to facilitate investors to get more accurate ESG information about companies and by encouraging and enforcing ESG-related investment activities by companies.

10.   Concluding Remarks

The relationship between ESG and corporate value during the COVID-19 pandemic remains significant. In this study, we investigated how the epidemic affects firms that adopt ESG practices compared to those that do not. We analysed the differences in firm value between ESG adopters and non-adopters during the outbreak, comparing these findings to their financial performance prior to the pandemic. Our results indicate that ESG firms outperform non-ESG firms in both financial and capital market performance during the pandemic, with the protective effect of ESG policies becoming more pronounced as the severity of the crisis increases. This resilience is attributed to ESG helping firms build stable relationships with investors, reducing the impact of the epidemic on their value. Our findings suggest that ESG is particularly appealing to investors in times of crisis.

These insights are valuable for corporate managers and industry practitioners. Given the heightened volatility in firm value during the pandemic, actively engaging in ESG can enhance investor interest and trust, thus improving firms' resilience in crises.

Policymakers should encourage ESG investments by aligning standards to help companies respond effectively to crises. Corporate decision-makers need to ensure that their ESG initiatives align with sustainable practices to achieve better returns. During economic downturns, ESG investments may face increased costs, so companies should adopt flexible ESG strategies to ensure future growth. Ultimately, ESG investments can create a win–win scenario for both society and businesses.

However, this study has limitations. Firstly, using a single scoring criterion to determine active ESG participation poses challenges that remain unresolved at this stage. Future research could benefit from unified ESG standards in China to further explore their impact. Secondly, we did not consider the conflicts of interest between corporate managers and stakeholders during COVID-19, which could be a focus for future studies. Finally, this research primarily addresses the negative impacts of the epidemic; future work could investigate market responses to positive events.

References

Akhtaruzzaman, M., Boubaker, S., & Umar, Z. (2022). COVID–19 media coverage and ESG leader indices. *Finance Research Letters, 45*, 102170.

Al Amosh, H., & Khatib, S. F. (2021). Corporate governance and voluntary disclosure of sustainability performance: The case of Jordan. *SN Business & Economics, 1*(12), 165.

Armani, A. M., Hurt, D. E., Hwang, D., McCarthy, M. C., & Scholtz, A. (2020). Low-tech solutions for the COVID-19 supply chain crisis. *Nature Reviews Materials, 5*(6), 403–406.

Asante Antwi, H., Zhou, L., Xu, X., & Mustafa, T. (2021, April). Beyond COVID-19 pandemic: an integrative review of global health crisis influencing the evolution and practice of corporate social responsibility. In *Healthcare* (Vol. 9, No. 4, p. 453). MDPI.

Bénabou, R., & Tirole, J. (2010). Individual and corporate social responsibility. *Economica, 77*(305), 1–19.

Berg, F., Koelbel, J. F., & Rigobon, R. (2022). Aggregate confusion: The divergence of ESG ratings. *Review of Finance, 26*(6), 1315–1344.

Boubaker, S., Cellier, A., Manita, R., & Saeed, A. (2020). Does corporate social responsibility reduce financial distress risk? *Economic Modelling, 91*, 835–851.

Boubaker, S., Liu, Z., & Zhan, Y. (2022). Customer relationships, corporate social responsibility, and stock price reaction: Lessons from China during health crisis times. *Finance Research Letters, 47*, 102699.

Breuer, W., Müller, T., Rosenbach, D., & Salzmann, A. (2018). Corporate social responsibility, investor protection, and cost of equity: A cross-country comparison. *Journal of Banking & Finance, 96*, 34–55.

Broadstock, D. C., Chan, K., Cheng, L. T., & Wang, X. (2021). The role of ESG performance during times of financial crisis: Evidence from COVID-19 in China. *Finance Research Letters, 38*, 101716..

Campbell, J. L. (2007). Why would corporations behave in socially responsible ways? An institutional theory of corporate social responsibility. *Academy of Management Review, 32*(3), 946–967.

Chetty, R., Friedman, J. N., Hendren, N., Stepner, M., & The Opportunity Insights Team. (2020). *How did COVID-19 and stabilization policies affect spending and employment? A new real-time economic tracker based on private sector data* (Vol. 91, pp. 1689–1699). National Bureau of Economic Research.

Coombs, T., & Holladay, S. (2015). CSR as crisis risk: Expanding how we conceptualize the relationship. *Corporate Communications: An International Journal, 20*(2), 144–162.

Cornell, B. (2021). ESG preferences, risk and return. *European Financial Management, 27*(1), 12–19.

Deng, X., & Cheng, X. (2019). Can ESG indices improve the enterprises' stock market performance? – An empirical study from China. *Sustainability, 11*(17), 4765.

Diab-Bahman, R., & Al-Enzi, A. (2020). The impact of COVID-19 pandemic on conventional work settings. *International Journal of Sociology and Social Policy, 40*(9/10), 909–927.

Drempetic, S., Klein, C., & Zwergel, B. (2020). The influence of firm size on the ESG score: Corporate sustainability ratings under review. *Journal of Business Ethics, 167*, 333–360.

Ferrell, A., Liang, H., & Renneboog, L. (2016). Socially responsible firms. *Journal of Financial Economics, 122*(3), 585–606.

Freeman, R. E. (1994). The politics of stakeholder theory: Some future directions. *Business Ethics Quarterly*, 409–421.

Freeman, R. E. (2010). *Strategic management: A stakeholder approach.* Cambridge University Press.

Friedman, A. L., & Miles, S. (2002). Developing stakeholder theory. *Journal of Management Studies, 39*(1), 1–21.

Friedman, M. (1970). A theoretical framework for monetary analysis. *Journal of Political Economy, 78*(2), 193–238.

Gao, S., Meng, F., Gu, Z., Liu, Z., & Farrukh, M. (2021). Mapping and clustering analysis on environmental, social and governance field a bibliometric analysis using Scopus. *Sustainability, 13*(13), 7304.

Garel, A., & Petit-Romec, A. (2021). The resilience of French companies to the COVID-19 crisis. *Finance, 42*(3), 99–137.

Gray, R., Kouhy, R., & Lavers, S. (1995). Corporate social and environmental reporting: A review of the literature and a longitudinal study of UK disclosure. *Accounting, Auditing & Accountability Journal, 8*(2), 47–77.

Griffin, D. W., Guedhami, O., Li, K., & Lu, G. (2020). National culture and the value implications of corporate social responsibility: A channel analysis. *SSRN Electronic Journal*, 3250222.

Gromis di Trana, M., Fiandrino, S., & Yahiaoui, D. (2022). Stakeholder engagement, flexible proactiveness and democratic durability as CSR strategic postures to overcome periods of crisis. *Management Decision, 60*(10), 2719–2742.

He, F., Du, H., & Yu, B. (2022). Corporate ESG performance and manager misconduct: Evidence from China. *International Review of Financial Analysis, 82*, 102201.

Hoque, A., Shikha, F. A., Hasanat, M. W., Arif, I., & Hamid, A. B. A. (2020). The effect of Coronavirus (COVID-19) in the tourism industry in China. *Asian Journal of Multidisciplinary Studies, 3*(1), 52–58.

Huang, W., Chen, S., & Nguyen, L. T. (2020). Corporate social responsibility and organizational resilience to COVID-19 crisis: An empirical study of Chinese firms. *Sustainability, 12*(21), 8970.

Jin, X., Zhang, M., Sun, G., & Cui, L. (2022). The impact of COVID-19 on firm innovation: Evidence from Chinese listed companies. *Finance Research Letters, 45*, 102133.

Khanchel, I., Lassoued, N., & Gargoury, R. (2023). CSR and firm value: Is CSR valuable during the COVID 19 crisis in the French market? *Journal of Management and Governance*, 1–27.

Kim, H. D., Kim, T., Kim, Y., & Park, K. (2019). Do long-term institutional investors promote corporate social responsibility activities? *Journal of Banking & Finance, 101*, 256–269.

Lanis, R., & Richardson, G. (2012). Corporate social responsibility and tax aggressiveness: A test of legitimacy theory. *Accounting, Auditing & Accountability Journal, 26*(1), 75–100.

Lassoued, N., & Khanchel, I. (2021). Impact of COVID-19 pandemic on earnings management: An evidence from financial reporting in European firms. *Global Business Review*, 09721509211053491.

Li, T. T., Wang, K., Sueyoshi, T., & Wang, D. D. (2021). ESG: Research progress and future prospects. *Sustainability*, *13*(21), 11663.

Lins, K. V., Servaes, H., & Tamayo, A. (2017). Social capital, trust, and firm performance: The value of corporate social responsibility during the financial crisis. *The Journal of Finance*, *72*(4), 1785–1824.

Miller, D., Tang, Z., Xu, X., & Le Breton-Miller, I. (2022). Are socially responsible firms associated with socially responsible citizens? A study of social distancing during the Covid-19 pandemic. *Journal of Business Ethics*, *179*(2), 387–410.

Ozili, P. K., & Arun, T. (2023). Spillover of COVID-19: Impact on the global economy. In *Managing inflation and supply chain disruptions in the global economy* (pp. 41–61). IGI Global.

Palma-Ruiz, J. M., Castillo-Apraiz, J., & Gómez-Martínez, R. (2020). Socially responsible investing as a competitive strategy for trading companies in times of upheaval amid COVID-19: Evidence from Spain. *International Journal of Financial Studies*, *8*(3), 41.

Patel, P. C., Pearce II, J. A., & Oghazi, P. (2021). Not so myopic: Investors lowering short-term growth expectations under high industry ESG-sales-related dynamism and predictability. *Journal of Business Research*, *128*, 551–563.

Phillips, R., Freeman, R. E., & Wicks, A. C. (2003). What stakeholder theory is not. *Business Ethics Quarterly*, *13*(4), 479–502.

Qiu, S. C., Jiang, J., Liu, X., Chen, M. H., & Yuan, X. (2021). Can corporate social responsibility protect firm value during the COVID-19 pandemic? *International Journal of Hospitality Management*, *93*, 102759.

Rahman, M. J., & Wu, J. (2023). M&A activity and ESG performance: Evidence from China. *Managerial Finance*, *50*(1), 179–197.

Rahman, M. J., Wu, Q., & Zhu, H. (2024). Corporate social responsibility in times of social distancing: Evidence from China. *Business Ethics, the Environment & Responsibility*.

Rahman, M. J., Zhu, H., & Chen, S. (2023). Does CSR reduce financial distress? Moderating effect of firm characteristics, auditor characteristics, and Covid-19. *International Journal of Accounting & Information Management*, *31*(5), 756–784.

Rahman, M. S., Hossain, M. A., Chowdhury, A. H., & Hoque, M. T. (2022). Role of enterprise information system management in enhancing firms competitive performance towards achieving SDGs during and after COVID-19 pandemic. *Journal of Enterprise Information Management*, *35*(1), 214–236.

Roubini, N. (2020, March 25). Coronavirus pandemic has delivered the fastest, deepest economic shock in history. *The Guardian*.

Shen, H., Fu, M., Pan, H., Yu, Z., & Chen, Y. (2020). The impact of the COVID-19 pandemic on firm performance. *Emerging Markets Finance and Trade*, *56*(10), 2213–2230.

Taleb, N. N. (2007). *The black swan: The impact of the highly improbable* (p. 2007). Random House.

Tang, S., He, L., Su, F., & Zhou, X. (2023). Does directors' and officers' liability insurance improve corporate ESG performance? Evidence from China. *International Journal of Finance & Economics*, 29, no. 3 (2024): 3713-3737.

Tian, J., Yu, L., Xue, R., Zhuang, S., & Shan, Y. (2022). Global low-carbon energy transition in the post-COVID-19 era. *Applied Energy*, *307*, 118205.

Wang, L. E., Tian, B., Filimonau, V., Ning, Z., & Yang, X. (2022). The impact of the COVID-19 pandemic on revenues of visitor attractions: An exploratory and preliminary study in China. *Tourism Economics*, *28*(1), 153–174.

Wang, Q., & Su, M. (2020). A preliminary assessment of the impact of COVID-19 on environment–A case study of China. *Science of the Total Environment*, *728*, 138915.

Yarovaya, L., Matkovskyy, R., & Jalan, A. (2021). The effects of a "black swan" event (COVID-19) on herding behavior in cryptocurrency markets. *Journal of International Financial Markets, Institutions and Money*, *75*(C), 101321.

Zhang, D., Wang, C., & Dong, Y. (2023). How does firm ESG performance impact financial constraints? An experimental exploration of the COVID-19 pandemic. *The European Journal of Development Research, 35*(1), 219–239.

Zhang, X., Zhao, X., & Qu, L. (2021). Do green policies catalyze green investment? Evidence from ESG investing developments in China. *Economics Letters, 207*, 110028.

Zhao, T., Xiao, X., & Dai, Q. (2021). Transportation infrastructure construction and high-quality development of enterprises: Evidence from the quasi-natural experiment of high-speed railway opening in China. *Sustainability, 13*(23), 13316.

Zhou, D., & Zhou, R. (2021). ESG performance and stock price volatility in public health crisis: Evidence from COVID-19 pandemic. *International Journal of Environmental Research and Public Health, 19*(1), 202.

Zhu, H., & Rahman, M. J. (2024). Ex-ante expected changes in ESG and future stock returns based on machine learning. *The British Accounting Review*, 101457.

Part 5

Carbon Accounting for Climate Change

14 The Relationship between Carbon Accounting, Carbon Dioxide Emissions, and Climate Change

Md Jahidur Rahman, Tarek Rana, Hongtao Zhu, and Xue Wenxin Shirley

1. Introduction

Climate change's cumulative impacts increasingly capture global attention. The signing of the Greenhouse Gas Protocol (GHGP) in 2001 and the establishment of the Climate Disclosure Project (CDP) in 2003 marked significant milestones in corporate transparency regarding carbon emissions. The Paris Agreement of 2016, signed by 178 countries, further underscores the global commitment to addressing climate change. Today, the CDP plays a critical role in promoting carbon disclosure, encouraging companies to report their emission data and pursue accountability.

Experts utilize tools from the GHGP and the CDP for carbon accounting. However, existing algorithms often fail to cover all relevant variables, prompting researchers to reconstruct these frameworks. For example, Judith et al. (2013) identified the impact of long-term carbon storage on atmospheric greenhouse gas (GHG) concentrations and developed a corresponding carbon accounting framework.

China's rapid economic growth and urbanization have led to significant environmental degradation, increasing pressure on its infrastructure and natural resources. The country's trade surplus has contributed to rising GHG emissions. In response, researcher Li and Colombier (2009) suggested redefining GHG responsibilities and enhancing energy conservation strategies. Effective coordination among local managers and stakeholders is essential to mitigate urban climate change. Several studies in China also investigated the relationship between environmental accounting and firm performance (Rahman & Wu, 2023; Rahman et al., 2023, 2024; Zhu & Rahman, 2024).

China primarily employs four methods for calculating carbon footprints: input–output analysis, life-cycle assessment, Intergovernmental Panel on Climate Change (IPCC) methodologies, and carbon footprint calculators. While many scholars focus on the benefits of carbon accounting for governance, few examine its actual impacts on emissions and policymaking. Understanding these dynamics is crucial for adapting to contemporary challenges.

The influence of carbon accounting on CO_2 emissions and temperature changes is significant. A clear presentation of carbon accounting's effects can deepen public

DOI: 10.4324/9781003488965-19"

understanding of climate change, prompting strategic actions. Positive impacts may encourage more companies to disclose carbon data, fostering transparency and public pressure for emission control. Conversely, highlighting negative impacts can draw attention to excessive emissions, motivating governmental policy interventions.

This study will employ comprehensive quantitative analysis methods, utilizing MATLAB to analyse the relationship among CO_2 emissions, carbon accounting, and average temperatures across China's provinces over nine years. The analysis will include two independent variables (carbon accounting emissions and CO_2 emissions) and one dependent variable (average temperature), generating three-dimensional maps and two-dimensional curves.

Expected outcomes include a continued rise in average annual temperatures in China, albeit at a decreasing rate, and an increase in carbon accounting values alongside rising CO_2 emissions, with a notable increase in the proportion of carbon accounting emissions.

The findings are anticipated to demonstrate that carbon accounting significantly influences China's climate trajectory, with potential for stabilizing or even reducing temperature trends in the future.

This research is valuable as it explores the relationship among carbon dioxide emissions, carbon accounting, and temperature changes – an area that has not been thoroughly investigated. The results are expected to inform national policies, corporate governance, and public awareness regarding climate action. However, the multifaceted nature of emissions and temperature changes necessitates careful consideration of external factors influencing the outcomes.

The remainder of this chapter is organized as follows: Section 2 presents the literature review and hypothesis development, Section 3 outlines the methodology, Section 4 details the variables, and Section 5 discusses the results and conclusions.

2.   Literature Review and Hypothesis Development

It is tacit for all countries to pay attention to climate change. For sustainable development, all countries are carrying out reforms and implementing new ways and methods to slow down the process of global climate change.

2.1   Carbon Accounting

Firstly, we need to clarify what carbon accounting is. "Carbon" is a term that refers to the main GHG carbon dioxide, or short for all GHGs (Bebbington et al., 2008). Carbon accounting is the process of CDP quantifying GHG emissions, which is roughly divided into three parts: emissions from direct activities of the company (range 1), emissions from purchased power consumed by the company (range 2), and emissions from other indirect sources in the company's supply chain (range 3) (Climate Disclosure Standards Board, 2009).

Companies control their supply chain and understand how their business pollution will influence the climate with the help of CDP and set goals to limit their carbon dioxide emission by quantifying their GHG emission.

Carbon accounting adopts the material balance method or stoichiometric method based on specific facilities or process flow to calculate emissions. The most common form of calculating GHG emissions is recorded emission factors (Ascui & Lovell, 2011). The emission factor is the calculated ratio between the activity level of emission sources and GHG emissions. When direct monitoring is impossible, or the cost is too high, accurate emission data can be calculated according to fuel consumption.

Auditors typically investigate the quantities, sources, overall trends, and projections of GHG emissions quantified in GHG reports, including ranges 1, 2, and 3, and here are the attestation/evaluation activities (see Table 14.1).

The emission is calculated at the site or the enterprise level. That is, the following formula is used in different steps:

$$Activity\ data \times Emission\ factor = GHG\ emissions$$

2.2 Carbon Dioxide Emissions and Carbon Accounting

Based on Foucault's concept of strategic allocation, the carbon accounting tools should focus on governance and effect. Therefore, Breton et al. (2020) combined the qualitative and comprehensive methods of gency for the Environment and the Energy Management's (ADEME) strategy and principal research in France. Breton explored and studied the company's strategy by rebuilding ADEME's weaving relationship network and observing its carbon footprint. According to this research, the author found that most companies prefer to use carbon accounting to control the cost and revenue during the business process. To obtain maximum benefit, with the help of carbon accounting tools, the company will effectively reduce carbon emissions, resulting in a slow-down trend on the improving amount of carbon dioxide emissions.

H1a: Carbon accounting leads to a slowing tread of carbon dioxide emissions.

Since range 3 is the very uncontrollable control of activities occurring upstream and downstream of the supply chain, nearly 70% or more of the company's carbon

Table 14.1 Categories of Carbon Accounting Auditing Method

	GHG Statement Assurance	*Compliance Carbon Audit*	*Carbon-Management Audit*	*Governmental Climate Change Audit*
Scope	Firm level	Firm and project level	Firm and project level	National or international level
Nature	Verification	Investigation	Evaluation	Evaluation
Users of audit report	External users	External users, internal managers	Internal managers	Public, governmental officials

accounting would happen in this period, and many leading companies have become pioneers and played a leading role. For example, Walmart's Gigaton Program plans to reduce GHG emissions throughout the company's supply chain by 1 billion tonnes by 2030. Smithfield promised to reduce 25% overall emissions from its supply chain by 2025 as part of the journey to carbon neutrality by 2050. Therefore, more and more companies and industries entered the CDP, which will lead to the rapid rise of carbon accounting emissions from scratch. Therefore,

H2a: The proportion of carbon accounting emissions in carbon dioxide emissions increases.

2.3 Carbon Dioxide Emission and Temperature Changes

The climate sensitivity of historical climate change simulation is less than that of long-term carbon dioxide increase, and the recorded energy budget change only slightly limits the climate sensitivity. By quantifying the mode effect related to the actual temperature change mode in the real world, which is equivalent to the simulation of long-term CO_2 increase by atmosphere-ocean general circulation model, scholars have obtained the point that the limitation of carbon dioxide on climate sensitivity will not give too low or excessive restriction (Andrew, 2018). Thus, the carbon dioxide emissions have a significant influence on the temperature changes, so the authors infer the following:

H3a: The temperature increases with the increase in carbon dioxide emissions.

China has launched the world's most extensive CO_2 emissions trading system (ETS) and implemented seven pilot emission trading projects in major cities and provinces since 2013. However, due to the universality of market transactions in China, the importance of state-owned enterprises in China has soared, and they account for a very high proportion in the natural gas and power industries., which may limit the cost of carbon dioxide emission reduction and reduce the efficiency of ETS due to monopoly behaviour and management pricing (Goulder & Hafstead, 2017).

The structural economic change implemented by China has proved to be the primary determinant of China's GHG emissions in the past 15 years – under the old growth model, it is the driving factor of emission growth. Under the new model, it is and will continue to be the driving force of emission reduction (Green & Stern, 2017).

At present, some scholars have predicted the peak of carbon dioxide emissions. According to the theoretical model of the carbon Kuznets curve (CKC), China's carbon dioxide emissions will peak in 2036 (Chen et al., 2020). If increased, China's energy CO_2 emissions may grow much slower than what grows under the old economic model, and it is assumed to peak in the first decade. Sane macroeconomic policies and planning are essential for decarbonization. China has chaired the G20, a key international economic institution actively engaged in the decarbonization process, particularly in financing infrastructure projects. Additionally, China plays a central role in both the Asian Infrastructure Investment Bank (AIIB) and the New Development Bank. Through initiatives like the Silk Road Fund, China aims to reduce carbon dioxide emissions (Fergus, 2017).

Experts speculate that China has entered a new economic model with the universal application of the CO_2 ETS. The structural economic reform implemented has become the driving factor of emission reduction, so more and more institutions have joined the ranks of emission reduction. Therefore,

H4a: The rising trend of carbon dioxide emissions has slowed down.

2.4   Carbon Accounting and Temperature Changes

Cordova et al. (2021) used logit and linear panel data models, along with the generalized method of moments, to examine how emerging and developing countries formulate carbon reports. The study also explores how local standard setters can encourage corporate actions to reduce carbon emissions through transparent disclosure practices. Corporate social responsibility (CSR) reporting is a key driver of low-carbon behaviour (Zhou et al., 2020), with more detailed carbon performance disclosures having a greater impact on addressing climate change (Giannarakis et al., 2017). By focusing on output performance, direct and indirect greenhouse gas (GHG) emissions, and environmental efforts to mitigate climate change – such as policies and emission reduction initiatives – both studies highlight the positive influence of environmental performance on climate change disclosure, using the Climate Performance Leadership Index as an indicator of disclosure levels.

H5a: Carbon accounting has a positive impact on temperature changes.

3.   Sample and Research Methodology

3.1   Sample and Data Selection

3.1.1   Provincial Annual Carbon Dioxide Emissions

Climate Watch provides data on China's carbon dioxide (CO_2) emissions, sourced from the Global Carbon Project (GCP). The data includes emissions generated by China's total fossil fuel consumption, as well as specific emissions from cement, oil, coal, gas, and cement production over the past nine years (2010–2018). For the analysis, only the total emissions from fossil fuels are used as one of the independent variables. However, since the annual average temperature (the dependent variable) is influenced by regional factors, the authors supplemented this with yearly carbon dioxide emissions data for each of China's provinces from 2010 to 2018. This provincial data was obtained from the Climate and Energy Project of the Beijing Representative Office of the World Resources Institute (WRI), which tracks and analyzes global environmental issues.

3.1.2 *Provincial Annual Carbon Accounting Emissions*

Carbon Emission Accounts and Datasets summarize the total emissions from raw coal, crude oil, and natural gas each year, based on the latest energy data provided by the China National Bureau of Statistics (He et al., 2020). As shown in Table 14.5 (referenced in the Appendix), the authors calculated the approximate carbon emissions, which are used as another independent variable in the analysis.

3.1.3 *Provincial Annual Average Temperature*

The average yearly temperature data of all provinces in China are provided by China Stock Market & Accounting Research (CSMAR), assisted by the National Tibetan Plateau Data Center. Data statistics of the climate of cities and counties in all provinces includes temperature, air pressure, precipitation, humidity, wind direction, and wind speed. Temperature is the dependent variable because temperature changes most directly among all carbon dioxide influencing factors, and is the indicator that can best reflect its impact and resonate with the masses.

3.2 **Methodology**

3.2.1 *Measurement of Carbon Dioxide Emission and Temperature Changes*

By using the carbon dioxide emissions and the temperature change rate, using the fitting curve method, the authors obtained the following expression:

$$f(x) = p_1 x^3 + p_2 x^2 + p_3 x + p_4, \tag{1}$$

where x is normalized by a mean of 351 and standard deviation of 121. Coefficients (with 95% confidence bounds) are given as follows:

$p_1 = 2.282\ (1.391,\ 3.173)$
$p_2 = -4.06\ (-6.356,\ -1.765)$
$p_3 = -6\ \text{(fixed at bound)}$
$p_4 = 16.85\ (14.66,\ 19.04).$

3.2.2 *Measurement of Carbon Accounting and Carbon Dioxide Emission*

There are different trading systems to calculate the carbon accounting emissions, and the authors use the financial method. To compare the proportion of carbon accounting in carbon dioxide emission, we should first divide carbon accounting by carbon emissions.

$$P = \frac{Carbon\ Accounting\ Emissions}{Carbon\ Dioxide\ Emissions} \tag{2}$$

To find out the relationship between temperature change and carbon accounting emissions in 2010–2018, the authors use MATLAB to fit and get the following formula:

$$f(x,y) = p_{00} + p_{10}x + p_{01}y + p_{20}x^2 + p_{11}xy \\ + p_{02}y^2 + p_{30}x^3 + p_{21}x^2y + p_{12}xy^2 + p_{03}y^3 ,$$

(3)

where x is normalized by mean 2014 and a standard deviation of 2.603 and y is normalized by a mean of 30,670 and a standard deviation of 9,536.

Coefficients (with 95% confidence bounds) are given as follows:

$$p_{00} = 14.57\ (11.76,\ 17.37)$$
$$p_{10} = -8.036\ (-12.52,\ -3.554)$$
$$p_{01} = 3.376\ (-1.097,\ 7.849)$$
$$p_{20} = 0.9315\ (-1.34,\ 3.203)$$
$$p_{11} = -0.5785\ (-2.315,\ 1.158)$$
$$p_{02} = 0.2496\ (-1.383,\ 1.883)$$
$$p_{30} = 5.693\ (2.812,\ 8.574)$$
$$p_{21} = -2.268\ (-4.531,\ -0.005257)$$
$$p_{12} = -0.4216\ (-2.17,\ 1.326)$$
$$p_{03} = -0.9841\ (-2.997,\ 1.029)$$

To find out the relationship among temperature change, caron accounting emissions, and carbon dioxide emissions in 2010–2018, the authors use MATLAB to fit and get the following formula:

$$f(x,y) = p_{00} + p_{10}x + p_{01}y + p_{20}x^2 + p_{11}xy \\ + p_{02}y^2 + p_{30}x^3 + p_{21}x^2y + p_{12}xy^2 + p_{03}y^3$$

(4)

where x is normalized by a mean of 351 and a standard deviation of 121 and y is normalized by a mean of 2014 and a standard deviation of 2.603.

Coefficients (with 95% confidence bounds) are given as:

$$p_{00} = 16.58\ (13.32,\ 19.83)$$
$$p_{10} = -2.155\ (-6.463,\ 2.153)$$
$$p_{01} = 0.4147\ (-4.08,\ 4.91)$$
$$p_{20} = -2.282\ (-6.081,\ 1.518)$$
$$p_{11} = 0.1673\ (-2.007,\ 2.342)$$
$$p_{02} = -0.02536\ (-1.977,\ 1.927)$$
$$p_{30} = 0.898\ (-1.149,\ 2.945)$$
$$p_{21} = 0.1091\ (-1.936,\ 2.154)$$
$$p_{12} = -0.3857\ (-3.231,\ 2.46)$$
$$p_{03} = -0.2003\ (-2.444,\ 2.044)$$

To find out the relationship between temperature change and carbon dioxide emissions in 2010–2018, the authors use MATLAB to fit and get the following formula:

$$f(x,y) = p_{00} + p_{10}x + p_{01}y + p_{20}x^2 + p_{11}xy \\ + p_{02}y^2 + p_{30}x^3 + p_{21}x^2y + p_{12}xy^2 + p_{03}y^3, \tag{5}$$

where x is normalized by a mean of 30,670 and a standard deviation of 953 and y is normalized by a mean of 351 and a standard deviation of 121. Coefficients (with 95% confidence bounds) are given as:

p_{00} = 8.494 (6.76, 10.23)
p_{10} = 4.947 (−0.681, 10.58)
p_{01} = −10.93 (−16.37, −5.487)
p_{20} = 8.468 (4.294, 12.64)
p_{11} = −16.78 (−26.15, −7.412)
p_{02} = 12.27 (6.834, 17.7)
p_{30} = 3.669 (−5.202, 12.54)
p_{21} = −4.362 (−26.68, 17.96)
p_{12} = 4.844 (−14.37, 24.06)
p_{03} = −2.625 (−8.517, 3.266)

4.   Variables Definition

4.1   Dependent Variables

According to the research hypothesis, two different dependent variables, carbon accounting and carbon dioxide emissions, are set. Nevertheless, we still need to consider the relationship between these two variables.

4.2   Control Variables

However, there are many provinces in China that significantly impact temperature because of longitude, latitude, and geographical location. To reasonably reduce this impact, the authors divide 30 provinces into seven regions: Central China, East China, North China, South China, Northeast China, Northwest China, and Southwest China. The annual average temperature of provinces within the scope with the average carbon dioxide emission and the average carbon accounting emission, respectively, are collected and the data input into MATLAB to obtain the connection.

4.3   Independent Variables

Temperature changes are the independent variables of this study. Carbon dioxide emissions influence it, and the relationship is clear: the more the carbon dioxide emissions, the higher the temperature. However, although carbon accounting will also cause temperature increases, its effect should be controlled to reduce carbon dioxide emissions. Therefore, carbon accounting emissions will lead to relatively adverse effects of temperature change.

5. Results

By displaying all data visually, mainly in the form of line chart and table, the authors obtain the following results according to the assumptions:

1) *The widespread use of carbon accounting has slowed down the upward trend of carbon dioxide emissions.*

Through the mapping function of Excel, the authors first counted and calculated the average annual carbon dioxide emissions, carbon accounting emissions, and temperature changes in seven regions from 2010 to 2018 (see in Appendix and Tables 14.6–14.8), as shown in Figures 14.1–14.3.

According to Figure 14.1, carbon dioxide emissions in South China have risen rapidly and remained the highest emissions, even reaching 662.4 mt per year. Carbon dioxide emissions in East China, South China, Southwest China, and Northwest China have increased slowly; that in Southwest and Northwest China has significantly improved; and that in Northeast and Central China has begun to decline.

Carbon accounting emissions have changed slightly, except in Southwest and Central China, wherein the emissions have increased. East China has surpassed Central China to become the first region in carbon accounting emissions. Southwest China has maintained a downward trend after a short rise. The growth rate in Northwest China was the fastest among all areas, rising from the lowest with

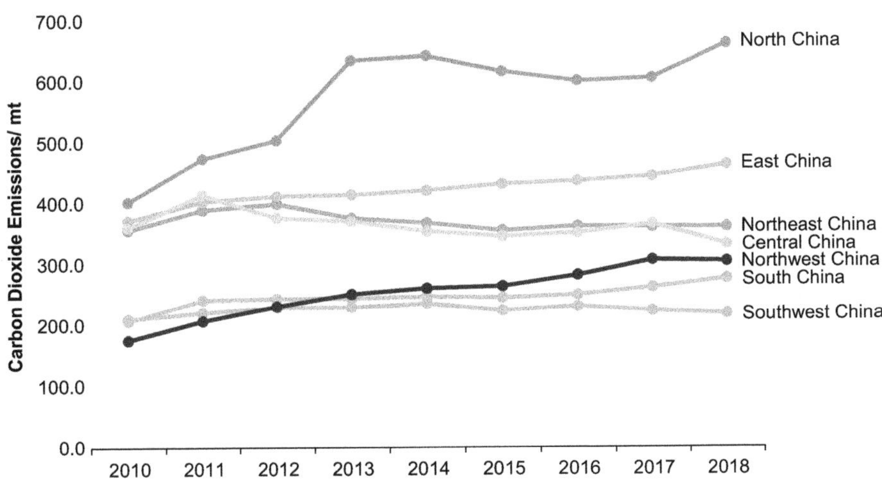

Figure 14.1 Carbon Dioxide Emissions of Seven Regions in China (2010–2018)

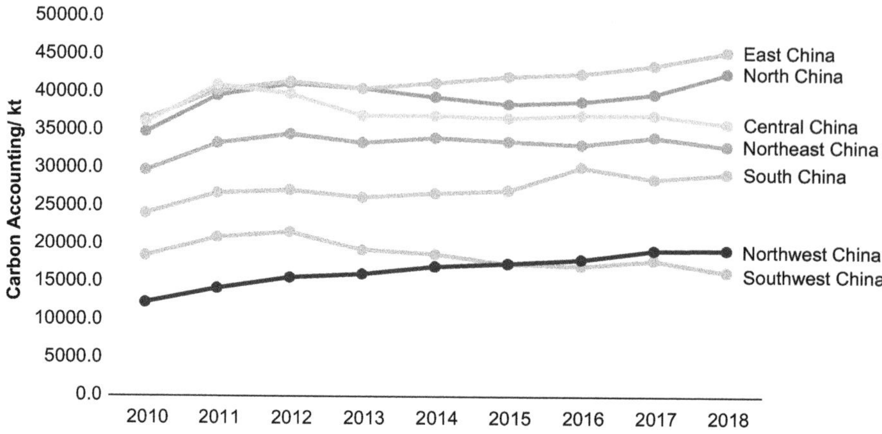

Figure 14.2 Carbon Accounting Emission of Seven Regions in China (2010–2018)

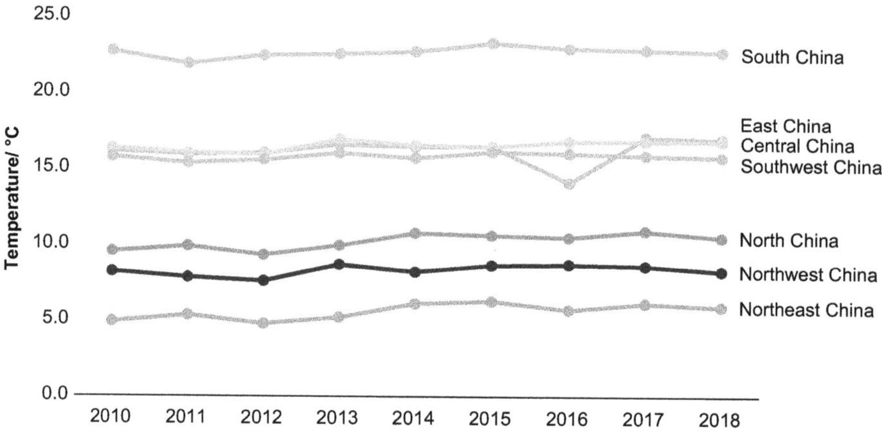

Figure 14.3 Temperature Changes of Seven Regions in China (2010–2018)

1234.5 kt to 19313.8 kt. The temperature changes in all regions are slight and relatively stable, except for a sudden drop in East China in 2016, and emissions in the rest of the regions keep rising steadily.

In summary, the curve trend in the chart shows that the carbon dioxide emission, carbon accounting emission, and temperature change in most regions have an upward trend, although the scope of change is different. For example, the range of temperature change is small, but carbon dioxide emission has a significant difference. At the same time, some regions also show a downward trend. For example, the carbon dioxide emission in Central China is almost stable, and the carbon accounting emission in South China and Northwest China is also in the opposite trend with other regions. By calculating the proportion of carbon accounting

Table 14.2 Proportion of Carbon Accounting Emission in Carbon Dioxide Emission

Region/CA/CO$_2$ (%)	2010	2011	2012	2013	2014	2015	2016	2017	2018
North China	8.64	9.84	10.23	10.07	9.79	9.56	9.65	9.89	10.57
Northeast China	7.39	8.27	8.57	8.29	8.45	8.33	8.23	8.48	8.18
East China	9.05	10.01	10.29	10.07	10.24	10.46	10.55	10.82	11.26
Central China	8.96	10.18	9.89	9.18	9.17	9.10	9.19	9.19	8.90
South China	5.99	6.65	6.75	6.51	6.64	6.74	7.50	7.11	7.28
Southwest China	4.60	5.21	5.38	4.79	4.65	4.33	4.26	4.45	4.07
Northwest China	3.07	3.53	3.89	4.00	4.24	4.34	4.48	4.78	4.79

Table 14.3 Year-on-Year Growth Rate of Carbon Dioxide Emission

Region/Growth Rate	2011	2012	2013	2014	2015	2016	2017	2018
North China	17.70%	6.35%	25.88%	1.17%	−4.02%	−2.55%	0.76%	9.29%
Northeast China	9.19%	2.69%	−5.93%	−2.17%	−3.35%	1.82%	−0.28%	−0.10%
East China	8.43%	2.17%	0.51%	1.63%	2.53%	1.12%	1.72%	4.31%
Central China	14.86%	−9.07%	−1.67%	−4.34%	−2.33%	1.53%	4.18%	−9.19%
South China	16.57%	0.84%	0.14%	1.11%	−0.87%	2.05%	4.82%	5.42%
Southwest China	5.04%	3.85%	0.01%	2.50%	−4.86%	2.92%	−3.15%	−2.21%
Northwest China	17.98%	11.32%	8.61%	3.76%	1.22%	6.88%	8.88%	−0.92%

emission in CO$_2$ emission, we obtained the following results, which are presented in Table 14.2.

It can be observed that the proportion of carbon accounting in East, South, Northeast, North, and Northwest China is increasing, while that in Central and Southwest China is decreasing.

Based on the calculation formula of year-on-year growth described in the method, the year-on-year growth rate of carbon dioxide emission (Table 14.3) is obtained. Through observation, the growth rate of all regions is very tortuous, where all show a downward trend and even have negative growth rates.

North, Northeast, East, South, Southwest, and Northwest China all have similar trends. Only Central China has noticeable differences from other regions in carbon dioxide emissions and carbon accounting proportion, which is the opposite trend.

2) *Carbon accounting has a positive impact on the rise of temperature.*

To judge the impact of carbon accounting on temperature change, the authors first calculate the change rate of temperature, then eliminate the effects of carbon dioxide by comparing the proportion of carbon accounting emissions in carbon dioxide emissions, and analyse the impact of carbon dioxide emission on temperature by obtaining the polynomial surface function, and the results are as follows (see Table 14.4, Figures 14.4–14.7).

Table 14.4 Year-on-Year Growth Rate of Temperature

Region/Growth Rate	2011	2012	2013	2014	2015	2016	2017	2018
North China	3.77	−5.86	6.65	8.25	−1.30	−1.32	4.20	−4.03
Northeast China	8.84	−10.63	9.09	17.31	2.73	−9.04	7.60	−3.80
East China	−1.33	0.54	3.48	−0.78	0.09	−14.41	21.20	−0.84
Central China	−1.84	−0.62	6.28	−2.36	−0.40	2.02	0.00	0.00
South China	−3.67	2.59	0.45	0.74	2.49	−1.5	−0.44	−0.58
Southwest China	−2.54	1.47	2.89	−2.03	2.55	−0.62	−0.78	−0.79
Northwest China	−4.63	−3.07	14.51	−5.53	5.37	0.69	−1.15	−4.19

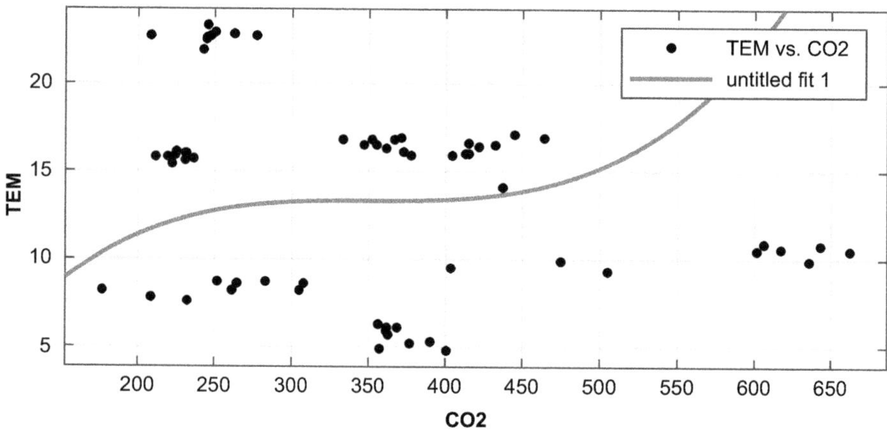

Figure 14.4 Temperature Change Trend vs. Carbon Accounting/Carbon Dioxide

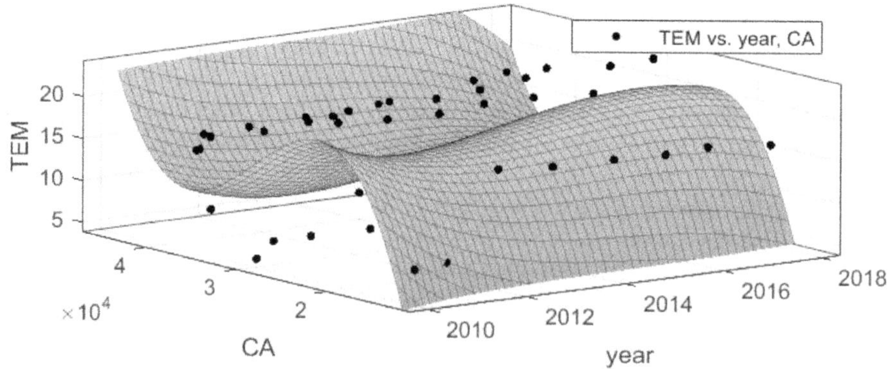

Figure 14.5 Temperature Change vs. Carbon Accounting in 2010–2018

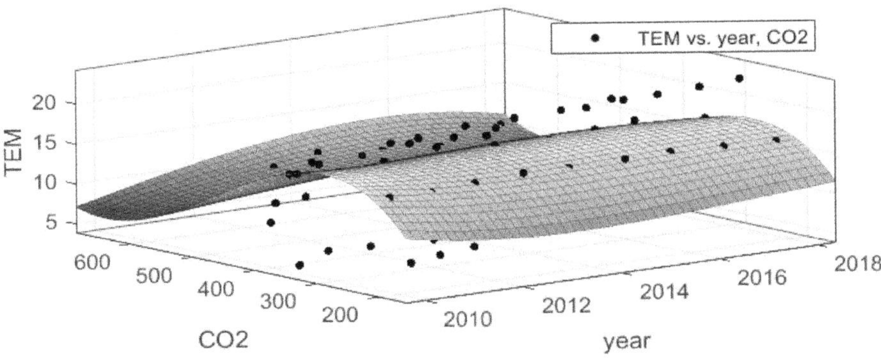

Figure 14.6 Temperature Change vs. Carbon Dioxide Emission in 2010–2018

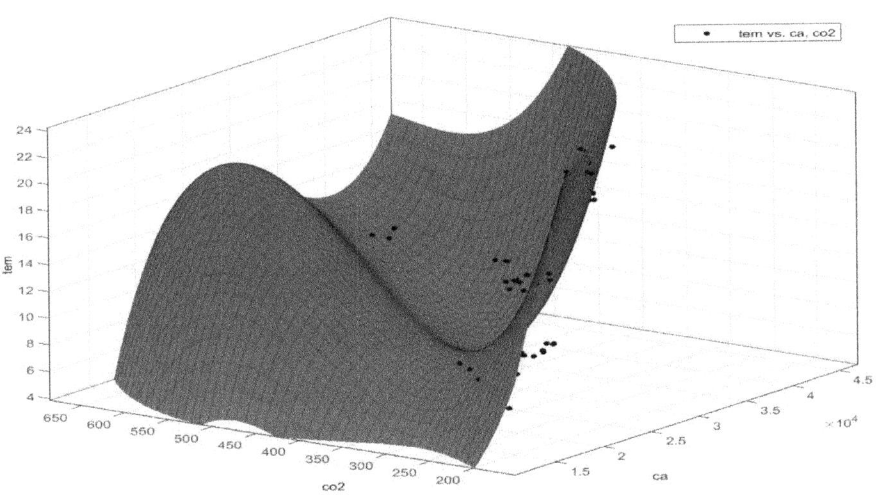

Figure 14.7 Temperature Change vs. Carbon Accounting, Carbon Dioxide Emission in 2010–2018

6. Discussion

The above consequences have proved that the hypothesis is tenable. Although some regions do not fit entirely, carbon dioxide emissions, carbon accounting emissions, and temperature change curves of various regions can roughly confirm that China's trend is developing in the positive direction.

A) *The rising trend of carbon dioxide emissions has begun to slow down, and emissions in some areas have decreased year by year*

The year-on-year growth rate of carbon dioxide emissions generally shows a downward trend and is almost less than 10%. In other words, the growth rate of carbon dioxide emissions in China has eased. If the situation remains stable, it will be possible to achieve negative growth in all regions and reduce temperature changes by affecting carbon dioxide emissions.

B) *The strange fact of average annual carbon dioxide emissions, carbon account-ing emissions, and temperature changes in seven regions from 2010 to 2018*

The most significant change is the sharp rise in carbon dioxide emissions in South China. South China includes Guangdong, Guangxi, and Hainan. As a first-tier city in China, the authors infer that Guangdong is China's first economic province. Its GDP, total retail sales of social consumer goods, total import and export volume, and deposits of financial institutions rank first in China. The advanced business environment and trade system will sharply increase Guangdong's carbon dioxide emissions. The Pearl River Delta is a manufacturing agglomeration. Likely, its developed economies have quickly produced a large amount of carbon dioxide, resulting in a sharp rise in emissions in South China.

C) *The proportion of carbon accounting emissions in carbon dioxide emissions increased*

As mentioned above, the increase in the proportion of carbon accounting emissions in carbon dioxide emissions will lead to a positive social cycle. More and more companies will disclose carbon accounting, actively optimize the company budget and its production chain, drive small and medium-sized enterprises with leading enterprises, attract the attention of the whole business circle to carbon emissions, and finally achieve the effect of reducing China's carbon emissions. Therefore, the authors reasonably infer that the wide use of carbon accounting will play an energetical role in carbon dioxide emissions and temperature changes. Thus, the hypothesis is tenable.

However, the proportion of carbon accounting emissions in carbon dioxide emissions still decreases in Central and Southwest China. According to the phenomenon, carbon dioxide emissions in Central China have begun to decline yearly. The proportion of carbon accounting is still rising, and the authors speculate that it may be because Central China pays more attention to other industries such as tourism and production. So, the growth of carbon dioxide emissions in other sectors exceeds the development of carbon accounting in business, so carbon accounting emissions account for a relatively small proportion of total carbon dioxide emissions. Nevertheless, the authors believe that all sectors of society should pay attention to carbon dioxide and climate issues. Generally, the CO_2 emission in Figure 14.7 has become a downward trend and can be expected to decrease in the future.

On the other hand, carbon accounting emissions in Southwest China have been decreasing, possibly because companies have taken steps to control their own emissions. However, due to emissions from other industries and activities, overall carbon dioxide emissions continue to rise, leading to a reduction in the proportion of emissions from these companies.

D) *Effect of carbon accounting on temperature change*

From Figure 14.4, we can find out when the carbon dioxide emission increases, the temperature will show an upward trend, and the same is for carbon accounting. Furthermore, we found that sometimes a single rise in carbon dioxide emissions and carbon accounting emissions will not directly lead to significant temperature changes, which means that emissions in different ranges have different effects on temperature. The combination of the two factors influences the temperature change. The slope of the curved surface represents the correlation affecting the temperature change. The trend brought by different numerical changes is entirely different. For example, when the CO_2 emission is 250 mt and the CA emission is 1.5 T, the impact on the temperature is the greatest. Any slight change will lead to a sharp rise in the temperature.

From Figure 14.1, we know that with the increase in the proportion of carbon accounting, the range of temperature change has a downward trend. From Figures 14.2 and 14.3, we can infer the carbon accounting emissions and the relationship between carbon dioxide emissions and temperature change from 2010 to 2018. The change range in Figure 14.2 is ultimately more incredible than that in Figure 14.3, which proves that carbon accounting plays a vital role in the impact of temperature change on carbon dioxide emissions. The discovery can also be verified in Figure 14.4. Due to different unit sizes of abscissa and ordinate scales, small changes in carbon accounting will greatly affect the temperature change range. From Figure 14.4, we can estimate future temperature changes based on carbon dioxide emissions and carbon accounting emissions.

7.   Limitations and Further Research

This study examines the relationship among carbon dioxide emissions, carbon accounting emissions, and temperature changes. Region is critical for ensuring the accuracy of the research, as temperature changes and carbon dioxide emissions are influenced not only by geographical environments but also by factors such as topography, longitude, latitude, and local economic conditions.

Abrupt temperature changes and interruptions in climate warming result from the interplay of multiple influencing factors. China's climate cooling and heating mechanisms are complex, shaped by various elements, leading to significant climatic differences across regions. Notable influences include the Atlantic Multidecadal Oscillation (AMO), El Niño Southern Oscillation (ENSO), Multivariate ENSO Index (MEI), solar radiation (SR), and Pacific Decadal Oscillation (PDO) (Chen et al., 2021). Additionally, different geographical locations prioritize different industries. Coastal provinces thrive in commerce and trade, northeastern provinces focus on industrial production, and central regions predominantly focus on agriculture. Consequently, the level of commercial development varies by province, complicating comparisons.

Future research should aim to minimize the impact of additional factors affecting climate and carbon accounting emissions, seeking a more precise relationship among the three elements. This could assist government and various sectors of society in guiding business development, controlling unchecked carbon emissions, and ultimately contributing positively to temperature changes. While it is evident that an increase in carbon dioxide emissions correlates with rising temperatures, this does not provide direct evidence of causation. Many factors influence temperature changes beyond just carbon dioxide emissions.

8. Conclusion

More and more countries have begun to pay great attention to climate change, especially the topic of rising temperature. This study discussed the effects and relationship among carbon dioxide emissions, carbon accounting emissions, and temperature changes. Through the study of these three elements, we hope to make all sectors of society pay more attention to carbon emissions; enable enterprises to strengthen the control of carbon emissions, that is, carbon disclosure, reduce carbon accounting emissions; and finally affect the temperature change, slow down the temperature change rate and even reduce the average annual temperature.

China's provinces are divided into seven regions according to the different geography. Through quantitative analysis and MATLAB, we found that carbon dioxide emission, carbon accounting emissions, and temperature change are connected, and the relationship between them is obtained and expressed by the formula.

By distinguishing carbon emission types with different attributes, auditors calculate carbon emission factors and carbon accounting emissions. The authors assumed that the proportion of carbon accounting emissions in carbon dioxide emissions increases and the growth rate of carbon dioxide emissions slows down. In addition, the author infers that the application of carbon accounting plays a positive role in temperature change.

The relationship curves and surfaces between several variables were obtained by fitting the data using the carbon dioxide emissions, carbon accounting emissions, and temperature changes of Chinese provinces from 2010 to 2018.

This study confirms that the use of carbon accounting has indeed alleviated the rise of carbon dioxide emissions and played a great role in it. Under the influence of the rising proportion of carbon accounting, the increasing temperature range also tends to decrease. However, we still cannot exclude the impact of regional and local economic conditions on temperature change and carbon dioxide emissions. Therefore, future research can try to exclude the effect of this variable.

References

Ajani, J. I., Keith, H., Blakers, M., Mackey, B. G., & King, H. P. (2013). Comprehensive carbon stock and flow accounting: A national framework to support climate change mitigation policy. *Ecological Economics, 89*, 61–72.

Andrew, R. M. (2018). Global CO 2 emissions from cement production. Earth System Science Data, 10(1), 195-217.

Ascui, F., & Lovell, H. (2011). As frames collide: making sense of carbon accounting. Accounting, Auditing & Accountability Journal, 24(8), 978-999.

Bebbington, J., Larrinaga, C., & Moneva, J. M. (2008). Corporate social reporting and reputation risk management. Accounting, Auditing & Accountability Journal, 21(3), 337-361.

Breton, M. D., Kanapka, L. G., Beck, R. W., Ekhlaspour, L., Forlenza, G. P., Cengiz, E., . . . & Wadwa, R. P. (2020). A randomized trial of closed-loop control in children with type 1 diabetes. New England Journal of Medicine, 383(9), 836-845.

Chen, C., Zhao, W., & Zhang, X. (2021). Pacific Decadal Oscillation-like variability at a millennial timescale during the Holocene. Global and Planetary Change, 199, 103448.

Chen, X., Shuai, C., Wu, Y., & Zhang, Y. (2020). Analysis on the carbon emission peaks of China's industrial, building, transport, and agricultural sectors. Science of the Total Environment, 709, 135768.

Climate Disclosure Standards Board. (2009). *The climate disclosure standards board (CDSB) reporting framework: Exposure draft*. Carbon Disclosure Project.

Cordova, C., Zorio-Grima, A., & Merello, P. (2021). Contextual and corporate governance effects on carbon accounting and carbon performance in emerging economies. *Corporate Governance, 21*(3), 536–550.

Fergus, E. (2017). The integration project among white teachers and racial/ethnic minority youth: Understanding bias in school practice. *Theory Into Practice, 56*(3), 169–177.

Giannarakis, G., Zafeiriou, E., & Sariannidis, N. (2017). The impact of carbon performance on climate change disclosure. *Business Strategy and the Environment, 26*, 1078–1094.

Goulder, L., & Hafstead, M. (2017). *Confronting the climate challenge: US policy options*. Columbia University Press.

Green, F., & Stern, N. (2017). China's changing economy: Implications for its carbon dioxide emissions. *Climate Policy, 17*(4), 423–442.

He, J., Yang, K., Tang, W., Lu, H., Qin, J., Chen, Y. Y., & Li, X. (2020). The first high-resolution meteorological forcing dataset for land process studies over China. *Scientific Data, 7*, 25.

Li, J., & Colombier, M. (2009). Managing carbon emissions in China through building energy efficiency. *Journal of Environmental Management, 90*(8), 2436–2447.

Rahman, M. J., & Wu, J. (2023). M&A activity and ESG performance: Evidence from China. *Managerial Finance, 50*(1), 179–197.

Rahman, M. J., Wu, Q., & Zhu, H. (2024). Corporate social responsibility in times of social distancing: Evidence from China. *Business Ethics, the Environment & Responsibility*.

Rahman, M. J., Zhu, H., & Chen, S. (2023). Does CSR reduce financial distress? Moderating effect of firm characteristics, auditor characteristics, and Covid-19. *International Journal of Accounting & Information Management, 31*(5), 756–784.

Zhou, Z., Nie, L., Ji, H., Zeng, H., & Chen, X. (2020). Does a firm's low-carbon awareness promote low-carbon behaviors? Empirical evidence from China. *Journal of Cleaner Production, 244*.

Zhu, H., & Rahman, M. J. (2024). Ex-ante expected changes in ESG and future stock returns based on machine learning. *The British Accounting Review*, 101457.

Appendix

Table 14.5 Provincial Carbon Dioxide Emissions (2010–2018)

Province	2010	2011	2012	2013	2014	2015	2016	2017	2018
Beijing	96.84	94.73	95.94	86.69	88.98	83.39	74.94	70.06	70.60
Tianjin	134.29	149.14	143.14	151.03	143.93	135.11	130.35	132.15	138.36
Hebei	569.37	623.46	642.31	657.72	624.59	639.37	614.57	541.90	595.29
Shanxi	654.05	766.12	854.67	1,499.06	1,552.01	1,474.50	1,433.11	1,521.41	1,650.24
Inner Mongolia	562.50	740.52	788.69	783.74	806.00	753.80	754.62	765.07	857.57
Liaoning	494.59	524.47	543.27	529.86	520.81	502.37	508.94	515.65	560.66
Jilin	225.77	264.67	265.30	237.29	234.97	218.85	214.18	214.52	193.87
Heilongjiang	351.54	381.24	393.33	363.48	350.35	347.80	365.37	355.33	329.91
Shanghai	161.42	170.60	168.52	181.78	156.55	161.65	158.23	156.58	151.47
Jiangsu	546.29	613.11	619.99	638.44	621.07	634.16	653.12	645.05	644.97
Zhejiang	375.77	398.62	383.59	387.56	380.79	381.50	379.18	401.44	401.66
Anhui	282.90	315.39	355.91	387.18	401.80	392.85	380.18	397.02	392.43
Fujian	179.64	210.15	208.53	203.57	229.16	234.47	217.15	232.92	254.14
Jiangxi	134.19	144.75	146.46	162.19	165.47	170.41	176.67	179.01	183.40
Shandong	929.12	976.56	1,007.56	944.49	997.83	1,052.18	1,096.72	1,101.80	1,220.03
Henan	573.13	654.10	545.87	594.42	557.53	537.07	536.84	557.62	498.13
Hubei	279.61	321.88	311.39	252.92	251.82	252.88	253.98	266.93	258.02
Hunan	231.70	269.56	275.30	266.33	255.97	250.52	265.51	275.97	243.25
Guangdong	445.06	501.36	486.70	493.09	500.65	497.95	506.86	533.20	557.28
Guangxi	134.28	173.34	192.85	189.77	183.50	173.23	182.48	193.85	209.37
Hainan	44.91	52.97	54.24	51.97	58.85	65.35	62.26	60.75	63.83
Chongqing	138.38	150.71	149.36	134.28	139.12	139.19	142.18	127.77	123.80
Sichuan	283.57	280.38	289.71	277.51	298.40	253.58	250.14	229.76	255.30
Guizhou	246.76	268.69	287.21	314.25	319.61	327.57	348.36	340.34	291.72
Yunnan	176.26	187.81	195.47	195.78	187.77	178.65	184.54	198.22	205.48
Shaanxi	308.25	344.52	405.13	482.84	502.04	529.36	590.05	637.83	576.35
Gansu	145.09	169.69	173.75	181.47	180.47	176.58	169.72	173.99	182.50
Qinghai	37.14	50.10	58.56	70.83	62.53	44.05	53.15	48.68	49.06
Ningxia	151.54	190.95	188.57	187.76	195.22	193.38	189.72	226.24	235.93
Xinjiang	240.82	286.31	333.44	336.34	366.26	379.09	410.85	452.26	480.95

Table 14.6 Carbon Dioxide Emissions of Seven Regions in China (2010–2018)

Mt/Year	2010	2011	2012	2013	2014	2015	2016	2017	2018
North China	403.4	474.8	504.9	635.7	643.1	617.2	601.5	606.1	662.4
Northeast China	357.3	390.1	400.6	376.9	368.7	356.3	362.8	361.8	361.5
East China	372.8	404.2	412.9	415.0	421.8	432.5	437.3	444.8	464.0
Central China	361.5	415.2	377.5	371.2	355.1	346.8	352.1	366.8	333.1
South China	208.1	242.6	244.6	244.9	247.7	245.5	250.5	262.6	276.8
Southwest China	211.2	221.9	230.4	230.5	236.2	224.7	231.3	224.0	219.1
Northwest China	176.6	208.3	231.9	251.8	261.3	264.5	282.7	307.8	305.0

Table 14.7 Carbon Accounting Emissions of Seven Regions in China (2010–2018)

Kt/Year	2010	2011	2012	2013	2014	2015	2016	2017	2018
North China	34,865.4	396,85.9	41,261.2	40,613.2	39,512.6	38,559.9	38,918.3	39,900.4	42,623.2
Northeast China	29,830.7	33,370.9	34,584.0	33,449.8	34,074.0	33,603.4	33,190.4	34,225.2	32,996.6
East China	36,524.3	40,361.6	41,494.4	40,603.5	41,324.2	42,180.5	42,550.2	43,654.2	45,424.1
Central China	36,155.7	41,073.5	39,883.9	37,020.9	36,981.2	36,693.3	37,080.8	37,069.0	35,916.5
South China	24,159.0	26,822.6	27,214.9	26,248.8	26,788.9	27,170.4	30,255.5	28,688.4	29,374.3
Southwest China	18,542.9	21,009.8	21,708.3	19,330.4	18,762.4	17,470.9	17,197.5	17,956.7	16,405.8
Northwest China	12,364.5	14,251.4	15,673.3	16,145.5	17,118.3	17,522.0	18,055.3	19,267.1	19,313.8

Table 14.8 Temperature Changes of Seven Regions in China (2010–2018)

Mt/Year	2010	2011	2012	2013	2014	2015	2016	2017	2018
North China	9.5	9.9	9.3	9.9	10.8	10.6	10.5	10.9	10.5
Northeast China	4.9	5.3	4.8	5.2	6.1	6.3	5.7	6.1	5.9
East China	16.1	15.9	16.0	16.6	16.4	16.5	14.1	17.1	16.9
Central China	16.3	16.0	15.9	16.9	16.5	16.5	16.8	16.8	16.8
South China	22.7	21.9	22.5	22.6	22.7	23.3	22.9	22.8	22.7
Southwest China	15.8	15.4	15.6	16.0	15.7	16.1	16.0	15.9	15.8
Northwest China	8.2	7.8	7.6	8.7	8.2	8.6	8.7	8.6	8.2

15 Carbon Accounting and Harvested Wood Products in China

Md Jahidur Rahman, Tarek Rana, Hongtao Zhu, and Zhang Sijie

1. Introduction

Forests are nature's "dispatchers". Forests can store carbon sinks and have the ability to solve the climate problem. The role of carbon sinks and storage is to transfer fixed carbon from the felled forest to harvested wood products (HWPs) when the forest is cut down. HWPs participate in the forest carbon cycle and is also a forest resource, so there is also a lag in carbon emissions.

According to the Carbon Emission Yearbook, the annual output and annual consumption of wood materials have increased. According to statistics from the Intergovernmental Panel on Climate Change (IPCC), carbon stocks in global HWPs increase about 26 million tonnes every year (Intergovernmental Panel on Climate Change, 2007). At the same time, other materials are estimated to increase annually between 40 million tonnes and 1.4 million tonnes. This huge amount of new carbon savings will be converted into carbon emissions and will have an impact on climate change to a certain extent.

A convention adopted by the United Nations (UN) General Assembly on 9 May 1992 and opened for signature during the UN Conference on Environment and Development held in Rio de Janeiro, Brazil, in June of the same year and the UN Framework Convention on Climate Change, which came into force on 21 March 1994, is effective and legally binding. The purpose of this convention is to reduce the concentration of greenhouse gases in the air. This low concentration level can eliminate and reduce the harm of human activities to the climate system. At the same time, the country's economic development also determines the obligations it undertakes. This convention stipulates that developed countries shall provide financial assistance to developing countries to fulfil their obligations. Every country has a certain number of indicators, which are not enough for developed and developing countries. Therefore, developed countries can purchase surplus indicators from developing countries. Countries with good economy can adopt more perfect methods and have the right to view the database. Countries with good economies usually have higher carbon budgets. Countries with underdeveloped economies are unable to formulate relevant measures on greenhouse gas emissions. At the same time, economically underdeveloped regions have no ability to control the climate and undertake related obligations. It can be seen that the signatories

DOI: 10.4324/9781003488965-20

of the convention are responsible for reporting changes in carbon stocks in many different areas, such as in agriculture, horticulture, and forestry. The IPCC introduced three methods such as the inventory change method (SCA), the production method (PA), and the atmospheric flow method to study the exposure and estimate the carbon flux to HWPs. The "Climate Convention" (Durban) stipulates that only domestic carbon emissions can be included in a country's calculations. Domestic output of HWP is also included in the national accounts. Therefore, the "Kyoto Protocol" adopts the production-based accounting (PA) method.

With China's reform, opening-up policies, and rapid economic development, the country's role in the global production and consumption of logging products has grown significantly. China now plays an important role in the trade of logging products and holds considerable influence. According to the *China Carbon Emissions Statistical Yearbook* and the *Table of Double Carbon Emissions from Industry and Agriculture in the Chinese Enterprises Database*, China's consumption of sawn wood surpassed that of the United States in 2016, making it the world's largest consumer. Additionally, China has maintained its position as the top producer of wood-based panels and paper products, the largest consumer of wood products, and the number one importer of logs and exporter of wood-based panels.

As a result, China generates a significant amount of carbon emissions from its manufacturing activities. In response to environmental concerns, national policies, such as logging bans and the State Council's plans for the import and export of logging products, aim to curb commercial logging. By 2020, China plans to reduce commercial logging of planted forests by about 20% and gradually halt commercial logging of natural forests. Consequently, the supply of domestic wood products is expected to decrease in the future. If the default IPCC estimation method (production-based accounting, or PA) is used, these changes will likely affect the contribution of the HWP carbon pool.

According to Food and Agriculture Organization of the United Nations Statistical Database (FAOSTAT) data, the carbon pool in China's HWP represents an ever-increasing carbon sink. China is a major importer of wood products. However, China's HWP carbon stock has not yet been reported, and the HWP carbon balance has not been included in the country's greenhouse gas inventory. Since the SCA method is no longer accepted by the reporting country, the purpose of this research is to analyse the basic data on carbon emissions from logging and carbon storage in China using the PA method. The study aims to estimate changes in carbon storage and waste from HWP every year from 2000 to 2020 (up to the Covid-19 pandemic) and to analyse the HWP carbon stock. This study also examines the structure of the carbon pool, discusses relevant policies affecting the HWP carbon pool in China, and proposes potential policy solutions that may be adopted in the near future.

2. Original Information and Solutions

2.1 *Changes in Carbon Storage in the Supplies Involved Every Year*

China's greenhouse gas list guidelines first described the methodology for the estimation of carbon removals and emissions in HWPs. This is a very common problem, and there are many ways to deal with it, including three main methods. The guide was revised in 2007 and enriched in 2013. As the only accepted method, PA is widely used in China. Therefore, next we need to improve the PA method. The other two methods, SCA and atmospheric flux accounting (AFA), are only used as recommendations rather than main force.

The carbon change in HWPs was estimated by the first-order attenuation method. Based on the domestic harvest yield, formulas (1) and (2) estimate the carbon reserves and changes in the HWP reservoir every year.

$$C(i+1) = e^{-k} \cdot C(i) + \left[(1 - e - k)k\right] \cdot Inflow(i) \tag{1}$$

$$\Delta C(i) = C(i+1) - C(i) \tag{2}$$

According to carbon-related regulations, the half-life is the number of years required for a substance to lose half its carbon reserves. $K = \ln(2)/HL$. Here, K represents the attenuation constant for first-order attenuation, in years and yr -1, respectively. The main purpose of this formula is to use the initial carbon reserves and obtain the cycle and year after half of the release.

According to the default value of IPCC, the half-lives of wood-based panels, sawn timber, and paper products are 25, 35, and 2 years, respectively.

$$Inflow(i) = P \times f_{DP}(i) \tag{3}$$

$$f_{DP}(i) = \begin{cases} f_{IRW}(i), \text{ for sawnwood and wood} - based\ panels \\ f_{IRW}(i) \cdot f_{PULP}(i), \text{ for paper and paperboard} \end{cases} \tag{4}$$

$$f_{IRW}(i) = \frac{IRW_P(i) - IRW_{EX}(i)}{IRW_P(i) + IRW_{IM}(i) + IRW_{EX}(i)} \tag{5}$$

$$f_{PULP}(i) = \frac{PULP_P(i) - PULP_{EX}(i)}{PULP_P(i) + PULP_{IM}(i) - PULP_{EX}(i)} \tag{6}$$

The amount of crystals (i) = the amount of liquid crystals in the number of HWP cells during the first year. $\Delta C(i)$ = the change in carbon storage in the first carbon year. From formula (7) and the data in China's Carbon Emission Yearbook, we get China's carbon expansion from 1900 to 1961, so as to estimate the carbon growth between 2000 and 2020. Using import and export data and related materials, where im = import, ex = export, and IRW = industrial log, the carbon in the product is estimated.

$$V_t = V_{1961} \cdot e^{(U \cdot (t - 1961))}, \tag{7}$$

where t represents the time of year.

This formula presents the multiplication of two sets of consumption annual rate of change and the number of years. The result obtained is the annual change. From this set of formulas, we find that the product of wood carbon emissions between 2000 and 2020 is -1, which also represents the annual rate of change.

2.2   Changes in Carbon Storage in Solid Waste Disposal Sites Every Year

Formula (8) reflects the annual carbon emission of wood products. According to the requirements of China's carbon emission planning, one-order attenuation method is used to calculate carbon attenuation. This requirement is of great help to determine the accuracy of the final data. The basis of China's basic carbon treatment method issued in 2007 shows that the quality of decomposable organic carbon (DDOC) is also a factor to be considered, which is reflected in formula (9) for calculating methane and carbon dioxide emission.

$$DDOC_m\left(i+1\right) = DDOC_m\left(i\right) \times e^{-kt} \tag{8}$$

$$DDOC_{mdecomp}\left(i+1\right) = DDOC_m\left(i\right) \times \left(1 - e^{-kt}\right). \tag{9}$$

The quality of degradable organic carbon (DOC) is reflected incisively and vividly in this formula. According to the distribution map of China's climate types, northern China is cold and needs to burn coal for heating in winter. According to the guidance on the combustion residual value of carbon products in China in 2007, taking waste paper with a residual value of 0.025 as an example, the K-value of waste wood is 0.05. This will allow the results to be calculated relatively accurately.

Formula (10) is the $DDOC_m$ calculation for the amount of waste to be disposed of.

$$DDOC_{md_T} = W_T \cdot DOC \cdot DOC_f \cdot MCF. \tag{10}$$

According to China's carbon emission standards, methane is also a carbon fuel. Methane and oxygen mix and burn after reaching a certain temperature to produce energy and emit carbon dioxide. Formula (11) describes the decomposition rate of DOC under anaerobic conditions. At the same time, through this formula, we can also calculate the proportion of degradable organic carbon. This is also a crucial point.

In this formula, 16 and 12 represent the period and total duration, respectively. This is also an important formula for calculating methane emissions.

$$CH_{4generated_T} = DDOC_m decomp_T \cdot F \cdot \frac{16}{12}. \tag{11}$$

According to Formula 12, we can calculate the annual change in long-term carbon storage. There are many ways and forms of carbon emissions, some of which be methane and carbon dioxide. These carbon emissions that do not decay will be treated in the following two ways. According to China's carbon emission treatment law in 2010, these non-decaying carbon emissions will be used as garbage, buried, and stored for a long time.

$$DOC_m long-term\,stored_T = W_T \times DOC\left(1-DOC_f\right) \cdot MCF. \tag{12}$$

In general, solid waste disposal (SWD) is still a very effective method. Although this residual value looks a bit distorted.

2.3　HWP's Helpful Way on Carbon Removal

The change in HWP carbon reserves every year can be attributed to annual contributions. The following formula (13) uses PA to estimate the removal and emissions of carbon pools in HWPs.

$$\Delta C_{PA} = \Delta C_{HWP\_DH} + \Delta C_{SWDS\_DH} \tag{13}$$

In the international trade of carbon products, community health departments and service centres will strictly grasp the import, export, and sales direction of carbon products. At the same time, they will also monitor the index within their ability.

2.4　Material Sources

HWP data are derived from forestry production and trade data in the FAO statistical database. It contains data on charcoal and wood residues, which is a very large collection. Table 15.1 covers all data of the HWP project and calorific value, as well as relevant names and attachment descriptions. However, due to the constraints of a long time, the data before 1980 are not accurate due to statistical reasons. Therefore, in order to reduce the error, we decided to focus on the analysis of the data between 2000 and 2020, which have high reliability. However, the data of 20 years may not be rich enough, so our research uses the data from 1980 to 2020, so the effect will be more clear and intuitive.

According to the China Circular Economy Yearbook, wood product waste can be treated in three ways: burning, landfill, and open dumping. According to Table 15.2, since 1990, open dumping has been decreasing, and since 2007, no waste has been regarded as open dumping. Since 2004, waste has been burned into an upward trend. From 1990 to 2007, landfills fluctuated slightly, but the overall trend was on the rise. It peaked in 2008 and has since kept falling.

3.　Outcomes

3.1　The Amount of Carbon Stored in HWPs and Its Changes Every Year

According to the China Forest Net, the National Greenhouse Gas Inventory report indicates that changes in the carbon storage of wood forest products are also part of

Table 15.1 Carbon Conversion Factors in HWPs Used in the Models

HWP Category	Conversion Factor (per Air Day Volume) (Mg C/m³)
Sawn wood (aggregate)	0.229
Coniferous sawn wood	0.225
Non-coniferous sawn wood	0.28
Wood-based panels (aggregate)	0.269
Hardboard	0.335
Other fibreboard	0.075
MDF/HDF[1]	0.315
Practice board	0.269
Plywood	0.267
Veneer sheets	0.253 (per air dry mass) [[Mg C/m]
Paper and paperboard (aggregate)	0.386
Charcoal	0.765
Roundwood (aggregate, industrial roundwood, pulpwood chips, particles, wood fuel, and wood residence[2])	0.26
Coniferous roundwood	0.225
Non-coniferous roundwood	0.295

[1] Assume half MDF (medium-density fibreboard) and half HDF (high-density fibreboard).
[2] Assume half temperature species and half topical species.

it. At the same time, as a component of the forest ecosystem's carbon cycle, HWPs play a vital role in the carbon balance between the forest ecosystem and the atmosphere. In order to estimate the carbon storage of wood forest products, three estimation frameworks were established at the Dakar Conference, namely: reserve change method, production method, and atmospheric flow method. The life cycle analysis method and stepwise recursion method were used to calculate the forest products in China. The results show that (1) The changes in wood products in China from 2000 to 2020 were estimated using the reserve change method, the production method, and the atmospheric flow method, confirming that China's wood forest products serve as a carbon pool, with the carbon storage in this pool continuously growing. Over time, the amount of carbon accumulation has increased significantly. The annual carbon storage changes of different HWP categories are different. (2) The three methods estimate the average annual increase in the carbon storage of wood forest products to be 11.72, 8.58, and 7.53 Tg.a^{-1}, respectively. (3) In 2000, the average annual increase in the carbon storage of wood forest products was 2.16 Tg.a^{-1}. (4) From the perspective of measurement and difficulty of estimation, measurement using the reserve change method is more beneficial to our country. This is a 20-fold increase over that in the 2000s. The small contribution and large fluctuation lead to the decline in carbon storage of paper, such as paperboard, from 2009. Between 2000 and 2020, the storage of SWDS increased steadily from 0.2 TgC per year to 3.5 TgC per year. Finally, as the proportion of HWP combustion increased, the annual carbon storage of SWDS dropped to 3.1 TgC per year in 2016. This is expected to reduce air pollution to a certain extent and improve the environment.

Table 15.2 Different Disposal Options of Waste HWP after Use

Year	Combustion	Unmanaged, Shallow	Unmanaged, Deep
1990	0%	67%	33%
1991	0%	64%	36%
1992	0%	60%	40%
1993	0%	57%	43%
1994	0%	44%	46%
1995	0%	41%	49%
1996	0%	37%	53%
1997	0%	34%	56%
1998	0%	31%	59%
1999	0%	28%	63%
2000	0%	24%	66%
2001	0%	19%	69%
2002	0%	13%	72%
2003	0%	7%	76%
2004	1%	1%	80%
2005	4%	4%	83%
2006	7%	7%	87%
2007	10%	0%	90%
2008	10%	0%	90%
2009	13%	0%	87%
2010	14%	0%	86%
2011	16%	0%	84%
2012	21%	0%	79%
2013	27%	0%	73%
2014	30%	0%	70%
2015	34%	0%	66%
2016	37%	0%	60%
2017	39%	0%	56%
2018	43%	0%	53%
2019	46%	0%	57%
2020	49%	0%	50%

3.2   *China's Carbon Flow in Round Timber Production, Import, and Export*

Forests are significant in mitigating climate change. As an extension of forest resources, the carbon storage of wood forest products has become an important issue in global climate change. Research on the carbon storage of wood forest products, the carbon flow of trade, and the requirements for reducing greenhouse gases is crucial for adjusting China's wood forest product trade structure, increasing carbon storage, and improving greenhouse gas emission inventories. Based on the wood forest product carbon accounting model under the IPCC guidelines, this paper examines China's wood forest products from 2000 to 2020 (Ainsworth, 2010), focusing on carbon storage, the contribution of carbon sinks, and the carbon flow of China's wood forest product trade. The study shows that, in terms of total amount, starting from the second decade of the 21st century, China's total carbon storage in wood forest products in use has reached 676 million tonnes of carbon. From the perspective of annual growth, the average increase in carbon storage of China's wood forest products, based on the reserve change method, production method, and atmospheric flow method, is 10.63 million tonnes, 7.61 million tonnes, and 2.62 million tonnes of carbon, respectively. The large imports of pulp and industrial roundwood have increased China's domestic carbon emissions. Using the atmospheric flow method for accounting reveals that logging and consumption of wood forest products in China may become a carbon source. In light of the current status of China's wood forest products trade, applying the reserve change method could help increase China's emission reduction potential.

3.3   *Carbon Emissions after HWP Use*

Carbon emissions after the use of HWP pools are greatly affected by the disposal methods. The main waste emitted by HWP combustion is methane, greenhouse gas emissions that cause the greenhouse effect and increase the global temperature. After 2000, the methane carbon released by papermaking waste was greater than that of wood waste, and its expansion rate was steady. The combustion emissions of HWPs in 2020 were 0.9 TgC (Biello, 2010). It can be seen that combustion is becoming increasingly popular, leading to higher carbon emissions and greenhouse gases. Among other methods, such as discarding, combustion is the treatment that produces the highest carbon emissions. In 2020 alone, carbon emissions from combustion reached 36 TgC.

4.   Discussion

4.1   *Impact of Assessment Methods and Policy*

In China, because the method of importing wood products is a contribution to the United Nations Framework Convention on Climate Change (UNFCCC) method, and the import of wood products increases the internal carbon storage, SCA is one of the good ways to calculate and report the UNFCCC contribution (Biello, 2010). China is a major timber importer. PA gives priority to using the country's HWP exports to increase carbon reserves, but trade is not considered by PA in the estimation method. If China's domestic harvest output decreases, the reported HWP contribution may weaken, regardless of whether the imported HWP supply increases in the future. However, this is unlikely to lead to a decrease in HWP carbon storage (Feng et al., 2014). If the consumption mode of exported HWP is the same as that of domestic HWP, and the assumption is invalid, the exporting country has the responsibility or obligation to report the carbon emissions in its national HWPs to a certain extent. Therefore, the reason for shortening the service life may be the unwillingness of importing countries to effectively manage imported HWPs. However, it may also lead to great errors or even errors in predicting China's carbon emissions and carbon liquidation. The advantages of distinguishing HWP sources include forest management, forest degradation, and deforestation. SCA considers the export and import of HWPs more comprehensively without considering the source of HWPs (Bebbington & Larrinaga-González, 2008). AFA is more suitable to consider carbon emissions. In addition, to improve the accuracy of estimation, in most cases, the carbon pool of HWPs acts as a carbon sink rather than a carbon source.

The country is not responsible for reporting imported HWPs based on production methods (Zhang et al., 2018). If China ignores the management of these wood goods via import and export, this practice may lead to more carbon leakage (Feng et al., 2014). According to China's forest and environmental protection policies, even if the demand for wood products increases, China will impose a logging ban and import wood. According to the Forest Products Report of the Republic of Congo, timber exports to China account for more than three-quarters of its total timber exports. Finally, the carbon storage potential of decommissioned HWPs is also worthy of attention.

4.2   *Difference from Previous Studies*

Data sources are different. There are database sources for China's statistical data, such as FAOSTAT and China Forestry Statistical Yearbook. Since the transaction data do not include wood, most studies use FAOSTAT as the source of original data, and it is also determined that the estimated carbon reserves in the study are small. The following two points are the reasons for this problem. Firstly, the calculation is affected by HWP parameters (Zhang et al., 2018). According to the HWP data in the China Forestry Statistical Yearbook, from 1950 to 2015, the carbon reserves of "direct use" industrial logs were 19–59% of HWPs. In this study, the existence of error is inevitable. Due to the different HWP data sources used, different studies may have different research results every time. Secondly, the 2013 supplement excluded some of these projects (Feng et al., 2014). According to the guidance scheme on carbon emissions from logging products released in 2007, carbon emissions are strictly controlled. However, even so, these data are not that accurate. By 2015, carbon emissions from wood-based panels decreased by 20%, while emissions from paper and paperboards decreased by 27%, respectively

4.3   *Limitations*

All research cannot guarantee meticulousness, errors are unavoidable, and therefore, errors may exist. The uncertainty of this research comes from the following three aspects. Firstly, there are few HWP data in China, and the flow of SWDS is not clear. Therefore, semi-finished products' carbon flow is not well-understood. Secondly, due to technological constraints, the transaction data from 2000 to 2003 are missing. The existing data are an estimate. The original data have errors and are not accurate, which may affect the results to a certain extent. Thirdly, due to the limited research conditions, the three dependent variables cannot be controlled at the same time, resulting in the neglect of recycled HWPs. Besides, this material could prolong the storage of carbon from products and cause data errors.

5.   Verdict

This research focused on the carbon sequestration, carbon emission, and removal mechanisms related to carbon flow in China's HWP pool, which is an important carbon sink. As consumption increases, carbon pool accumulation will rise sharply in the estimation period, and carbon accumulation will continue to increase. According to China Forestry Net and FAOSTAT, by 2020, China's HWPs in use will have 746.2 TgC of carbon storage and 92.1 TgC of disposal. The increase in the accumulation of carbon pools indicates to a certain extent that the country can better respond to climate change, and to a certain extent delay global warming, so as to buy time for the earth's climate change. At the same time, the study also found that the biggest influence on Chinese HWP carbon storage is the use of wood-based panels. The burning of wood-based panels produces a large amount of greenhouse gases, which has led to a logging ban in China. At the same time, China's demand for wood is increasing, and the promulgation of this ban has also led to an inevitable increase in imported carbon flows. It is worth noting that PA may have errors on carbon sinks and is not aware of the huge carbon sinks in China's HWP pool. In summary, the management of wood-based panels should be strengthened to reduce the amount of HWP combustion and promote the mitigation of HWPs in China to optimize the environment to the greatest extent.

6.   Results

Since the adoption of the "Kyoto Protocol" in 1997, countries around the world have carried out a series of emission reduction measures to deal with climate change brought about by industrialization. However, different countries, different regions, and different companies need to rely on scientific data to clarify carbon reduction targets and measure carbon reduction effects. Carbon accounting is a method to measure the direct and indirect emissions of carbon dioxide and its equivalent gas into the earth's biosphere by industrial activities. It can be seen that in terms of accounting objects, carbon accounting needs to include at least the following two conditions: one is to delineate the gases that cause the greenhouse effect, and the other is to determine the main body of industrial activities.

Greenhouse gases are natural and man-made gaseous components that absorb and re-emit infrared radiation into the atmosphere, including carbon dioxide (CO_2), methane (CH_4), nitrous oxide (N_2O), hydrofluorocarbons (HFCs), perfluorocarbons (PFCs), sulphur hexafluoride (SF_6), and nitrogen trifluoride (NF_3). Since different gases have different degrees of impact on the greenhouse effect, the IPCC proposed the concept of carbon dioxide equivalent (CO_2e) to uniformly measure the impact of these gas emissions on the environment. Based on the global warming potential (GWP), we can see the degree of influence of different gases on the greenhouse effect relative to carbon dioxide.

In addition, only for energy activities and industrial production processes, according to the "Provincial Greenhouse Gas Inventory Compilation Guidelines", HFCs, PFCs, and SF_6 mainly involve a few industrial wastes such as aluminium and magnesium, while N_2O has long been included in the scope of air pollution monitoring. Therefore, the main objects of carbon accounting for most companies are CO_2 and CH_4. According to the "China Greenhouse Gas Bulletin 2017", carbon dioxide (CO_2) and methane (CH_4) are the primary and secondary long-lived greenhouse gases that affect the Earth's radiation balance, respectively. The total radiative forcing resulting from the increase in the concentration of all long-lived greenhouse gases contributes about 66% and 17%, respectively. China's carbon trading market is the world's largest. According to the "2020 China Carbon Price Survey" report jointly issued by the China Carbon Forum and ICF International Consulting Company, the price of carbon allowances in the carbon ETS is estimated to continue to rise to 71 yuan/tonne, and the market size will reach 284 billion yuan (approximately 44 billion yuan). China proposed a national carbon emissions trading plan in 2011 and confirmed in the Sino-US Joint Statement on Climate Change before the Paris Climate Summit in 2015 that the national carbon trading market has been piloted for a long time before the launch of the national carbon trading market on 16 July 2021.

The main body of China's carbon trading system includes three parts: Shanghai's National Carbon Emissions Trading Center, Wuhan's Carbon Quota Registration System, and Beijing's National Greenhouse Gas Resource Emission Reduction Management and Trading Center. The first batch of 2,225 power companies to enter the market has a total annual carbon dioxide emissions of more than 4 billion tonnes. The International Energy Agency (IEA) data show that these companies account for one-seventh of global fossil fuel carbon emissions. Energy-intensive and high-emission industries such as petrochemicals, chemicals, building materials, steel, non-ferrous metals, papermaking, and domestic aviation will be gradually included in the carbon trading system in the next few years.

The Ministry of Ecology and Environment of China is responsible for monitoring the trading platform, verifying the data declared by the local government, and ensuring the smooth operation of the system. The carbon trading system functions in several ways: listing agreement trading, block agreement trading, or one-way bidding. The transaction price of a listed agreement transaction is determined within ±10% of the closing price of the previous trading day. The minimum declaration quantity for a single transaction of a block agreement transaction is at least 100,000 tonnes of carbon dioxide equivalent, and the transaction price is determined within ±30% of the closing price of the previous trading day. Applying for one-way bidding is to purchase carbon emission allowances from potential sellers based on information released by trading institutions.

The Chinese carbon trading market was launched on 16 July 2021, and the first day carbon emission allowance trading volume was 4.104 million tonnes, and the closing price was 51.23 yuan ($7.91), which was similar to a trading mechanism in the United States. The current EU carbon market price is 59–70 US dollars per tonne, and that of the United Kingdom is 55–69 dollars.

Founded in 1956, the National Building Materials Exhibition and Trade Center is an institution directly under the State-owned Assets Supervision and Administration Commission of the State Council. It is managed by the China Building

Materials Federation (formerly the National Building Materials Bureau). It is an international exchange and cooperation platform for the Chinese building materials industry and national development and reform. The specific executive unit of the China Building Materials International Production Capacity Cooperation Enterprise Alliance approved by the committee is the initiator of the China-ASEAN Building Materials Industry Cooperation Committee.

The current main business of the International Capacity Cooperation Department is independent research and development and design of international assembly housing systems and exports; international bulk import and export trade; international capacity cooperation funds and overseas investment, park construction; and international certification and testing services. The International Production Capacity Cooperation Department not only participates in international production capacity cooperation itself but also actively leads more high-quality building materials and electric power Chinese enterprises to enter the international market, and integrates with international standards through diversified cooperation models.

References

Ainsworth, R. T. (2010, January 7). *CO2 MTIC fraud – technologically exploiting the EU VAT (again)*. Boston University School of Law Working Paper No. 10-01, School of Law, Boston University. Retrieved January 18, 2010, from www.bu.edu/law/faculty/scholarship/workingpapers/2010.html

Bebbington, J., & Larrinaga-González, C. (2008). Carbon trading: Accounting and reporting issues. *European Accounting Review, 17*(4), 697–717.

Biello, D. (2010). Negating climate gate. *Scientific American, 302*(2), 16.

Feng, K., Hubacek, K., Pfister, S., Yu, Y., & Sun, L. (2014). Virtual scarce water in China. *Environmental Science & Technology, 48*(14), 7704–7713.

IPCC (Intergovernmental Panel on Climate Change). (2007). Climate change 2007. The physical science basis. In *Contribution of working group I to the fourth assessment report of the intergovernmental panel on climate change*. Cambridge University Press.

Zhang, X., Yang, H., & Chen, J. (2018). Life-cycle carbon budget of China's harvested wood products in 1900–2015. *Forest Policy and Economics, 92*, 181–192.

16 The Effects of Ownership Structure, Corporate Governance, and External Supervision on the Quality of Environmental Accounting Information Disclosure

Md Jahidur Rahman, Tarek Rana, Hongtao Zhu, and Zhao Jiayi

1. Introduction

Since its reform and opening, China has achieved rapid development in many fields, which has greatly enhanced its international status. However, as in many developing countries, China's economic development has been accompanied by increasingly serious environmental problems (Qi et al., 2021). The quality of environmental accounting information disclosure (EAIDQ) reveals the utilization of environmental resources and the treatment of environmental pollution, which are inevitable requirements for the management of serious environmental problems (Wang et al., 2014). On the one hand, environmental accounting arises from the need for the public and stakeholders to know information related to environmental accounting of enterprises, mainly from the regulatory role of the government and the public from outside the company (Wei & Zhou, 2020). On the other hand, environmental accounting arises from the need for principals to check the behaviour of agents and the need for internal decision-making (Lin, 2017). So, studying the influence factors of EAIDQ comes from the need of both internal ownership structure and corporate governance. In the case of insufficient supervision by the external sector, proper ownership structure and corporate governance can effectively prevent moral hazard and adverse selection behaviour of managers and ensure that the environmental accounting information disclosure (EAID) is not distorted to a large extent (Wei & Zhou, 2020). Therefore, both intrinsic needs and public pressure force companies to issue EAIDQ (Wei & Zhou, 2020). Based on the above, this study aims to analyse the influence of ownership structure, corporate governance, and external supervision on EAIDQ to help mining companies disclose environmental accounting information in a more standardized way. And in this study, ownership structure is measured by ownership concentration, corporate governance is measured by the proportion of independent directors, and external supervision is measured by government supervision and public supervision.

EAIDQ refers to the disclosure of environmental accounting information reflected in financial reports and financial information related to the environment by actors related to the environment, in accordance with the requirements of EAID

DOI: 10.4324/9781003488965-21

regulations, using specific disclosure methods to provide environmental accounting information to the demanders of environmental information on the use of environmental resources and the fulfilment of fiduciary responsibilities, in order to fully meet their needs for information and decision-making (Lin, 2017). Environmental accounting information can be divided into two categories: monetary accounting information and non-monetary accounting information (Liu & Bai, 2022). Monetary accounting information includes data related to accounts such as environmental assets, liabilities, and expenses, while non-monetary information includes environmental policies, environmental certifications, and environmental logos (Liu & Bai, 2022).

This study was carried out for several reasons. Firstly, socio-economic development has led to an increasing demand for natural resources by society and businesses (Ge, 2017). However, the exploitation of natural resources is often accompanied by huge environmental pollution (Santosa et al., 2008). In 2010, the Ministry of Environmental Protection (MEP) issued the Guidelines for Environmental Information Disclosure of Listed Companies, which announced 16 categories of industries as heavy-polluting industries, and mining is one of them (Guo & Liu, 2018). Mining activities come at the cost of large-scale ecological damage and serious environmental pollution, such as water pollution and heavy metal pollution (Wang et al., 2021). Therefore, it is essential to pay attention to the EAIDQ for mining companies. Studying the influence factors of the EAIDQ can help mining companies to better deal with environmental issues, which is important for both companies and society (Agyemang et al., 2021).

Secondly, at present, the state of EAID of enterprises in China is still not optimistic (Wei & Zhou, 2020). There are also problems of incomplete information disclosure, low EAIDQ, and large differences in the EAIDQ of different industries (Wang et al., 2014). Many studies have shown that although the EAID by companies is increasing, few companies will voluntarily disclose more environmental accounting information except for mandatory disclosures (Xu & Shen, 2014). According to the content of environmental accounting information disclosed by mining companies, it can be found that some mining companies disclose a large amount of environmental accounting information, but the quality is not high (Qi et al., 2021). This is mainly because the companies disclose some general environmental accounting information but not substantive environmental accounting information (Cheng & Gong, 2022). From the content of disclosed environmental accounting information in China, enterprises disclose more qualitative information but less quantitative information, and the information lacks relevance and comparability (Lin, 2017).

Thirdly, most listed mining companies choose to disclose some positive and beneficial environmental accounting information, and few companies disclose negative environmental accounting information (Moroney et al., 2012). This leads to the lack of comprehensiveness and authenticity of EAID in the mining industry. Li et al. (2015) supposed that mining enterprises in China are paying more and more attention to EAID, but the overall willingness to disclose is low. Li and Zhai (2018) revealed that companies are more willing to disclose non-monetary environmental accounting information compared with monetary environmental accounting information.

Fourthly, it should be noted that, as of now, China does not have a complete regulatory system for EAID (Liu & Liu, 2021). In general, China's laws and regulations regarding EAIDQ are still inadequate. Although the state has issued a series of resource management regulations and environmental protection laws and regulations, coverage of these laws and regulations is incomplete (Lin, 2017). And because of poor enforcement by regulators, these laws and regulations are often difficult to function as it should. Without the pressure from environmental risks, enterprises will lack the motivation to disclose environmental accounting information actively (Xu, 2011). In the absence of mandatory laws and regulations and lack of social supervision, most enterprises are reluctant to disclose environmental accounting information to the society to a certain extent because of their environmental image. Therefore, the depth, breadth, and enforcement of environmental information disclosure are not sufficient. In addition, it is also difficult for the government to create a deterrent and high pressure on not disclosing environmental accounting information as required (Liu, 2016).

Finally, previous studies have not reached a uniform conclusion on this issue. And according to the survey, it is found that no scholars have previously divided the EAID of the mining industry into monetary EAID and non-monetary EAID specifically to study. Therefore, it is important to understand the influencing factors of EAIDQ to improve the EAID problem in the mining industry.

Based on previous studies, for this study, the mining industry is selected as the research object, and the effects of ownership structure, corporate governance, and external supervision on EAIDQ from inside and outside the company are investigated. In addition, based on whether the disclosed environmental accounting information is expressed in monetary form, EAIDQ is divided into quality of monetary environmental accounting information (M-EAIDQ) and quality of non-monetary environmental accounting information (NM-EAIDQ). And the effects of ownership structure, corporate governance, and external supervision on them are discussed. Finally, based on the relevant findings of this study, some pertinent suggestions are made.

By observing the sample data of 71 Chinese listed mining companies from 2012 to 2021, we have drawn the following findings. The higher ownership concentration in mining companies helps improve the EAIDQ, mainly because the controlling shareholder supervises the management to disclose more environmental accounting information in order to protect the interests of the company and itself, so as to prevent the management from the act of concealing environmental accounting information. In addition, the higher the proportion of independent directors in the board of directors, the more likely the mining companies will disclose non-monetary environmental accounting information. This is mainly because independent directors are often able to make more objective and reasonable corporate decisions. In order to maintain corporate image and corporate interests, independent directors usually only require the management to disclose positive and operable qualitative environmental accounting information. From the perspective of external supervision, environmental supervision does not seem to have a significant impact on the EAIDQ of mining companies. However, through hysteresis detection, this study found that the impact of environmental supervision on EAIDQ has a certain hysteresis. If the research time is moved back one year, it can be found that environmental supervision can effectively prompt mining companies to disclose more non-monetary environmental accounting information. This is mainly due to the incompleteness and slow development of China's environmental supervision system. In addition to environmental supervision from the government, supervision from the public also has an impact on corporate environmental information disclosure. But the difference is that the public pays more attention to the M-EAIDQ of mining enterprises. This is mainly because for high-pollution industries, monetary environmental accounting information such as environmental protection investments disclosed directly by enterprises is more convincing to the public than vague qualitative information. Through the analysis of robustness test and endogeneity test, we found that the above findings are robust, so it can provide a certain reference for EAID of mining companies.

The current study seeks to make the following contributions to the existing literature: Firstly, most previous studies on EAIDQ in the mining industry have studied all environmental information disclosed by enterprises as a whole. In this study, according to whether it can be expressed in monetary terms, environmental accounting information is divided into monetary environmental accounting information and non-monetary environmental accounting information, which can fill the gap in the existing literature. Moreover, few scholars have studied the factors influencing EAIDQ in the Chinese mining industry based on the classification of environmental accounting information into monetary and non-monetary environmental accounting information. The results of this study can provide some theoretical support for domestic and foreign mining companies in their environmental accounting disclosure decisions.

Secondly, this study specifically investigates the effects of ownership structure, corporate governance, and external supervision on monetary and non-monetary EAID, which leads to more specific conclusions and improves the usability and comparability of the findings. In addition, this study helps companies understand the importance and necessity of environmental information disclosure and improves the directionality and accuracy of corporate environmental accounting disclosure. By studying the impact of different factors on the EAIDQ of mining companies, mining companies with low EAIDQ can clearly understand and pinpoint the problem points, so that they can quickly and accurately solve the problems and fundamentally improve the EAIDQ of the entire mining industry.

Thirdly, for high-polluting enterprises such as the mining industry, it is their social responsibility to disclose environmental accounting information. The disclosure quality of environmental accounting information affects the corporate image of the mining enterprises to a certain extent and is an important standard for the public and investors to evaluate enterprises. The research results of influencing factors of EAIDQ in this study can help enterprise operators and managers better understand the environmental problems that may be caused by business decisions in different links and realize the importance of EAIDQ for enterprise operation and development. In addition, the conclusions and suggestions of this study will also provide business operators and managers with operational and management guidance. Enterprise management authorities can check the environmental performance of enterprises through comparative analysis of environmental information disclosure, adjust corporate strategy in time, reduce the risks of excessive environmental pollution brought by government punishment, reduce the suspension of business and other risks, and establish a good corporate image.

Fourthly, the purpose of environmental accounting disclosure by companies is for higher corporate performance, in addition to fulfilling their corporate social responsibility (CSR). Based on the shareholder theory and other relevant statements, this study focuses on the reasons why the ownership structures and internal governance of a firm have an impact on EAIDQ. In the long run, corporate performance is the ultimate goal of firms to regulate their ownership structure and internal governance. Therefore, studying the relationship between EAIDQ and its influencing factors can provide behavioural and decision-making guidance for companies to improve their corporate performance.

Finally, based on the results of the empirical study, this work puts forward specific and feasible suggestions for completing EAIDQ from the perspectives of the company's ownership structure and internal governance, as well as the local government and the public, that are conducive to the development of a more effective environmental information disclosure mechanism for mining companies and are of great significance for the improvement of EAIDQ for mining enterprises.

The remainder of this chapter is organized as follows: Section 2 presents the background of environmental information disclosure of the mining industry in China. Section 3 presents the literature review and hypothesis development. Section 4 describes the methodology and sample. Section 5 presents the robustness tests. Section 6 presents the endogeneity test. Section 7 presents the discussion. Section 8 presents the conclusions and suggestions.

2. Background

2.1 Environmental Information Disclosure

EAID mechanism is divided into voluntary disclosure mechanism and mandatory disclosure mechanism. China adopts a disclosure mechanism that combines mandatory and voluntary disclosure – the legal system mandates listed companies in heavily polluting industries to disclose environmental accounting information, while a voluntary disclosure mechanism that encourages disclosure is adopted for other companies (Tian, 2021). However, due to the overall low level of development of CSR in China, the EAID system is not perfect, which makes it difficult to mobilize the initiative and enthusiasm of listed companies to disclose environmental accounting information and seriously restricts the disclosure of environmental accounting information of listed companies in China (Liu, 2016).

Xu (2011) proposed that the strong support and cooperation from the government and related environmental management agencies are prerequisites for the development of environmental accounting. Exploring the factors affecting corporate environmental information disclosure can help provide more theoretical basis for the government to make better policies. As an industry with a large impact on the environment, it is of great practical significance to explore the factors influencing environmental information disclosure in the mining industry (Fu, 2016). Several studies in China also investigated the relationship between environmental accounting and firm performance (Rahman & Wu, 2023; Rahman et al., 2023, 2024; Zhu & Rahman, 2024).

2.2 *Institutional Background of the Mining Industry in China*

In recent years, with the establishment of the environmental information disclosure system and the awakening of the public's awareness of being informed, the society has become more and more vocal about the disclosure of environmental accounting information (Liu, 2012). The related national departments have also introduced a series of environmental regulatory policies and environmental protection rules one after another (Chang, 2013).

The MEP and the National Development and Reform Commission jointly issued the "Guidance on Strengthening the Construction of Enterprise Environmental Credit System" at the end of 2015, which clearly stipulates that in the next five years, key emission enterprises should disclose environmental information according to the law (Zhang, 2018). The fourth ministerial meeting of the Ministry of Ecology and Environment on 26 November 2021 considered and adopted the Administrative Measures for the Legal Disclosure of Enterprise Environmental Information, which was formally implemented on 8 February 2022. This bill aims to regulate the activities of legal disclosure of environmental information of mining enterprises, strengthen social supervision of enterprises, and organize and guide enterprises to disclose environmental information in a timely, accurate, and complete manner and establish a good standard of environmental information disclosure.

3. Literature Review and Hypothesis Development

For Chinese listed mining companies, the factors affecting their EAIDQ can be divided into internal and external factors (Liu, 2022). In terms of internal factors, ownership structure (Liu & Bai, 2022) and corporate governance (Sun & Ye, 2020) have a greater impact on EAIDQ. The ownership structure is studied specifically through ownership concentration; the corporate governance aspect focuses on the impact of the proportion of independent directors on corporate EAIDQ. External factors mainly include environmental supervision from the government (Liu & Bai, 2022) and public opinion pressure from society (Dai, 2020). This study reviews the effects of ownership structure and corporate governance on corporate EAIDQ, and the effects of environmental supervision and public opinion monitoring on corporate EAIDQ.

3.1 Ownership Concentration and EAIDQ

Deng (2018) studied the influence of ownership structure and capital structure on EAIDQ in the mining industry and concluded that ownership concentration is significantly positively related to the content of EAIDQ. With the increase in ownership concentration, the company's decision-making and management power are concentrated in the hands of some major shareholders, and in order to satisfy the minority shareholders' interest in environmental management, listed companies take the initiative to disclose more environmental information. Huang (2014) studied the relevant data of 619 A-share listed companies in 20 heavily polluting industries from 2008 to 2012, and the results showed that the effect of ownership concentration on EAIDQ of listed companies is significantly positively related. Wang et al. (2014) selected A-share listed companies in Shanghai Stock Exchange and Shenzhen Stock Exchange that disclosed social responsibility reports due to the implementation of circular economy in steel, cement, chemical, and petroleum industries, and used regression models to study the data of 218 samples. The results of the study indicate that the circular economy accounting disclosure index is significantly and positively related to ownership concentration. Zheng and Zheng (2018) suggested that, according to the shareholder theory, as ownership concentration increases, shareholder control over management also increases, and in order to better supervise management and to avoid management from harming shareholders' interests, large shareholders will demand more information disclosure from the company.

However, some researchers have shown that ownership concentration has no significant relationship or has a negative relationship with EAIDQ. Lu et al. (2016) reached the conclusion that ownership concentration is not significantly correlated with EAIDQ by studying the influencing factors of accounting information disclosure in the ferrous metal smelting industry. Qi et al. (2021) adopted the fixed-effect regression model to analyse the relevant data of listed mining companies in Shanghai and Shenzhen from 2014 to 2016, and the results showed that the concentration of equity is significantly and negatively correlated with the level of EAID. Wang (2015) proposed that when the shareholding is concentrated in a small number of hands, the major shareholders will reduce the disclosure cost and disclose less information that is unfavourable to them. Conversely, the more decentralized the shareholding is, the more shareholders will demand more disclosure of environmental accounting information in order to maintain their own interests and promote the disclosure of environmental information.

Based on the above literature review, and according to the "monitoring hypothesis", the increase in ownership concentration will enhance shareholders' monitoring of the company and has the effect of limiting the management's behaviour of pursuing its own interests at the expense of shareholders (Cheng et al., 2011). Therefore, the first hypothesis of this study is proposed as follows:

H1: Ownership concentration is positively correlated with the EAIDQ.

3.2   *Proportion of Independent Directors and NM-EAIDQ*

Improving the governance structure of a company helps clarify responsibilities and authority at all levels of the company and helps improve the credibility and transparency of environmental information disclosure (Hua, 2020). The proportion of independent directors is an important factor affecting corporate governance. Kong (2017) conducted an empirical study on the data of 67 mining companies for five consecutive years, and the results showed that the proportion of independent directors was significantly correlated with the level of EAID. Hua (2020) analysed the effect of the proportion of independent directors on the degree of carbon information disclosure by establishing a multiple linear regression model. The experimental results show that independent directors play a supervisory function to a certain extent in the company's management decisions. The proportion of independent directors is positively correlated with the level of carbon information disclosure. Agyemang (2020) studied the effect of board characteristics on EAID in listed companies. The study showed that independent directors play a role in controlling the management's interest in seeking private benefits for the company. Therefore, the greater the proportion of independent directors, the higher the degree of voluntary disclosure of company information. There is a significant positive relationship between the proportion of independent directors and EAID. However, Fu (2016) studied the mining industry and concluded that the effect of the proportion of independent directors on environmental information disclosure is not significant.

Xiong and Cheng (2013) argued that as the overall level of culture and skills of corporate managers increases, the ability of independent directors to guide them diminishes. On the other hand, the monitoring influence of independent directors may decrease as the majority shareholder's holdings increase. However, Huang and Zhou (2020) suggested that independent directors are able to monitor and limit the plans and decisions of company management, have a high degree of autonomy and high level of cognitive ability, and can solve problems from the perspective of long-term corporate development and overall effectiveness. Therefore, the higher the proportion of independent directors, the more the companies focus on external effects and governance, and the higher the level of environmental information disclosure. However, since true monetary information is usually not beneficial to the overall effectiveness of the enterprise (Liu & Bai, 2022), enterprises are more willing to disclose more non-monetary environmental accounting information. Based on the above, the third hypothesis of this study is proposed:

H2: The proportion of independent directors is positively related to the NM-EAIDQ.

3.3   *Government Supervision and NM-EAIDQ*

From the government's point of view, the current EAID in China lacks regulation, compulsion, and supervision by relevant departments (Lin, 2017). Normative and mandatory pressure has a positive promotion effect on the disclosure of environmental accounting information by enterprises (Liu & Liu, 2021). Nguyen et al.

(2020) used Cronbach's alpha analysis, exploratory factor analysis, and multiple regression analysis with mining enterprises in Binh Dinh Province, Vietnam, to show that the coercive pressure from government agencies has a significant effect on environmental accounting disclosure. Wang (2019) took an empirical test to study the data related to A-share listed companies in the chemical industry from 2014 to 2018. The results showed that the higher the level of government regulation, the higher the level of EAID. Yan et al. (2018) argued that strong government regulation can play a strong supervisory role on corporate environmental information disclosure and improve the level of corporate EAID, while establishing a good public image will also enhance corporate value and bring long-term economic benefits to the company.

Nie and Wang (2022) studied the external pressure of resource-based companies based on relevance theories such as reputation theory. The results showed that government regulation has a positive moderating effect on environmental information disclosure, and EAID can significantly improve the innovation level of resource-based enterprises, and "hard" disclosure has a stronger effect on the innovation level than "soft" disclosure (Nie & Wang, 2022). Usually, government documents require companies to disclose mostly non-monetary environmental accounting information, and few companies are willing to voluntarily disclose monetary environmental accounting information (Tian, 2021). Based on this, the fourth hypothesis of this study is proposed:

H3: Government supervision is positively related to the NM-EAIDQ.

3.4   Supervision by Public Opinions and M-EAIDQ

With the improvement of living standards, the public is increasingly concerned about environmental issues. In the process of disclosure of environmental information by listed companies, supervision of the public is a very effective driving force (Wang, 2015). Zhang (2017) studied the sample data of listed mining companies from 2012 to 2014, and the results showed that public supervision had a significant positive impact on the disclosure level of environmental information of mining enterprises. Shen and Feng (2012), using a sample of listed companies in China's heavily polluting industries, found that social media publishing reports on corporate environmental performance can enhance the role of public opinion monitoring, thus significantly contributing to the level of corporate environmental information disclosure.

Social media is one of the ways for the public to monitor the public opinion of enterprises and for enterprises to disclose environmental accounting information to the public. When listed companies get higher media attention, they will be more proactive in releasing better quality environmental accounting information. Especially when the number of positive media reports increases, it will prompt companies to pay more attention to improving the EAIDQ (Xue et al., 2021). According to the legitimacy theory, by disclosing environmental accounting information, enterprises can demonstrate to the public that their actions are legal and reflect positive values such as sustainable development, indicating that they are able to achieve long-term economic development and reduce the social pressure on them (Dowling & Pfeffer, 1975). However, due to the uncontrollability of monetary environmental accounting information (Liu & Bai, 2022), it is more conducive to shareholders and other stakeholders to obtain the authenticity of information. Based on the above, the fifth hypothesis of this study is proposed:

H4: Supervision by public opinion is positively related to the M-EAIDQ.

4. Research Methodology

4.1 Sample Selection and Data Sources

In recent years, as the state pays more and more attention to high-consumption and high-pollution enterprises, the quality of corporate EAID has also received much attention, especially in the mining industry, which is a key high-pollution industry (Kong, 2017). In this context, we selected data of 86 listed companies in China's mining industry from 2012 to 2021 as the research sample. Among them, environmental accounting disclosure data are obtained from manually collected annual reports, social responsibility reports, and sustainability reports of listed companies. Most of the data of this study including shareholding structure and corporate governance are mainly obtained from the China Stock Market & Accounting Research (CSMAR) database. The data related to government monitoring are measured according to the Pollution Information Transparency Index (PITI), and the data related to public opinion monitoring are mainly from the China Research Data Service Platform (CNRDS) database. In the process of selecting listed companies, this study screened the sample excluding (1) listed companies with special treatment (ST) and *ST from 2012 to 2021 and (2) listed companies with abnormal data. Finally, a total of 71 listed companies were selected in this study as the sample subjects.

4.2 Dependent Variables

Referring to Nei and Wang (2022), this study uses the "content analysis method" to evaluate the EAIDQ in the mining industry. For the classification of environmental accounting information, there are relatively uniform standards in China. Firstly, environmental accounting information can be classified into monetary environmental accounting information and non-monetary environmental accounting information according to whether the environmental information of the enterprise is expressed in money or not. In addition, environmental accounting information can be classified into "hard" disclosure and "soft" disclosure according to whether the environmental accounting information disclosed by the enterprise is mandatorily required by the relevant legal documents. Among them, environmental accounting information that is mandatorily required to be disclosed by relevant legal documents is "hard" disclosure, while the rest of the items that are voluntarily disclosed by enterprises are "soft" disclosure (Liu & Bai, 2022).

In this study, with reference to the first classification method, the environmental accounting information disclosure index (EAIDI), monetary environmental accounting information disclosure index (M-EAIDI), and non-monetary environmental accounting information disclosure index (NM-EAIDI) are constructed, and the specific evaluation indexes are shown in Table 16.1 (Nei & Wang, 2022).

M-EAIDI consists of 6 indicators, and NM-EAIDI consists of 7 indicators, and the indicators are assigned with values. Among them, whether there is an independent social responsibility report, sustainability report, and independent environmental report: "yes" = 2, "no" = 0. For other indicators: "non-disclosure" = 0, "qualitative disclosure" = 1, and "combination of qualitative and quantitative" = 2. According to the calculation, the optimal score of EAIDI, M-EAIDI, and

Table 16.1 Rating System of Environmental Information Disclosure Indicators

Classification	Index Set	Index Score		
		Non-disclosure	Qualitative Disclosure	Combine Qualitative and Quantitative Disclosure
M-EAIDI	Pollutant discharge fee expenditure (environmental tax)	0	1	2
	Emergency expenditures for major environmental problems	0	1	2
	Green investment spending or borrowing	0	1	2
	Pollution reduction benefits	0	1	2
	Income from waste utilization	0	1	2
	Environmental grants, grant reduction and exemption of incentive income	0	1	2
NM-EAIDI	Environmental information disclosure system	0	1	2
	Environmental management objectives	0	1	2
	Environmental measures and improvements	0	1	2
	Environmental certification and implementation	0	1	2
	Energy-saving measures and results	0	1	2
	Types and quantities of pollutants and the discharge standards	0	1	2
	Whether there is an independent social responsibility report, sustainability report, and independent environmental report	If it is issued, it is 2 If it is not issued, it is 0		

NM-EAIDI are 18, 10, and 9, respectively. The calculation formula of the EAIDI is as follows (Liu & Bai, 2022):

$$EAIDI = \frac{EAIDI_i}{18} \tag{1}$$

$$M{-}EAIDI = \frac{M{-}EAIDI_i}{10} \tag{2}$$

$$NM{-}EAIDI = \frac{NM{-}EAIDI_i}{9} \tag{3}$$

4.3   Independent Variables

To estimate the determinants of EAIDI, M-EAIDI, and NM-EAIDI, we use the following four sets of independent variables: ownership structure, corporate governance, government monitoring, and public opinion monitoring. The ownership structure of the firm is measured by the ownership concentration (Liu & Bai, 2022). Corporate governance is mainly reflected by the proportion of independent directors (Habib, 2008). For government monitoring, in this study, we selected the environmental monitoring index (PITI) published by the China Environmental Research Center as the data sample. The PITI index quantifies the daily supervision, self-monitoring, report response, emission data, and environmental assessment information of pollution source management in four dimensions: systematic, timely, complete, and user-friendly. It also evaluates the status of environmental information disclosure from eight aspects, for example, enterprise exceedance violations, complaints, and environmental assessment (Nei & Wang, 2022). Public opinion supervision, on the other hand, is reflected by the number of media reports (Li & Zeng, 2020).

4.4   Control Variables

In this study, board size (BSIZE), number of board meetings (BOME), supervisory board scale (SUPSIZE), company size (SIZE), return on equity (ROE), operating capacity (TAT), solvency (LEV), Big Four (AUDIT), growth capacity (GROWTH), and annual disclosure (YEAR) are selected as control variables to attenuate the interactions among factors affecting environmental accounting disclosure and to ensure that the empirical results are truly presented. The indicators of each variable are shown in Table 16.2 (Liu & Bai, 2022; Nei & Wang, 2022).

4.5   Multiple Linear Regression Model

The main purpose of this empirical study is to explore the influencing factors of EAIDQ, M-EAIDQ, and NM-EAIDQ in the mining industry in recent years and analyse the data mainly from the aspects of ownership structure (ownership concentration) and corporate governance (proportion of independent directors), as well as government supervision and public opinion supervision. Since EAIDQ, M-EAIDQ, and NM-EAIDQ are affected by many important factors, the following

Table 16.2 Variable Definition

Types	Name	Variable Definition	Data Source
Dependent variables	EAIDI	Environmental information disclosure index	CSMAR
	M-EAIDI	Monetary environmental information disclosure index	CSMAR
	NM-EAIDI	Non-monetary environmental information disclosure index	CSMAR
Independent variables	FIRST	Top one shareholders holding proportion	CSMAR
	IDR	Proportion of independent directors to the total number of directors	CSMAR
	PITI	Index of the PITI	IPE
	MEDI	Natural log of (number of media reports +1)	CNRDS
Control variables	BSIZE	Board size	CSMAR
	BOME	Number of board meetings	CSMAR
	SUPSIZE	Scale of board of supervisors	CSMAR
	STATE	Ownership nature. State = 1. Non-state = 0	CSMAR
	ROE	Net profit/shareholders' equity balance	CSMAR
	TAT	Net operating income/average total assets	CSMAR
	LEV	Leverage measured as the latest year-end long-term debt over total assets	CSMAR
	AUDIT	Big Four. Yes equals 1; no equals 0	CSMAR
	GROWTH	Total growth of operating revenue/ total operating revenue of the previous year	CSMAR
	YEAR	Virtual variable, controlling for year effects	

multiple linear regression models are constructed to test the main hypotheses of this study (Liu & Bai, 2022).

4.5.1   Test for Hypothesis 1

This model is used to test the relationship between EAIDQ and the ownership concentration in the ownership structure. In this model, FIRST represents the shareholding ratio of the largest shareholder. If the regression coefficient of FIRST is positive, it indicates that the degree of ownership concentration is positively correlated with EAIDQ.

$$
\begin{aligned}
\boldsymbol{EAIDI}_i = {} & \alpha_0 + \alpha_1 \boldsymbol{FIRST}_i + \alpha_2 IDR_i + \alpha_3 PITI_i + \alpha_4 MEDI_i \\
& + \alpha_5 BSIZE_i + \alpha_6 BOME_i + \alpha_7 SUPSIZE_i \\
& + \alpha_8 STATE_i + \alpha_9 ROE_i + \alpha_{10} TAT_i + \alpha_{11} LEV_i \\
& + \alpha_{12} AUDIT_i + \alpha_{13} GROWTH_i + \alpha_{14} YEAR_i + \varepsilon_i
\end{aligned} \tag{4}
$$

4.5.2   Test for Hypotheses 2 and 3

This model is used to test the relationship between NM-EAIDQ and the proportion of independent directors, as well as the relationship between NM-EAIDQ and external environmental supervision. In this model, IDR stands for the proportion of independent directors, and PITI stands for Environmental Disclosure Index. If the regression coefficients of IDR and PITI are positive, it indicates that both corporate governance and external environmental supervision are positively correlated with NM-EAIDQ.

$$\begin{aligned}
NM_EAIDI_i = {}& \alpha_0 + \alpha_1 FIRST_i + \alpha_2 IDR_i + \alpha_3 PITI_i + \alpha_4 MEDI_i \\
& + \alpha_5 BSIZE_i + \alpha_6 BOME_i + \alpha_7 SUPSIZE_i + \alpha_8 STATE_i \\
& + \alpha_9 ROE_i + \alpha_{10} TAT_i + \alpha_{11} LEV_i + \alpha_{12} AUDIT_i \\
& + \alpha_{13} GROWTH_i + \alpha_{14} YEAR_i + \varepsilon_i
\end{aligned} \tag{5}$$

4.5.3   Test for Hypothesis 4

This model is used to test the relationship between M-EAIDQ and external supervision by public opinion. In this model, MEDI represents the number of public media reports. If the regression coefficient of MEDI is positive, it indicates that supervision by public opinion is positively correlated with M-EAIDQ.

$$\begin{aligned}
M_EAIDI_i = {}& \alpha_0 + {-}_1 FIRST_i + \alpha_2 IDR_i + \alpha_3 PITI_i + \alpha_4 MEDI_i \\
& + \alpha_5 BSIZE_i + \alpha_6 BOME_i + \alpha_7 SUPSIZE_i \\
& + \alpha_8 STATE_i + \alpha_9 ROE_i + \alpha_{10} TAT_i + \alpha_{11} LEV_i \\
& + \alpha_{12} AUDIT_i + \alpha_{13} GROWTH_i + \alpha_{14} YEAR_i + \varepsilon_i
\end{aligned} \tag{6}$$

The relevant variables involved in the model are shown in Table 16.2. α_0 is the constant term, $\alpha_1 \sim \alpha_4$ is the regression coefficient of the explanatory variable, $\alpha_5 \sim \alpha_{14}$ is the regression coefficient of the control variable, and ε_i is the model residual term.

4.6   Descriptive Statistics

The descriptive statistical analysis of annual overall EADI, M-EAIDI, and NM-EAIDI of 71 sample listed companies is shown in Table 16.3. In addition to the traditional annual reports of listed companies as a carrier of environmental information disclosure, independent environmental reports and liability reports are also increasingly valued by listed companies. The independent environmental report is the carrier with the most comprehensive data, the most detailed content, and the highest quality, while the social responsibility report is the simplest disclosure carrier. In addition, when compiling the data, it can be found that nearly half of the listed mining companies in China have issued independent environmental responsibility reports and social responsibility reports. And the EAIDQ of listed companies that have issued social responsibility reports is higher than that of listed companies that have not issued social responsibility reports. The more comprehensive the environmental information disclosure vehicle is, the higher the EAIDQ is. As in Table 16.3, some companies publish independent environmental reports and social responsibility reports at the same time, so the maximum value of NM-EAIDI is 1.00.

As can be seen from Table 16.3, during the period from 2012 to 2021, EAIDI increased from 0.30 to 0.42, M-EAIDI increased from 0.20 to 0.29, and NM-EAIDI increased from 0.39 to 0.51, indicating that EAIDQ in China has increased year by year, and environmental transparency has improved. The sample means of NM-EAIDI are higher than that of EAIDI and M-EAIDI in all years, indicating that NM-EAIDQ is higher than overall EAIDQ and M-EAIDQ in the process of annual environmental information disclosure.

However, it can be seen from Table 16.4 that the sample averages of EAIDI, M-EAIDI, and NM-EAIDI are 0.33, 0.20, and 0.44, respectively, which are all

Table 16.3 Annual Descriptive Statistics of EAIDI, M-EAIDI, and NM-EAIDI

Annual_Variable	Sample Size	Max	Min	Mean	Median	Std. Dev.
2012_EAIDI	71	0.89	0.06	0.30	0.22	0.24
2013_EAIDI	71	0.89	0.06	0.34	0.28	0.24
2014_EAIDI	71	0.89	0.00	0.34	0.33	0.25
2015_EAIDI	71	1.00	0.00	0.36	0.33	0.26
2016_EAIDI	71	1.00	0.06	0.38	0.33	0.27
2017_EAIDI	71	1.00	0.06	0.38	0.28	0.27
2018_EAIDI	71	0.94	0.06	0.40	0.33	0.26
2019_EAIDI	71	0.94	0.06	0.41	0.39	0.26
2020_EAIDI	71	1.00	0.06	0.34	0.33	0.25
2021_EAIDI	71	1.00	0.06	0.42	0.39	0.27
2012_M-EAIDI	71	0.80	0.00	0.20	0.10	0.22
2013_M-EAIDI	71	0.80	0.00	0.24	0.20	0.22
2014_M-EAIDI	71	0.80	0.00	0.25	0.20	0.24
2015_M-EAIDI	71	1.00	0.00	0.25	0.20	0.24
2016_M-EAIDI	71	1.00	0.00	0.28	0.20	0.26
2017_M-EAIDI	71	1.00	0.00	0.29	0.30	0.28
2018_M-EAIDI	71	0.90	0.00	0.30	0.30	0.26
2019_M-EAIDI	71	1.00	0.00	0.30	0.20	0.27
2020_M-EAIDI	71	1.00	0.00	0.22	0.10	0.26
2021_M-EAIDI	71	1.00	0.00	0.29	0.30	0.27
2012_NM-EAIDI	71	1.00	0.11	0.39	0.33	0.28
2013_NM-EAIDI	71	1.00	0.11	0.41	0.33	0.27
2014_NM-EAIDI	71	0.89	0.00	0.41	0.33	0.27
2015_NM-EAIDI	71	1.00	0.00	0.44	0.44	0.28
2016_NM-EAIDI	71	1.00	0.11	0.45	0.44	0.27
2017_NM-EAIDI	71	1.00	0.11	0.43	0.33	0.26
2018_NM-EAIDI	71	1.00	0.11	0.47	0.44	0.26
2019_NM-EAIDI	71	1.00	0.11	0.49	0.44	0.26
2020_NM-EAIDI	71	0.89	0.11	0.43	0.44	0.25
2021_NM-EAIDI	71	1.00	0.11	0.51	0.44	0.28

Table 16.4 Descriptive Statistics of Major Variables

Variable	Sample Size	Max	Min	Mean	Median	Std. Dev.
EAIDI	710	1.00	0.00	0.33	0.37	0.26
M-EAIDI	710	1.00	0.00	0.20	0.26	0.25
NM-EAIDI	710	1.00	0.00	0.44	0.44	0.27
FIRST	710	0.87	0.01	0.43	0.42	0.18
IDR	710	0.67	0.29	0.36	0.37	0.05
PITI	710	82.40	8.30	52.10	51.03	17.83
MEDI	710	9.81	2.30	5.29	5.42	1.17
BSIZE	710	18.00	3.00	9.00	9.20	2.24
BOME	710	37.00	3.00	9.00	9.76	4.50
SUPSIZE	710	11.00	1.00	3.00	4.17	1.67
STATE	710	1.00	0.00	1.00	0.65	0.48
ROE	710	1.39	−3.34	0.06	0.05	0.25
TAT	710	4.61	0.04	0.48	0.63	0.52
LEV	710	0.36	0.00	0.04	0.06	0.07
AUDIT	710	1.00	0.00	0.00	0.11	0.32
GROWTH	710	56.17	−0.96	0.07	0.43	3.00

lower than 0.5, indicating that the EAIDQ of listed companies in heavily polluting industries in China is still relatively low overall. On the basis that the difference between the maximum value (1) and the minimum value (0) of EAIDI, M-EAIDI and NM-EAIDI are 1, and the overall EAIDQ of listed companies is quite different. Moreover, there is a big difference between the quality of monetary environmental information disclosure and the quality of non-monetary environmental information disclosure of listed mining companies.

The difference between the maximum value (0.87) and the minimum value (0.01) of the shareholding ratio of the controlling shareholder is 0.86, the average value is 0.43, and the standard deviation is as high as 0.18. It indicates that some of the mining sample companies studied in this work have a high degree of ownership concentration and a high degree of dispersion of ownership structure data. For the ratio of independent directors, the difference between the maximum value (0.67) and the minimum value (0.29) is 0.38, the average value is 0.36, and the standard deviation is 0.05. It indicates that the proportion of independent directors in some sample companies is relatively high, but the proportion of independent directors in most sample companies is not much different, and the proportion of independent directors is generally low in mining companies. The average of PITI is 52.10, which does not meet the passing line. According to the region where the sample companies are located, the maximum and minimum values of PITI are 82.40 and 8.30, respectively, and the standard deviation is as high as 17.83. It can be seen that the PITI of different sample companies is quite different and highly dispersed, indicating that the overall level of environmental supervision in China is relatively low, and the level of supervision varies greatly among regions.

4.7 Pearson Correlation Analysis

In this study, Pearson correlation analysis was conducted on the main variables, and the analysis results are shown in Table 16.5. As can be seen from the table below, the correlation between variables shows that FIRST and MEDI are significantly correlated with EAIDI, M-EAIDI, and NM-EAIDI; PITI is significantly correlated with NM-EAIDI; and FIRST is significantly correlated with IDR and MEDI. There is a strong correlation between some variables, so it is necessary to test the multicollinearity among the variables.

In this study, the variance inflation factor (VIF) is used for multicollinearity test to extract the VIF value of each explanatory variable. It can be seen from Table 16.6 that the VIF values of all explanatory variables are far less than 5, and the average VIF value is less than 2, indicating that there is no multicollinearity problem in the multiple regression model, which can be further analysed.

4.8 Reliability Test

Considering that the unit root test and cointegration test require a long period, the data from 2012 to 2021 that we use are short panel data (N > T). From the micro level, there was no need to conduct the unit root test and cointegration test in this study.

4.9 Regression Analysis

How ownership structure, corporate governance, and external environmental supervision affect EAIDQ is shown in Table 16.7. The results of the regression analysis show that there is a significant correlation between ownership structure and EAIDQ, M-EAIDQ, and NM-EAIDQ. At the 0.1% level, ownership concentration is significantly and positively correlated with EAIDQ. This result verifies H1. It indicates that controlling shareholders play a significant role in the environmental accounting disclosure of mining companies. The more concentrated the ownership structure of the company, the greater the role played by controlling shareholders and the higher the EAIDQ of the company.

At the 1% level, the proportion of independent directors is positively and strongly correlated with EAIDQ and NM-EAIDQ; therefore, H2 is verified. This indicates that the percentage of independent directors in the board of directors of mining companies has an impact on the quality of EAID. In particular, in terms of NM-EAIDQ, the higher the percentage of independent directors, the higher the quality of non-monetary environmental information disclosure.

According to the regression results of PITI, PITI has a positive relationship with NM-EAIDQ, but the regression coefficient of PITI is not significant at the 5% level. It indicates that environmental monitoring does not have a significant effect on NM-EAIDQ. Therefore, H3 is not established. This is mainly because in China, there is no complete environmental supervision system, the government's supervision is not strong enough, and the environmental supervision system is not perfect.

Table 16.5 Pearson Correlation Coefficient between Variables

	EAIDI	M-EAIDI	NM-EAIDI	FIRST	IDR	PITI	MEDI	BSIZE	BOME	SUPSIZE	STATE	ROE	TAT	LEV	AUDIT	GROWTH
EAIDI	1															
M-EAIDI	.943**	1														
NM-EAIDI	.937**	.767**	1													
FIRST	.377**	.376**	.333**	1												
IDR	0.038	0.026	0.046	−.076*	1											
PITI	.079*	0.063	.086*	0.006	0.025	1										
MEDI	.387**	.400**	.328**	.400**	0.052	−0.022	1									
BSIZE	.258**	.235**	.250**	.369**	−.377**	−.125**	.208**	1								
BOME	−0.040	−.080*	0.007	−.327**	−0.041	.129**	0.039	−0.038	1							
SUPSIZE	.287**	.285**	.254**	.514**	−.131**	−.194**	.344**	.557**	−.118**	1						
STATE	.506**	.473**	.479**	.560**	−0.021	−0.049	.301**	.306**	−.175**	.424**	1					
ROE	0.035	0.056	0.010	.086*	−0.045	−0.051	0.070	0.066	−0.006	−0.036	−0.046	1				
TAT	0.034	0.041	0.023	.099**	0.000	0.042	.152**	0.014	0.024	.152**	.137**	0.072	1			
LEV	.414**	.384**	.396**	.108**	−0.057	0.005	.250**	.101**	0.040	.127**	.233**	0.025	−.131**	1		
AUDIT	.379**	.394**	.317**	.283**	.121**	.209**	.516**	0.010	0.003	.137**	.264**	0.040	0.066	.180**	1	
GROWTH	−0.049	−0.020	−0.074	0.041	−0.036	0.026	−0.045	−0.020	0.052	−0.020	−0.014	.117**	.170**	−0.030	−0.034	1

Note: * and ** indicate significance levels of 0.10 and 0.05, respectively.

Table 16.6 Result of the VIF Model

Variable	VIF	1/VIF
FIRST	2.14	0.467355
SUPSIZE	1.95	0.512561
BSIZE	1.77	0.564114
MEDI	1.74	0.576059
STATE	1.67	0.600364
AUDIT	1.54	0.64967
IDR	1.22	0.819394
BOME	1.22	0.820316
LEV	1.17	0.854388
PITI	1.17	0.854635
TAT	1.15	0.8695
ROE	1.07	0.934044
GROWTH	1.06	0.940608
Mean	1.45	

Therefore, the crippled environmental regulatory system leads to the uneven level of environmental regulation in different places, which cannot have a significant effect on EDIAQ, M-EAIDQ, and NM-EAIDQ.

The regression coefficients of MEDI passed the significance test. There is a significant positive effect of public opinion monitoring on M-EAIDI at the 1% level. It indicates that the higher the level of public regulation of mining companies, the more willing the companies are to disclose more monetary environmental accounting information. This result can verify H5.

The regression results of the control variables show that board size has a significant positive effect on EAIDQ, M-EAIDQ, and NM-EAIDQ at the 0.1% level. This indicates that increasing the size of the board of directors can increase the role of the board of directors in corporate decision-making, which can significantly increase the overall EAIDQ, M-EAIDQ, and NM-EAIDQ. The number of board meetings and the supervisory board size have no significant effect on EAIDQ, M-EAIDQ, and NM-EAIDQ. In addition, solvency and whether the audit unit is a Big Four have significant effects on EAIDQ, M-EAIDQ, and NM-EAIDQ at the 0.1% level. In contrast, operating capacity, profitability, and development capacity have no significant effects on EAIDQ, M-EAIDQ, and NM-EAIDQ.

5.　Robustness Test

In order to further verify the reliability of the above conclusion, in this study the shareholding ratio of the top three shareholders (CRIO) is used as a surrogate variable for the shareholding ratio of the largest shareholder (FIRST); the female proportion of independent directors (FIDR) is used as a substitution variable for the proportion of independent directors (IDR); and the marketization index (MI) is used as a substitution variable of the PITI to regress the former models (Liu & Bai, 2022).

Table 16.7 Results of Multiple Regression Analysis

Variable	Model 4 EAIDI	Model 5 NM-EAIDI	Model 6 M-EAIDI
FIRST	0.30***(4.93)	0.33*** (5.09)	**0.24*** (3.93)**
IDR	0.45**	0.56**	0.30
	(2.69)	(3.13)	(1.81)
PITI	0.00	0.00	0.00
	(1.78)	(1.84)	(1.42)
MEDI	0.02*	0.01	0.02**
	(2.07)	(1.05)	(2.70)
BSIZE	0.02***	0.02***	0.01**
	(3.88)	(4.14)	(2.97)
BOME	0.00	0.00	−0.00
	(0.19)	(1.79)	(−1.40)
SUPSIZE	0.00	−0.00	0.00
	(0.10)	(−0.38)	(0.55)
ROE	0.01	−0.01	0.02
	(0.23)	(−0.35)	(0.76)
TAT	0.02	0.02	0.01
	(1.09)	(1.09)	(0.90)
LEV	1.24***	1.29***	1.07***
	(9.97)	(9.59)	(8.62)
AUDIT	0.15***	0.11***	0.16***
	(4.70)	(3.38)	(5.18)
GROWTH	−0.00	−0.01*	−0.00
	(−1.35)	(−2.17)	(−0.32)
STATE	control	control	control
YEAR	control	control	control
_cons	−0.34***	−0.32**	−0.32***
	(−3.54)	(−3.11)	(−3.36)
N	710	710	710
Adj. R^2	0.35	0.30	0.33

t-statistics in parentheses.
* $p < 0.05$, ** $p < 0.01$, *** $p < 0.001$.

5.1 Test for Hypothesis 1

In this model, the sum of the equity ratios of the top three shareholders is used
to replace the equity ratio of the largest shareholder to test the relationship between ownership concentration and EAIDQ. If CRIO is significantly correlated for
EAIDQ, we can prove the model to be robust.

$$\begin{aligned}
EAIDI_i = {} & \alpha_0 + \alpha_1 CRIO_i + \alpha_2 FIDR_i + \alpha_3 MI_i + \alpha_4 MEDI_i \\
& + \alpha_5 BSIZE_i + \alpha_6 BOME_i + \alpha_7 SUPSIZE_i \\
& + \alpha_8 STATE_i + \alpha_9 ROE_i + \alpha_{10} TAT_i + \alpha_{11} LEV_i \\
& + \alpha_{12} AUDIT_i + \alpha_{13} GROWTH_i + \alpha_{14} YEAR_i + \varepsilon_i
\end{aligned} \tag{7}$$

5.2   Test for Hypotheses 2 and 3

In this model, the proportion of independent directors is replaced by the proportion of female independent directors to test the relationship between corporate governance and NM-EAIDQ. And the MI is used instead of the environmental disclosure index to test the relationship between environmental supervision and NM-EAIDQ. Significantly correlation of both FIDR and MI for NM-EAIDQ can prove the model to be robust.

$$
\begin{aligned}
NM\_EAIDI_i = {}& \alpha_0 + \alpha_1 CRIO_i + \alpha_2 \mathbf{FIDR_i} + \alpha_3 \mathbf{MI_i} \\
& + \alpha_4 MEDI_i + \alpha_5 BSIZE_i + \alpha_6 BOME_i + \alpha_7 SUPSIZE_i \\
& + \alpha_8 STATE_i + \alpha_9 ROE_i + \alpha_{10} TAT_i + \alpha_{11} LEV_i + \alpha_{12} AUDIT_i \quad (8) \\
& + \alpha_{13} GROWTH_i + \alpha_{14} YEAR_i + \varepsilon_i
\end{aligned}
$$

5.3   Test for Hypothesis 4

This model is still used to test the relationship between M-EAIDQ and external supervision by public opinion. Although there is no suitable replacement variable for MEDI, this work still retains this model to study the effect of replacing other variables on the relationship between MEDI and M-EAIDQ. If the relationship between MEDI and M-EAIDQ is the same as the previous regression results, it proves that the model is robust.

$$
\begin{aligned}
M\_EAIDI_i = {}& \alpha_0 + \alpha_1 CRIO_i + \alpha_2 FIDR_i + \alpha_3 MI_i + \alpha_4 \mathbf{MEDI_i} \\
& + \alpha_5 BSIZE_i + \alpha_6 BOME_i + \alpha_7 SUPSIZE_i \\
& + \alpha_8 STATE_i + \alpha_9 ROE_i + \alpha_{10} TAT_i + \alpha_{11} LEV_i \quad (9) \\
& + \alpha_{12} AUDIT_i + \alpha_{13} GROWTH_i + \alpha_{14} YEAR_i + \varepsilon_i
\end{aligned}
$$

The robustness results are shown in Table 16.8. It can be seen from the table that the shareholding ratio of the top three shareholders is significantly positively correlated with EAIDQ, M-EAIDQ, and NM-EAIDI at the 0.1% level. There is a significant positive correlation between the proportion of female independent directors and NM-EAIDQ at the 1% level. MI has no significant correlation with EAIDQ, M-EAIDQ, and NM-EAIDI at the 5% level. In summary, the robustness results are consistent with the original results, which can indicate that the conclusions of this study are robust and reliable.

6.   Mitigating Endogeneity Issue

6.1   Propensity Score Matching

Propensity score matching (PSM) is a non-parametric analysis method for counterfactual inference. By analysing and processing non-experimental data and

Table 16.8 Results of the Robustness Test

Variable	Model 7 EAIDI	Model 8 NM-EAIDI	Model 9 M-EAIDI
CRIO	0.28*** (7.20)	0.32*** (7.57)	0.22*** (5.60)
FIDR	0.41* (2.45)	0.54** (2.98)	0.26 (1.52)
MI	0.00 (0.22)	0.00 (0.68)	−0.00 (−0.26)
MEDI	0.02* (2.34)	0.01 (1.35)	0.03** (2.89)
BSIZE	0.02*** (3.72)	0.02*** (4.02)	0.01** (2.79)
BOME	−0.00 (−0.09)	0.00 (1.60)	−0.00 (−1.70)
SUPSIZE	0.00 (0.10)	−0.00 (−0.41)	0.00 (0.57)
ROE	0.01 (0.25)	−0.01 (−0.30)	0.02 (0.74)
TAT	0.02 (1.43)	0.02 (1.40)	0.02 (1.21)
LEV	1.11*** (9.02)	1.15*** (8.65)	0.97*** (7.80)
AUDIT	0.09** (2.76)	0.05 (1.29)	0.12*** (3.70)
GROWTH	0.00 (0.20)	−0.00 (−0.56)	0.00 (0.89)
STATE	control	control	Control
YEAR	control	control	Control
_cons	−0.28** (−2.82)	−0.28** (−2.59)	−0.25* (−2.54)
N	710	710	710

t-statistics in parentheses.
* $p < 0.05$, ** $p < 0.01$, *** $p < 0.001$.

observation data, PSM can effectively reduce selection bias and endogeneity (Tang et al., 2022). The research object of this study is listed companies in China's mining industry, and the research industry is relatively single. The EAID of mining companies has certain self-selection characteristics. If only the ordinary least squares (OLS) method is used to estimate the impact of ownership structure, corporate governance, and external supervision on EAIDQ, M-EAIDQ, and NM-EAIDQ, the result error may be caused by sample selection bias, so the PSM method is used to solve this problem, which can make the conclusion more robust.

Since the EAIDQ is mainly reflected in high-pollution industries, it is not meaningful to study the influencing factors of EAID in other non-pollution industries, so in this study non-mining companies in high-pollution industries are selected as the control group. The PSM method is used to match the treatment group (mining industry) and the control group (non-mining industry) and find the control group with similar characteristics, so that it can simulate the "counterfactual" state of the treatment group, so as to compare and analyse the difference in influencing factors of EAIDQ between the mining industry and non-mining industry in high-polluting industry.

6.2 Lagged Test

Based on time series, some data and information have a certain lag (Tan & Xiong, 2021). For listed mining companies, the EAIDQ is affected by uncertain factors such as time. When collecting data, it was found that most companies did not invest in the environment on a year-to-year basis. Some sample companies made a lot of investment in environmental governance and environmental protection in a certain year but did not make any environmental investment in the previous year and the next year. This situation may lead to hysteresis bias in the regression results. Secondly, the external market environment and social environment are changing rapidly. The issuance of environmental policies in a certain year or the occurrence of major environmental protection events may lead to year-to-year fluctuations in the company's EAID. In addition, we consider the slow pace of development of environmental supervision system. Therefore, in this study a lag test is conducted on the main regression results.

Table 16.9 shows the regression results with the data shifted one year forward. From the results, we can see that at the 1% level, ownership concentration and the proportion of independent directors have a significant positive correlation with EAIDI, M-EAIDI, and NM-EAIDI. There was a significant positive correlation between MEDI and NM-EAIDI at the 5% level. Slightly different from the original regression results, PITI is significantly positively correlated with NM-EAIDI at the 10% level. This result is consistent with H3. It can be proved that PITI has a significant impact on NM-EAIDI, but this impact is affected by time lag and is not always significant.

Table 16.10 presents the regression results of the data moved backward by one year as a whole. From the results, we can see that at the 1% level, ownership concentration and the proportion of independent directors have a significant positive correlation with EAIDI, M-EAIDI, and NM-EAIDI. There is a significant positive correlation between MEDI and NM-EAIDI at the 5% level. Slightly different from the original regression results, there is a significant positive correlation between PITI and EAIDI at the 10% level, and a significant positive correlation between PITI and NM-EAIDI at the 5% level. This result is consistent with H3. It can be proved that PITI has a significant impact on NM-EAIDI, but this impact is affected by hysteresis and is not always significant.

In general, although increased environmental regulation would lead to an increase in NM-EAIDQ, this relationship may not be immediately apparent. It may take some time to cause changes in NM-EAIDQ with the strengthening of environmental supervision.

7. Discussions

7.1 Relationships with Previous Literature

With the intensification of global environmental problems and the strengthening of people's awareness of environmental protection, many scholars have studied the internal and external factors that affect the EAIDQ. However, the theoretical evidence and research results provided by relevant scholars are not consistent with

Table 16.9 Result of Lagged Test (F)

Variable	EAIDI	NM-EAIDI	M-EAIDI
F.FIRST	0.29*** (4.82)	0.30*** (4.55)	**0.26*** (4.25)**
F.IDR	0.59*** (3.52)	0.71*** (3.91)	0.42** (2.53)
F.PITI	0.00 (1.63)	0.00* (1.91)	0.00 (1.06)
F.MEDI	0.02* (1.74)	0.01 (0.79)	0.02** (2.38)
F.BSIZE	0.02*** (4.42)	0.02*** (4.78)	0.02*** (3.27)
F.BOME	0.00 (0.34)	0.00* (1.85)	−0.00 (−1.20)
F.SUPSIZE	0.00 (0.50)	0.00 (0.16)	0.01 (0.76)
F.ROE	0.01 (0.40)	−0.01 (−0.25)	0.03 (0.95)
F.TAT	0.03* (1.78)	0.04* (1.89)	0.02 (1.37)
F.LEV	1.28*** (10.22)	1.28*** (9.45)	1.16*** (9.24)
F.AUDIT	0.16*** (5.19)	0.14*** (4.02)	0.17*** (5.44)
F.GROWTH	−0.00 (−1.39)	−0.01** (−2.15)	−0.00 (−0.41)
F.STATE	control	control	control
F.YEAR	control	control	control
_cons	−0.42*** (−4.36)	−0.41*** (−3.95)	−0.39*** (−4.00)
N	639.00	639.00	639.00
R^2	0.39	0.33	0.37

t-statistics in parentheses.

* $p < 0.1$, ** $p < 0.05$, *** $p < 0.01$.

the impact of ownership structure, corporate governance, and external supervision on the EAIDQ.

Similar to the conclusion of this study, Deng (2018) believed that the concentration of ownership improves the level of environmental information disclosure. Zheng and Zheng (2018) believed that the increase in the proportion of controlling shareholders' equity will help improve their control over the management and can effectively prevent the management from harming the interests of major shareholders for their own interests, thereby effectively improving the EAIDQ. However, Wang (2015) proposed just the opposite of the conclusion of this study. He believed that when the equity is concentrated in the hands of a few people, the controlling shareholder will reduce the disclosure of environmental accounting information in order to avoid the loss of interests caused by the disclosure of negative environmental accounting information. Some scholars' research results show that the relationship between ownership concentration and EAIDQ is not significant. Liu and Bai (2022) believed that large shareholders did not play a role in improving EAIDQ, which may be because the ownership structure of enterprises is mainly to improve enterprise value and profitability, but there is no more development in fulfilling environmental responsibilities.

Wu and Dong (2021) believed that independent directors do not have close business contacts with management, and it is easier to make objective

Table 16.10 Result of Lagged Test (L)

Variable	L.EAIDI	L.NM-EAIDI	L.M-EAIDI
L.FIRST	0.26*** (4.25)	0.28*** (4.28)	**0.21*** (3.51)**
L.IDR	0.57*** (3.39)	0.74*** (4.11)	0.36** (2.13)
L.PITI	0.00* (1.90)	0.00** (2.09)	0.00(1.37)
L.MEDI	0.02* (1.92)	0.01 (1.16)	0.02** (2.35)
L.BSIZE	0.02*** (4.61)	0.03*** (5.25)	0.02*** (3.20)
L.BOME	0.00 (0.17)	0.00* (1.67)	−0.00 (−1.32)
L.SUPSIZE	0.00 (0.23)	−0.00 (−0.27)	0.00 (0.68)
L.ROE	−0.02 (−0.65)	−0.05 (−1.38)	0.00 (0.14)
L.TAT	0.02 (1.19)	0.02 (1.23)	0.02 (0.96)
L.LEV	1.30*** (10.17)	1.32*** (9.64)	1.15*** (8.97)
L.AUDIT	0.15*** (4.93)	0.13*** (3.78)	0.16*** (5.21)
L.GROWTH	−0.00 (−1.04)	−0.01* (−1.89)	−0.00 (−0.04)
L.STATE	control	control	Control
L.YEAR	control	control	Control
_cons	−0.42*** (−4.34)	−0.45*** (−4.33)	−0.35*** (−3.60)
N	639.00	639.00	639.00
R^2	0.38	0.34	0.35

t-statistics in parentheses.
* $p < 0.1$, ** $p < 0.05$, *** $p < 0.01$.

judgements on corporate decisions. Therefore, he proposed that in listed companies, the higher the proportion of independent directors, the more effective their supervision and restraint on management behaviour, thereby enhancing the awareness of corporate EAID and promoting corporate disclosure of more environmental accounting information. However, Xu and Shen (2014) mentioned that in addition to the voluntary disclosure of environmental accounting information, few companies will voluntarily disclose non-mandatory environmental accounting information, especially in heavily polluting industries like the mining industry. The above two points of view are consistent with the conclusions of this study. Independent directors can promote mining companies to raise awareness of EAID, but to a large extent, independent directors will only require management to meet mandatory. Therefore, the quality of non-monetary environmental accounting information of mining companies with a high proportion of independent directors is usually higher.

However, the conclusions in this study on environmental supervision are not consistent with the research conclusions of most scholars. Liu and Bai (2022) believed that when the level of environmental supervision is improved, enterprises will actively disclose non-monetary environmental accounting information in order to establish a good corporate image. However, the research results of this study show that there is no significant correlation between environmental supervision and NM-EAIDQ of mining companies. This is mainly because the time spans of different studies are different, and the impact of environmental supervision on NM-EAIDQ has a certain lag.

We also found that public scrutiny has a significantly positive effect on M-EAIDQ. This is similar to the results of many previous studies. Zhang (2017) believed that for mining industries, social media's role of public opinion supervision can effectively improve the level of EAID. Liu and Bai (2022) mentioned that due to the uncontrollability of monetary environmental accounting information, it is more conducive to the authenticity of information obtained by stakeholders such as high shareholders. Therefore, we believed that supervision from the public can effectively improve the M-EAIDQ of mining companies.

Wang et al. (2021) emphasized that the joint effect of internal and external governance factors is very important for heavy-polluting enterprises to improve the level of environmental information disclosure. This is consistent with the research purpose of this study. This work studied the influence of ownership structure, corporate governance and external supervision on EAIDQ, M-EAIDQ, and NM-EAIDQ from internal and external factors and provides theoretical and practical suggestions for EAID for enterprises and governments.

7.2 *Theoretical Implications*

This study verifies the influence of ownership structure, corporate governance and external supervision on the EAIDQ and proves the relationship between them and three internal or external influencing factors from M-EAIDQ and NM-EAIDQ, respectively. This study fills the gaps in the mining industry in terms of M-EAIDQ and NM-EAIDQ influencing factors and also contributes to shareholder theory and supervisory theory.

Our research found that there is a significant positive correlation between ownership concentration and EAIDQ. The results showed that the largest shareholder has absolute influence on the company's decision-making and plays a huge role in improving the EAIDQ of the company. Our research also shows that the company's independent directors further promote the improvement of NM-EAIDQ and disclose as much high-quality non-monetary environmental accounting information as possible without affecting the company's economic interests.

The results of this work's research on external supervision showed that environmental supervision is positively correlated with NM-EAIDQ, but the impact is not significant. According to the hysteresis test results, there is a certain delay in the significance of environmental supervision on NM-EAIDQ. The results of this study indicated that the pressure from external environmental supervision forces companies to disclose more environmental protection information to a certain extent. However, due to weak government supervision, such pressure has not played a significant role in improving NM-EAIDQ at present. Our survey also shows that public supervision generally has higher requirements for corporate M-EAIDQ, which prompts companies to increase the M-EAIDQ and present more real and effective environmental accounting information to the public.

7.3 Practical Implications

Our research results provide valuable enlightenment for decision-makers and managers of enterprises and provide reasonable reference for environmental policymakers. Firstly, overall, the empirical results show that the firm's ownership structure is an important factor affecting EAIDQ. Improving the ownership structure and increasing the shareholding ratio of major shareholders is one of the important ways for enterprises to improve EAIDQ and increase enterprise value.

Secondly, the research in this study gives specific conclusions from two aspects of M-EAIDQ and NM-EAIDQ, which improves the usability and comparability of empirical results, and provides key insights for enterprises and governments to solve environmental disclosure problems. For example, corporate governance is one of the important influencing factors of mining enterprises NM-EAIDQ. For the pursuit of corporate value and the protection of self-interest, increasing the proportion of independent directors in the board of directors mainly plays a role in improving NM-EAIDQ and has little effect on improving M-EAIDQ. This kind of targeted discovery can help companies adjust corporate governance policies in a timely manner, so as to make relevant decisions more accurately.

Thirdly, this study empirically proved the impact of ownership structure, corporate governance, and external supervision on EAIDQ, M-EAIDQ, and NM-EAIDQ; emphasized the importance and necessity of environmental information disclosure; and enhanced the awareness of environmental disclosure of enterprises and the government's awareness of environmental regulation. The empirical research results of this study will help business operators and managers provide guidance on behaviour and decision-making to establish a good corporate image; at the same time, it will also provide effective support for the government and other relevant agencies to improve the environmental disclosure mechanism, which will help increase environmental protection. It is of positive significance to strengthen supervision and improve the environmental supervision system.

Finally, based on the development goals of enterprises, the research results of this study are helpful to improve the performance of companies. The ultimate goal of standardizing company ownership structure and internal governance to improve EAIDQ is to improve corporate performance, which realizes the common development of social economy and environmental protection, and is one of the development goals pursued by China at this stage. Thus, our findings should pique the curiosity of corporate decision-makers and regulators, who are interested in improving firm performance in relation to environmental accounting disclosure.

8. Conclusions and Suggestions

Starting from the factors that affect China's EAIDQ, this study focused on the impact of ownership structure, corporate governance, and external supervision on the EAIDQ of China's A-share listed companies in the mining industry by constructing a multiple linear regression equation. EAIDQ is further subdivided into M-EAIDQ and NM-EAIDQ, and the impacts of ownership structure, corporate governance, and external supervision on the two were studied, respectively. The final basic conclusions are as follows:

(1) The EAIDQ of Chinese listed mining companies is improving, but the overall disclosure level is still not high. We believed that there may be two main reasons for this. One reason may be because the awareness of EAID in some enterprises still needs to be improved. From the data collected in this study, it can be found that many Chinese listed mining companies have invested very little in environmental protection and governance, indicating that the awareness and actions of companies on environmental protection and governance are still insufficient. Companies that do well in greening their environment will naturally disclose higher quality environmental accounting information, as this information can bring positive impact to the company. Another reason may be due to the lack of complete environmental accounting disclosure standards in China and the still inadequate environmental monitoring system. Many companies do not have standard disclosure rules, and many companies will look for environmental supervision loopholes and only selectively disclose environmental accounting information that is mandatory by law, and usually, this information is monetary environmental accounting information, which is highly operable, and its truthfulness and accuracy are yet to be examined, so it leads to the low quality of EAID.

Therefore, this study suggests that relevant government departments need to establish a complete set of EAID guidelines as soon as possible, improve the environmental disclosure supervision system, and standardize the content of EAID of enterprises to urge enterprises to pay more attention to environmental protection and governance, and improve their environmental protection and governance awareness.

(2) Equity ownership structure, corporate governance, and external oversight all have significant effects on environmental accounting disclosure quality, but the effects of different factors differ. Equity shareholding structure has a significant effect on overall EAIDQ, corporate governance and external government oversight have a significant effect on NM-EAIDQ, while external public oversight has a significant effect on M-EAIDQ. Thus, it can be seen that companies and different stakeholders have different effects on environmental information disclosure. Therefore, disclosing comprehensive environmental accounting information is very important for enterprises to meet their own development needs and to satisfy the needs of different stakeholders. According to business management theory, the ultimate purpose of disclosing environmental accounting information by enterprises is to improve the performance of enterprises. We can find that many enterprises disclose a lot of non-monetary environmental information in order to build up their corporate image, but this has little impact on the social public groups. Ordinary stakeholders, such as the general stockholders, judge a company's corporate value and its ability to continue as a going concern mainly based on quantifiable monetary environmental accounting information.

Therefore, this study suggests that companies should disclose more true and effective monetary environmental accounting information to increase corporate value and thus attract more investors. In addition, it is necessary for the government to make clear requirements on the quality of monetary environmental information disclosure by enterprises. The supervision and law enforcement should be increased, and enterprises that can conceal relevant environmental issues should be severely punished. At the same time, the government should also give some praise and reward to enterprises that disclose both monetary and non-monetary environmental accounting information well and encourage enterprises to disclose more environmental accounting information.

This study has the following limitations. First of all, because there are very few mining companies listed on the Beijing Stock Exchange, but the data are incomplete, in this work we only studied the listed companies in the mining industry announced by the Shanghai Stock Exchange and the Shenzhen Stock Exchange. In addition, the PITI used in this article was compiled by the Institute of Public and Environmental Affairs (IPE) release. PITI is missing in some years in some areas, and the average value method is used to replace it. Finally, in this study we used content analysis to evaluate the quality of EAID, which has a certain subjectivity.

References

Agyemang, J. K. (2020). The relationship between audit committee characteristics and financial performance of listed banks in Ghana. *Research Journal of Finance and Accounting, 11*(2), 145–166.

Agyemang, A. O., Yusheng, K., Twum, A. K., Ayamba, E. C., Kongkuah, M., & Musah, M. (2021). Trend and relationship between environmental accounting disclosure and environmental performance for mining companies listed in China. *Environment, Development and Sustainability, 23*(8), 12192–12216.

Chang, K. (2013). The effects of ownership and capital structure on environmental information disclosure: Empirical evidence from Chinese listed electric firms. *WSEAS Transactions on Systems, 12*(12), 637–649.

Cheng, L., Li, Z., & Ma, L. (2011). Analysis of influencing factors of enterprise environmental information disclosure. *Economics and Management Research, 11*, 83–90.

Cheng, Y., & Gong, Q. L. (2022). Problems and countermeasures of enterprise environmental accounting information disclosure. *Hebei Enterprises, 9*, 8–10.

Dai, R. (2020). *Research on the influence of public opinion supervision and manager characteristics on enterprise environmental accounting information disclosure* [Master's thesis, Chengdu University].

Deng, X. (2018). *Research on influencing factors of environmental accounting information disclosure of listed mining companies* [Master's thesis, Hunan University of Science and Technology].

Dowling, J., & Pfeffer, J. (1975). Organizational legitimacy: Social values and organizational behavior. *Pacific Sociological Review, 18*(1), 122–136.

Fu, M. (2016). Analysis on influencing factors of environmental information disclosure of listed mining companies. *Cooperative Economy and Science and Technology, 19*, 68–71.

Fu, Z. (2017). *What factors determine the success and failure of innovation in China? A systemic study of the Chinese mining industry* (Doctoral dissertation).

Ge, C. X. (2017). An empirical study on the impact of corporate social responsibility fulfillment on environmental information disclosure: A case study of listed mining enterprises. *Green Accounting, 2,* 30–36.

Guo, S., & Liu, X. (2018). Research on environmental accounting information disclosure of listed companies in the mining industry in China. *The Collective Economy of China, 32,* 144–145.

Habib, A., & Azim, I. (2008). Corporate governance and the value-relevance of accounting information: Evidence from Australia. *Accounting Research Journal, 21*(2), 167–194.

Hua, Y. F. (2020). Analysis of influencing factors of carbon information disclosure of listed companies under the background of low carbon economy – Statistical analysis of the annual reports of the Top 100 companies in Shanghai Social Responsibility Index from 2015 to 2017. *Modern Marketing (Next Issue), 1,* 58–60.

Huang, B. L., & Zhou, Y. P. (2020). Research on influencing factors of environmental information disclosure – taking listed companies in the building materials industry as an example. *Ecological Economy, 8,* 175–180.

Huang, Q. (2014). Analysis on influence factors of public company's environmental information disclosure. *Journal of Chongqing Liberal Arts College (Social Science Edition), 3,* 99–104.

Kong, Y. (2017). *Research on influencing factors of environmental accounting information disclosure of listed companies in mining industry* [Master's thesis, Xiangtan University].

Li, R. J., & Zhai, X. (2018). Research on public pressure and environmental information disclosure – based on empirical evidence of listed companies in heavy pollution industries. *Friends Account, 23,* 76–83.

Li, S. X., & Zeng, L. Y. (2020). The influence of media reports on environmental information disclosure of pharmaceutical enterprises. *Journal of Shenyang Agricultural University (Social Science Edition).*

Li, Z. P., Ban, H. F., & Yu, H. (2015). Exploring the disclosure of environmental accounting information of listed companies in mining industry. *Friends of Accounting, 20,* 21–25.

Lin, L. (2017). Discussion on China's environmental accounting information disclosure. *Journal of Sichuan Agricultural University, 32*(2), 242–246.

Liu, J. B. (2016). Analysis on the status quo and constraints of environmental accounting information disclosure of listed companies – based on the evidence of listed companies in Sichuan Province. *Reform of the Economic System, 4,* 121–126.

Liu, J. B. (2022). *Under the background of carbon neutral carbon accounting information disclosed the question to explore* [Master's degree thesis, Northeast University of Finance and Economics].

Liu, S. (2012). Legal research on environmental information disclosure of listed companies – from the perspective of the supervision of environmental protection departments. *Environmental Science and Management, 37*(4), 5.

Liu, Z., & Bai, Y. (2022). The impact of ownership structure and environmental supervision on the environmental accounting information disclosure quality of high-polluting enterprises in China. *Environmental Science and Pollution Research, 29*(15), 21348–21364.

Liu, Z., & Liu, M. (2021). Quality evaluation of enterprise environmental accounting information disclosure based on projection pursuit model. *Journal of Cleaner Production, 279,* 123679.

Lu, X. P., Wei, Y. Z., & Xu, S. T. (2016). Research on environmental accounting information Disclosure of listed companies in ferrous metal smelting and rolling processing. *Journal of Commercial Accounting, 10,* 44–47.

Moroney, R., Windsor, C., & Aw, Y. T. (2012). Evidence of assurance enhancing the quality of voluntary environmental disclosures: An empirical analysis. *Accounting & Finance, 52*(3), 903–939.

Nguyen, H. T., Do, B. N., Pham, K. M., Kim, G. B., Dam, H. T., Nguyen, T. T., . . . & Duong, T. V. (2020). Fear of COVID-19 scale – Associations of its scores with health literacy and health-related behaviors among medical students. *International Journal of Environmental Research and Public Health, 17*(11), 4164.

Nie, Z. P., & Wang, Y. W. (2022). The impact of environmental information disclosure on innovation of resource-based enterprises: from the perspective of external pressure. *Resources and Industry*, 1–29.

Qi, Y. A., Yao, J., & Liu, L. (2021). Research on evolutionary game of environmental accounting information disclosure from the perspective of multi-agent. *PLoS ONE, 16*(8), e0256046.

Rahman, M. J., & Wu, J. (2023). M&A activity and ESG performance: Evidence from China. *Managerial Finance, 50*(1), 179–197.

Rahman, M. J., Wu, Q., & Zhu, H. (2024). Corporate social responsibility in times of social distancing: Evidence from China. *Business Ethics, the Environment & Responsibility.*

Rahman, M. J., Zhu, H., & Chen, S. (2023). Does CSR reduce financial distress? Moderating effect of firm characteristics, auditor characteristics, and Covid-19. *International Journal of Accounting & Information Management, 31*(5), 756–784.

Santosa, S. J., Okuda, T., & Tanaka, S. (2008). Air pollution and urban air quality management in Indonesia. *CLEAN–Soil, Air, Water, 36*(5–6), 466–475.

Shen, H. T., & Feng, J. (2012). Supervision by public opinion, Government supervision and enterprise environmental information disclosure. *Journal of Accounting Research*, 72–78.

Sun, J. H., & Ye, L. Y. (2020). Research on the impact of corporate governance structure on the level of environmental accounting information Disclosure – A case study of listed companies in the steel industry. *Jiang Reading Journal, 4*, 66–78 + 122–123.

Tan, C. H., & Xiong, M. Y. (2021). Research lag between the same subject areas at home and abroad: A case study of data mining. *Journal of Information Technology, 6*, 590–602.

Tang, X. B., Bin, J. Y., & Tan, X. L. (2022). Retesting the effect of household income on poverty risk identification: A "counterfactual" estimation based on Propensity score Matching (PSM). *Journal of Natural Sciences*. Hunan Normal University.

Tian, Y. (2021). *Research on environmental accountability and enterprise capital cost based on legitimacy theory* [PhD dissertation, Dong Bei University of Finance and Economics].

Wang, P., Che, F., Fan, S., & Gu, C. (2014). Ownership governance, institutional pressures and circular economy accounting information disclosure: An institutional theory and corporate governance theory perspective. *Chinese Management Studies.*

Wang, X. (2019). *Government regulation, environmental accounting information disclosure and debt financing cost research* [Master's thesis, Hebei University].

Wang, X., Gao, Y., Jiang, X., Zhang, Q., & Liu, W. (2021). Analysis on the characteristics of water pollution caused by underground mining and research progress of treatment technology. *Advances in Civil Engineering, 2021.*

Wang, Y. (2015). Analysis on influencing factors of environmental information disclosure of listed companies in steel industry. *Accounting for Commerce, 14*, 44–47.

Wei, F., & Zhou, L. (2020). Multiple large shareholders and corporate environmental protection investment: Evidence from the Chinese listed companies. *China Journal of Accounting Research, 13*(4), 387–404.

Wu, Y., & Dong, B. (2021). The value of independent directors: Evidence from China. *Emerging Markets Review, 49*, 100763.

Xiong, T., & Cheng, B. (2013). Analysis on influencing factors of environmental information disclosure – Empirical data from listed steel companies in China. *Accounting Communication, 21*, 113–116.

Xu, G. L. (2011). Foreign environmental accounting research: Review, characteristics and enlightenment to our country. *Accounting Communication, 30*, 19–22 + 161.

Xu, S., & Shen, H. (2014). Problems and environmental countermeasures of enterprise accounting information disclosure. *China Market, 40*, 58–59.

Xue, J., He, Y., Liu, M., Tang, Y., & Xu, H. (2021). Incentives for corporate environmental information disclosure in China: Public media pressure, local government supervision and interactive effects. *Sustainability, 13*(18), 10016.

Yan, J. X., Xuan, F. G., & Qin, Y. M. (2018). Impact of environmental accounting information disclosure on enterprise value under government supervision. *The Collective Economy of China, 14*, 135–136.

Zhang, L. (2018). Research on Enterprise environmental accounting information disclosure from the perspective of environmental protection – taking Sinopec as an example. *IOP Conference Series: Materials Science and Engineering, 452*(3), 032022.

Zhang, N. (2017). *An empirical study on the relationship between ownership structure and the quality of carbon accounting information disclosure* [Master's thesis, Capital University of Economics and Business].

Zheng, F. H., & Zheng, L. X. (2018). Influencing factors and mode selection of environmental information disclosure of listed companies. *Statistics and Decision, 21*, 175–178.

Zhu, H., & Rahman, M. J. (2024). Ex-ante expected changes in ESG and future stock returns based on machine learning. *The British Accounting Review*, 101457.

17 Management Control, Performance Measurement, and Climate Risk Management Perspectives for Carbon Accounting

Tarek Rana, Md Jahidur Rahman, and Peter Öhman

1. Introduction

In an era characterized by escalating environmental challenges and increasing regulatory pressures, the role of carbon accounting has gained prominence within the realm of corporate governance and sustainability practices. This chapter delves into the multifaceted dimensions of management control performance measurement and climate risk management, particularly through the lens of carbon accounting. The integration of carbon accounting within management control systems (MCS) and performance measurement frameworks is not merely a technical endeavour but a strategic imperative that underscores the commitment of organizations to sustainable development and environmental stewardship.

The global urgency to address climate change has catalysed significant shifts in corporate strategies, necessitating a robust framework for measuring and managing carbon emissions. This chapter aims to explore the theoretical underpinnings, practical applications, and strategic implications of carbon accounting in fostering sustainable business practices. By examining the intersection of management control, performance measurement, and climate risk management, it provides a comprehensive perspective on how carbon accounting can enhance corporate accountability, transparency, and overall sustainability performance.

The significance of carbon accounting lies in its ability to quantify and report greenhouse gas (GHG) emissions, thereby providing critical data for decision-making processes aimed at reducing environmental impacts. This chapter emphasizes the integration of carbon accounting into MCS, highlighting how it supports strategic planning, operational control, and performance evaluation. Through case studies and empirical analyses, practical benefits of incorporating carbon metrics into financial reporting and managerial decision-making are illustrated.

Moreover, the chapter addresses the regulatory landscape governing carbon accounting practices, emphasizing the need for clear and comprehensive regulations to standardize and institutionalize these practices. It underscores the importance of regulatory compliance and voluntary reporting standards, such as the Global

DOI: 10.4324/9781003488965-22

Reporting Initiative (GRI) and the Carbon Disclosure Project (CDP), in enhancing transparency and accountability.

In addition to regulatory considerations, the chapter explores the role of professional development in advancing carbon accounting practices. It highlights the critical need for dedicated training programmes and interdisciplinary collaboration to cultivate a cohort of skilled professionals equipped to navigate the complexities of green accounting. It further advocates for the integration of carbon accounting principles into academic curricula and professional development initiatives to bridge the gap between theory and practice.

By examining the correlation between the adoption of green accounting practices and enhanced environmental performance, it presents empirical evidence demonstrating that companies with robust carbon accounting practices tend to exhibit superior sustainability outcomes.

The chapter also explores the intricate relationship between carbon accounting and performance management, challenging the traditional notion that environmental initiatives are merely cost centres. It argues that green accounting practices can drive financial value by fostering innovation, improving operational efficiencies, and enhancing corporate reputation. By integrating carbon metrics into performance measurement systems, organizations can achieve a more holistic view of their performance, aligning sustainability goals with strategic and operational objectives.

Overall, the chapter provides a comprehensive exploration of carbon accounting within the broader context of management control, performance measurement, and climate risk management. It underscores the strategic importance of carbon accounting in fostering sustainable business practices and achieving long-term environmental and financial performance. Through a combination of theoretical insights, practical applications, and empirical evidence, the chapter offers valuable guidance for organizations seeking to enhance their sustainability efforts and contribute to the global agenda for climate action and sustainable development.

2. Key Themes

2.1 *Theoretical Development*

In the realm of green accounting, China finds itself at a crucial juncture, navigating the complexities of theoretical development against the backdrop of its evolving socio-economic landscape. While strides have been made, the chapter underscores the pressing need for a robust theoretical framework tailored to China's unique context. At present, China's theoretical groundwork in green accounting is a work in progress. While the country benefits from a foundation built upon traditional accounting principles and ecological understanding, there exists a noticeable gap in the adaptation of these theories to the realm of green accounting. This disconnect poses a significant challenge, as it inhibits the effective integration of green accounting practices into China's economic fabric. The sections in this chapter sheds light on the current shortcomings in theoretical research, emphasizing the imperative of aligning green accounting principles with China's socio-economic realities. One of the challenges lies in reconciling traditional accounting methodologies with the multifaceted considerations inherent in green accounting. Unlike conventional accounting practices, which primarily focus on financial metrics, green accounting necessitates a holistic approach that incorporates environmental and social factors alongside economic indicators. The interdisciplinary nature of green accounting further complicates theoretical development. It requires a synthesis of accounting principles with insights from fields such as ecology, environmental science, and sustainable development. This interdisciplinary approach demands innovative thinking and collaboration across diverse academic disciplines, underscoring concerted efforts to bridge disciplinary divides. To address these challenges, the chapter advocates for collaborative research endeavours that leverage insights from domestic and international sources. By drawing upon established theories from developed nations and tailoring them to suit China's unique context, researchers can enrich the theoretical landscape of green accounting. Additionally, the chapter emphasizes the importance of practical application in driving theoretical advancements. Small-scale pilot projects and real-world case studies serve as learning opportunities, enabling researchers to refine theoretical frameworks based on empirical evidence. Thus, we call for a concerted effort to cultivate a robust theoretical foundation for green accounting in China. By aligning principles with socio-economic realities, and fostering interdisciplinary collaboration, China can pave the way for the effective integration of green accounting practices into its economic framework. This endeavour not only holds the promise of enhancing environmental stewardship but also contributes to the sustainable development goals of the nation.

2.2 Legal and Regulatory Framework

Within the discourse surrounding green accounting in China, a prominent theme emerges: the need for a well-defined legal and regulatory framework. The chapter underscores the role of clear and comprehensive regulations in facilitating the effective implementation of green accounting practices, emphasizing the urgency of addressing existing gaps in environmental legislation. At present, China's legal landscape pertaining to environmental accounting is marked by ambiguity and inconsistency. While the country has made strides in enacting environmental protection laws, there remains a notable dearth of specific regulations governing green accounting practices. This regulatory void poses significant challenges, as it hampers efforts to standardize and institutionalize green accounting within the national framework. The chapter highlights the imperative of filling these regulatory gaps through targeted legislative measures. It calls for the formulation of specific laws and regulations tailored to the intricacies of green accounting, encompassing aspects such as environmental asset valuation, liability assessment, and disclosure requirements. By delineating clear guidelines and standards, such regulations can provide much-needed clarity and direction to businesses and accounting professionals engaged in green accounting practices. Furthermore, the chapter advocates for the integration of green accounting provisions into existing accounting laws and environmental regulations. This harmonization ensures coherence between different legal frameworks, fostering synergy and alignment in the pursuit of environmental stewardship. By embedding green accounting principles within established regulatory structures, China can leverage existing institutional mechanisms to drive sustainable practices across sectors. However, the chapter also highlights the challenges inherent in translating regulatory intent into effective enforcement mechanisms. Despite the existence of environmental laws, lax enforcement and lenient penalties for non-compliance have undermined their efficacy. This underscores the need for robust enforcement mechanisms and stringent penalties to deter environmental violations and incentivize adherence to green accounting standards. The role of regulatory bodies in overseeing and monitoring compliance with green accounting regulations, calls for enhanced regulatory oversight and enforcement mechanisms to ensure accountability and transparency in environmental reporting practices. By empowering regulatory authorities with the resources and mandate to enforce green accounting standards, China can instil greater confidence in the integrity of environmental reporting processes. This underscores the role of a clear and comprehensive legal and regulatory framework in advancing green accounting practices in this country. By addressing existing regulatory gaps, integrating green accounting provisions into existing laws, and strengthening enforcement mechanisms, China can foster an environment conducive to sustainable development and environmental stewardship. Concerted effort towards regulatory enhancement not only enhances accountability and transparency but also signals China's commitment to advancing the global sustainability agenda.

2.3 *Professional Development*

Amid discussions surrounding green accounting in China, a critical theme emerges: the shortage of skilled professionals and the imperative of cultivating talent in this burgeoning field. The chapter elucidates the pressing need for dedicated training programmes in universities to equip aspiring professionals with the requisite knowledge and skills to navigate the interdisciplinary landscape of green accounting effectively. At present, China grapples with a dearth of professionals well-versed in green accounting principles and practices. The nascent nature of the field, coupled with its interdisciplinary nature, poses significant challenges to talent acquisition and retention. Traditional accounting professionals often lack the specialized expertise required to engage with environmental accounting intricacies, while environmental scientists may possess limited accounting acumen. This underscores the need for targeted training programmes that bridge disciplinary divides and foster interdisciplinary competencies, and the importance of collaborative efforts between academic institutions and industry stakeholders in designing and implementing tailored training programmes. Universities can play a pivotal role in nurturing the next generation of green accountants by offering comprehensive curricula that integrate accounting, environmental science, and sustainability principles. By fostering collaboration between accounting and environmental science disciplines, universities can cultivate a cohort of professionals equipped with the diverse skill set necessary to address multifaceted environmental challenges. Furthermore, the chapter advocates for the establishment of specialized green accounting disciplines or tracks within accounting and business programmes to cater to the growing demand for expertise in this field. These specialized programmes would offer targeted coursework covering topics such as environmental economics, carbon accounting, and sustainability reporting, providing students with a robust foundation in green accounting principles and practices. Internships, industry partnerships, and case study analyses could enable students to apply theoretical concepts to real-world scenarios, fostering critical thinking and problem-solving skills essential for success in the field.

The chapter also acknowledges the challenges associated with talent cultivation in green accounting, including faculty shortages, resource constraints, and curriculum development complexities. Addressing these challenges necessitates concerted efforts from policymakers, academic institutions, and industry stakeholders to invest in faculty development, infrastructure, and curriculum innovation. This chapter underscores the critical importance of talent cultivation in advancing green accounting practices in China to address the complex environmental challenges of the 21st century. Through collaborative efforts and strategic investments, China can build a robust talent pipeline capable of driving sustainable development and environmental stewardship for generations to come.

2.4 *Impact of Green Accounting on Corporate Sustainability*

The chapter delves deeply into the impact of green accounting on corporate sustainability within Chinese firms. The central argument posited by the authors is that the integration of environmental costs into financial reporting mechanisms enables companies to prioritize sustainability. This shift not only leads to enhanced environmental performance but also fosters long-term corporate viability. Empirical findings presented in the chapter underscore a significant positive relationship between the implementation of green accounting practices and corporate sustainability, illustrating that companies adopting these practices tend to exhibit superior sustainable development outcomes. Green accounting is designed to aggregate a corporation's environmental costs into its financial reports. This approach aims to reflect the environmental expenses required to generate profits. Traditionally, financial reports have been criticized for overlooking environmental costs, leading to an incomplete picture of a company's true impact. By incorporating these costs, green accounting encourages companies to be more transparent about their environmental responsibilities and the associated costs.

The sections in this chapter highlight several mechanisms through which green accounting enhances corporate sustainability. Firstly, by internalizing environmental costs, companies are motivated to adopt more sustainable practices. This includes investing in cleaner technologies, improving resource efficiency, and reducing waste. Such initiatives not only mitigate environmental harm but also often result in cost savings and operational efficiencies, further reinforcing the business case for sustainability. Secondly, green accounting fosters a culture of sustainability within organizations. When environmental costs are explicitly accounted for, they become a visible part of managerial decision-making processes. This visibility encourages managers to consider the long-term environmental impacts of their actions, leading to more sustainable strategic choices. It also aligns the interests of various stakeholders, including investors, customers, and regulators, who are increasingly demanding greater environmental accountability from businesses. Empirical results indicate that companies with green accounting practices in place show significantly better performance in sustainability measures compared to those that do not. This includes metrics such as resource efficiency, emissions reductions, and overall environmental impact. Moreover, the book's findings suggest that the benefits of green accounting extend beyond mere compliance with regulatory requirements. Companies that proactively adopt these practices are seen to gain a competitive advantage by building a positive reputation, and securing long-term financial performance. This aligns with the growing body of literature suggesting that sustainability and profitability are not mutually exclusive but can be mutually reinforcing. By integrating environmental costs into financial reporting, companies are better equipped to make informed decisions that balance economic and environmental objectives. Accordingly, the adoption of green accounting practices should be encouraged not only as a means of regulatory compliance but also as a strategic approach to achieving long-term sustainability and business success.

2.5   *Green Accounting and Performance Management*

Another critical theme explores whether the implementation of green accounting practices enhances or detracts from a company's financial outcomes. Findings presented in this book suggest a positive correlation between green accounting and financial performance, challenging the traditional notion that environmental initiatives are merely cost centres, instead demonstrating that they can, in fact, create financial value. Green accounting is designed to incorporate environmental costs into a company's financial reports, offering a more comprehensive view of its financial health. By including these costs, green accounting provides a clearer picture of a company's financial status and encourages more sustainable and financially prudent decision-making. One of the primary ways green accounting can enhance financial performance is through cost savings and efficiency improvements. By identifying and accounting for environmental costs, companies can pinpoint areas where resources are being wasted and implement measures to improve efficiency. For instance, reducing energy consumption not only lowers utility bills but also decreases a company's carbon footprint, leading to both financial and environmental benefits. Similarly, waste reduction initiatives can cut disposal costs and material expenses, directly impacting the bottom line, and green accounting can lead to innovation and new revenue streams.

Companies that adopt green practices often find opportunities to develop new products or services that cater to the growing demand for environmentally friendly options. This can open up new markets and enhance a company's competitive edge.

A company that invests in sustainable product design may attract eco-conscious consumers, driving sales and boosting profitability. As consumers become more concerned with sustainability, companies that demonstrate a commitment to environmental responsibility are likely to enjoy better reputations and stronger stakeholder support. This can translate into financial benefits related to increased investor confidence, better access to capital, and enhanced customer loyalty. For instance, companies with strong sustainability records may find it easier to attract investment from funds that prioritize environmental, social, and governance (ESG) criteria. Additionally, green accounting helps companies mitigate risks associated with environmental regulations and potential liabilities. By proactively managing environmental impacts, companies can avoid fines, sanctions, and costly clean-up operations. This risk management aspect is crucial for maintaining financial stability and protecting shareholder value. For example, a company that invests in pollution control technologies not only complies with regulations but also minimizes the risk of legal action and associated costs.

The positive relationship between green accounting and financial performance provides compelling evidence that sustainability and profitability can go hand in hand. Companies that integrate environmental considerations into their financial strategies are better positioned to achieve long-term success, benefiting both their bottom line and the broader community. Thus, the chapter demonstrates that green accounting practices positively impact financial performance, refuting the notion that environmental initiatives are merely expenses. Instead, they are strategic investments that can yield significant financial returns.

2.6 *Environmental Accounting and Organizational Performance*

One of the key themes of this chapter is the intricate relationship between environmental accounting and firm performance, with a particular focus on listed Chinese oil companies. This delves into whether the implementation of environmental accounting practices, especially the disclosure of environmental costs, significantly impacts the financial performance of these firms. The theme is explored through various theoretical frameworks and empirical analyses, providing a comprehensive understanding of the motivations, practices, and outcomes associated with environmental accounting, especially in industries with substantial environmental impacts like the oil sector.

The emphasis on "green growth", "low-carbon development", and "sustainable development" has intensified, with stakeholders such as governments, non-governmental organizations, and communities urging companies to minimize their environmental footprint. This societal demand for greater environmental responsibility underscores the relevance of studying the link between environmental accounting and firm performance. In environmental accounting, this involves addressing environmental costs in financial statements, thereby providing a more comprehensive view of operations and enhancing transparency and accountability. Effective stakeholder management, facilitated by environmental accounting, can lead to improved long-term financial performance by mitigating risks, enhancing corporate reputation, and ensuring regulatory compliance. By aligning their operations and disclosures with societal expectations regarding environmental stewardship, companies can mitigate potential legitimacy gaps that may arise from negative environmental impacts or public scrutiny. This strategic disclosure helps companies sustain their legitimacy over time, particularly in industries perceived as having significant environmental impacts.

However, we call for improved regulatory frameworks and corporate governance to enhance environmental reporting by emphasizing the need for the Chinese government to mandate the disclosure of environmental costs to improve transparency and accountability. By doing so, organizations can better align their operations with sustainable development goals, contributing to broader societal objectives while potentially reaping long-term financial benefits. This dual focus on economic growth and environmental responsibility is critical in shaping the future trajectory of China's petrochemical sector and ensuring sustainable corporate practices.

2.7   *ESG and Organizational Performance*

Another key theme revolves around the relationship between ESG performance and corporate financial performance, specifically examining the moderating effect of firm ownership on this relationship. This exploration is rooted in the context of Chinese A-share listed companies, providing a detailed empirical analysis over the period from 2014 to 2020. By utilizing advanced econometric techniques, an aim is to offer robust evidence on how ESG initiatives influence financial outcomes, with a particular focus on the distinct impacts observed in state-owned enterprises (SOEs) versus non-state-owned enterprises (non-SOEs).

ESG performance has emerged as a critical factor in modern corporate governance, reflecting a company's commitment to sustainable practices and social responsibility. These metrics go beyond traditional financial indicators to assess how well companies manage environmental risks, contribute to social welfare, and uphold governance standards. The chapter underscores the importance of ESG as a multi-dimensional framework that can potentially drive long-term corporate value. It posits that integrating ESG considerations into business strategies can lead to enhanced operational efficiencies, improved stakeholder relations, and ultimately, superior financial performance. The empirical results of Chapter 7 reveal a significant positive relationship between ESG performance and corporate financial performance. Companies with higher ESG scores demonstrate better financial performance, suggesting that ESG initiatives contribute to economic value. This finding supports the hypothesis that ESG activities, through various mechanisms such as risk mitigation, reputation enhancement, and operational improvements, can drive financial success. This also underscores the dual benefits of ESG investments, highlighting their role in fostering both social and environmental sustainability and enhancing corporate financial performance. This dual benefit is particularly pronounced in non-SOEs, making a compelling case for the widespread adoption of ESG practices in corporate strategies.

3.   Research Propositions

3.1   Integration of Carbon Accounting into Management Control Systems

This theme integrates carbon accounting within the broader framework of management accounting and control to enhance corporate sustainability and environmental performance. MCS are pivotal in aligning corporate strategies with sustainability goals, and carbon accounting serves as a critical tool in this alignment, providing quantifiable data to manage and reduce GHG emissions. Carbon accounting, as a subset of environmental accounting, involves measuring and reporting the carbon footprint of an organization's activities (Schaltegger & Burritt, 2010). It is essential for internal decision-making processes aimed at reducing emissions and improving sustainability. The integration of carbon accounting into MCS allows organizations to track carbon emissions alongside financial performance, fostering a comprehensive approach to sustainability management (Hopwood et al., 2010).

A primary function of MCS in this context is to provide relevant information that supports strategic planning and operational control. Carbon accounting data can inform various management control activities, such as budgeting, performance evaluation, and strategic planning. For instance, incorporating carbon costs into budgeting processes encourages managers to consider the environmental impacts of their financial decisions (Henri & Journeault, 2010). By doing so, organizations can set more realistic and environmentally conscious financial targets. Furthermore, performance measurement systems (PMS) within MCS can be adapted to include carbon metrics, enabling organizations to monitor their progress towards sustainability goals (Epstein & Buhovac, 2014). These metrics can include absolute emissions, emission intensity per unit of output, or progress towards emission reduction targets. By integrating these measures into regular performance reviews, companies can ensure that environmental objectives are given due importance alongside financial targets. Incentive systems are another crucial element of MCS that can be aligned with carbon accounting. Linking managerial compensation to the achievement of carbon reduction targets can motivate managers to prioritize sustainability initiatives (Merchant & Van der Stede, 2017). This alignment of incentives ensures that efforts to reduce carbon emissions are not seen as ancillary to the primary business activities but are integrated into the core operational objectives.

Carbon accounting supports environmental management by identifying areas where emissions can be reduced cost-effectively. This aligns with the concept of eco-efficiency, which aims to create more value with less environmental impact (Schaltegger & Wagner, 2006). By analysing carbon data, managers can identify high-emission activities and processes, evaluate alternatives, and implement changes that reduce emissions while maintaining or improving financial performance. This can lead to the adoption of cleaner technologies, process improvements, and more sustainable supply chain practices. Additionally, carbon accounting facilitates compliance with environmental regulations and reporting standards, such as the Greenhouse Gas Protocol and the CDP (Wiedmann & Minx, 2008). Adhering to these standards not only helps organizations avoid legal penalties but also enhances their reputation among stakeholders who increasingly demand transparency and accountability in environmental performance.

The integration of carbon accounting into MCS is vital for advancing corporate sustainability and environmental management. By providing critical data for decision-making, aligning incentives with sustainability goals, and supporting compliance with regulatory standards, carbon accounting helps organizations manage their environmental impact more effectively. This holistic approach ensures that environmental considerations are embedded in the strategic and operational fabric of the organization, fostering long-term sustainability and financial viability.

3.2 Integration of Carbon Accounting and Sustainability Performance Measurement

The integration of carbon accounting with sustainability performance measurement (SPM) represents a significant advancement in how organizations manage and report their environmental impacts. This integration ensures that environmental considerations, particularly carbon emissions, are systematically measured, managed, and reported alongside traditional financial metrics. By incorporating carbon accounting into SPM frameworks, organizations can achieve a more holistic view of their performance, aligning sustainability goals with strategic and operational objectives. Carbon accounting involves quantifying and reporting an organization's carbon emissions, providing critical data for managing and reducing these emissions (Schaltegger & Burritt, 2010).

When integrated into SPM systems, carbon accounting offers a clear and measurable way to track environmental performance, facilitating better decision-making and enhanced transparency. This integration is vital as it aligns environmental initiatives with the broader business strategy, ensuring that sustainability is embedded into the organizational culture and operations. One of the key benefits of integrating carbon accounting with SPM is the ability to set, monitor, and achieve sustainability targets. PMS traditionally focus on financial metrics, but the inclusion of carbon metrics enables organizations to track their progress towards reducing GHG emissions (Henri & Journeault, 2010). This can involve metrics such as total carbon emissions, emissions intensity per unit of output, and progress towards specific reduction targets. By incorporating these metrics, organizations can ensure that their sustainability goals are specific, measurable, achievable, relevant, and time-bound (SMART), thus enhancing accountability and performance. The integration also supports strategic planning by providing a comprehensive understanding of the environmental impacts associated with different business activities.

Carbon accounting data can highlight high-emission areas, enabling managers to identify opportunities for improvement and prioritize investments in low-carbon technologies and processes (Epstein & Buhovac, 2014). For instance, data from carbon accounting can inform decisions on energy efficiency projects, renewable energy adoption, and supply chain optimizations, all of which contribute to reduced emissions and improved sustainability performance. Furthermore, the integration of carbon accounting into SPM enhances stakeholder communication and reporting. As stakeholders, including investors, customers, and regulators, increasingly demand transparency and accountability in corporate sustainability, having robust carbon accounting data integrated into performance reports becomes crucial (Wiedmann & Minx, 2008). This transparency not only builds trust but also meets regulatory requirements and voluntary reporting standards such as the GRI and the CDP. Effective communication of carbon performance can improve a company's reputation, attract socially responsible investors, and satisfy customer expectations.

Integrating carbon accounting with SPM can drive continuous improvement through benchmarking and performance evaluation. Organizations can compare their carbon performance against industry standards or best practices, identify gaps, and implement corrective actions. This benchmarking process is essential for fostering a culture of continuous improvement and innovation in sustainability practices (Hopwood et al., 2010). For example, a company can use benchmarking data to evaluate the effectiveness of its emission reduction initiatives, adjust its strategies, and implement more effective measures based on industry leaders' practices. The integration also aligns with the broader management accounting practices by linking environmental performance with financial outcomes. By demonstrating the financial benefits of carbon reduction initiatives, such as cost savings from energy efficiency or increased revenue from green products, companies can justify sustainability investments and align them with their financial goals (Schaltegger & Wagner, 2006). This alignment ensures that sustainability is not seen as a cost burden but as an integral part of the value creation process. Hence, the integration of carbon accounting with SPM is a critical step towards achieving comprehensive and effective sustainability management. By embedding carbon metrics into PMS, organizations can better manage their environmental impacts, enhance transparency, and drive continuous improvement. This integration supports strategic planning, stakeholder communication, and financial justification for sustainability initiatives, ensuring that environmental performance is aligned with broader business objectives.

3.3 *Carbon Accounting and Reporting for Climate Risk Management*

The integration of carbon accounting and reporting within climate risk management frameworks is an emerging imperative for organizations aiming to mitigate the impacts of climate change. Carbon accounting involves the measurement and reporting of GHG emissions generated by a company's operations (Schaltegger & Burritt, 2010). This data forms the foundation for effective climate risk management, allowing organizations to identify, assess, and mitigate the risks associated with their carbon footprint. Climate risk management encompasses strategies and actions taken by organizations to address the risks posed by climate change. These risks can be broadly categorized into physical risks, such as extreme weather events and sea-level rise, and transition risks, including policy changes, technological advancements, and market shifts towards a low-carbon economy (Task Force on Climate-related Financial Disclosures [TCFD], 2017).

Integrating carbon accounting into climate risk management enables companies to quantify their exposure to these risks and develop appropriate mitigation strategies. One of the primary benefits of carbon accounting in climate risk management is the ability to identify high-risk areas within an organization's operations. By quantifying GHG emissions, companies can pinpoint processes, activities, and assets that contribute significantly to their carbon footprint. This identification is crucial for developing targeted mitigation strategies, such as investing in energy-efficient technologies, switching to renewable energy sources, and optimizing supply chain logistics to reduce emissions (Epstein & Buhovac, 2014). For instance, a company that identifies its manufacturing process as a significant source of emissions can focus its efforts on implementing cleaner production techniques and energy-saving measures.

Carbon accounting supports compliance with regulatory requirements and voluntary reporting standards. Governments and regulatory bodies worldwide are increasingly mandating the disclosure of carbon emissions and climate-related risks (Wiedmann & Minx, 2008). Adhering to these regulations not only helps companies avoid legal penalties but also enhances their reputation and credibility with stakeholders. Voluntary frameworks like the GRI and the CDP encourage companies to disclose their carbon footprint and climate risk management strategies, fostering greater transparency and accountability. Integrating carbon accounting into climate risk management also aligns with the recommendations of the Task Force on Climate-related Financial Disclosures (TCFD). The TCFD advocates for the disclosure of climate-related risks and opportunities in financial reports, emphasizing the importance of considering climate impacts in financial planning and decision-making (TCFD, 2017). By incorporating carbon accounting data into TCFD-aligned reports, companies can provide stakeholders with a comprehensive understanding of their climate-related exposures and the measures being taken to mitigate these risks. This transparency is particularly valued by investors who are increasingly considering ESG factors in their investment decisions.

Additionally, carbon accounting enhances an organization's resilience to climate-related risks by informing scenario analysis and stress testing. Scenario analysis involves evaluating the potential impacts of different climate scenarios on a company's operations and financial performance (Bowen & Wittneben, 2011). Carbon accounting data provide the necessary inputs for these analyses, enabling companies to assess how various emission reduction pathways and climate policies might affect their business. This forward-looking approach helps organizations prepare for a range of possible futures, ensuring they are better equipped to navigate the uncertainties of climate change. Carbon accounting also supports the identification and pursuit of climate-related opportunities. As the global economy transitions to a low-carbon model, companies that proactively manage their carbon footprint can gain competitive advantages, such as access to new markets, enhanced brand reputation, and cost savings from improved efficiency (Schaltegger & Wagner, 2006). For example, a company that invests in renewable energy not only reduces its emissions but may also benefit from lower energy costs and increased appeal to environmentally conscious consumers. Therefore, the integration of carbon accounting and reporting within climate risk management frameworks is essential for organizations seeking to address the challenges posed by climate change.

By providing accurate and comprehensive data on GHG emissions, carbon accounting enables companies to identify high-risk areas, comply with regulatory requirements, enhance transparency, and build resilience to climate-related risks. This approach supports strategic decision-making and positions companies to capitalize on the opportunities arising from the transition to a low-carbon economy. Effective carbon accounting and reporting are thus pivotal in managing climate risks and ensuring sustainable business practices.

3.4   *Carbon Accounting and Ecological Management Control*

The integration of carbon accounting into ecological management control represents a transformative approach in the way organizations manage their environmental impacts. Ecological management control (EMC) encompasses systems and processes that guide and evaluate an organization's environmental performance, aligning it with broader ecological and sustainability goals (Burritt & Schaltegger, 2014). By incorporating carbon accounting into EMC, companies can more effectively measure, manage, and mitigate their GHG emissions, contributing to improved ecological outcomes and sustainable business practices. Carbon accounting involves the systematic measurement and reporting of an organization's carbon emissions. This process provides a detailed understanding of the carbon footprint associated with various operations and activities (Wiedmann & Minx, 2008). When integrated into EMC systems, carbon accounting serves as a tool for tracking environmental performance and identifying areas for improvement. This integration enables organizations to set precise targets for emission reductions, monitor progress, and implement strategies that enhance their overall environmental stewardship.

A primary function of EMC is to provide accurate and relevant information that supports decision-making processes aimed at reducing environmental impacts. Carbon accounting data, when utilized within EMC frameworks, informs various aspects of environmental management, such as resource allocation, operational adjustments, and strategic planning (Henri & Journeault, 2010). By identifying the most carbon-intensive processes, managers can prioritize interventions that yield the greatest ecological benefits. This might include investing in energy-efficient technologies, optimizing logistics to reduce transportation emissions, or transitioning to renewable energy sources.

Performance measurement is a key component of EMC, and the inclusion of carbon metrics enhances the robustness of environmental performance evaluations. Traditional performance measurement systems often focus on financial indicators, but integrating carbon metrics allows for a comprehensive assessment of an organization's ecological footprint (Epstein & Buhovac, 2014). Metrics such as total GHG emissions, emissions intensity per unit of output, and reductions achieved relative to targets provide tangible measures of ecological performance. These metrics facilitate continuous monitoring and benchmarking, driving improvements in environmental management practices.

Incentive systems within EMC can also benefit from the integration of carbon accounting. Aligning managerial incentives with environmental performance goals ensures that sustainability initiatives receive the necessary attention and resources (Merchant & Van der Stede, 2017). By linking compensation to the achievement of carbon reduction targets, organizations can motivate managers to prioritize actions that enhance ecological outcomes. This alignment reinforces the importance of sustainability within the corporate culture and operational strategies. Moreover, carbon accounting supports regulatory compliance and adherence to environmental standards, which are integral to EMC. As governments and international bodies implement stricter regulations on carbon emissions, having robust carbon accounting practices helps organizations stay compliant and avoid legal penalties (Wiedmann & Minx, 2008). Additionally, voluntary frameworks such as the GRI and the CDP encourage transparency in reporting environmental impacts, further embedding carbon accountability into organizational practices. The integration of carbon accounting into EMC also facilitates stakeholder engagement and communication. As stakeholders, including investors, customers, and regulators, demand greater transparency and accountability in environmental performance, carbon accounting provides the necessary data to meet these expectations (Hopwood et al., 2010). Transparent reporting of carbon emissions and reduction efforts enhances an organization's reputation and credibility, attracting investment from socially responsible investors and building trust with consumers.

Carbon accounting further aids in the identification of opportunities for ecological innovation and efficiency. By analysing carbon data, organizations can uncover areas where resource use can be optimized, leading to both environmental and economic benefits. For example, reducing energy consumption through efficiency measures not only lowers carbon emissions but also reduces operational costs, creating a win–win scenario for sustainability and profitability (Schaltegger & Wagner, 2006). Hence, the integration of carbon accounting into ecological management control systems is essential for advancing sustainable business practices and achieving ecological goals. By providing detailed insights into carbon emissions, supporting regulatory compliance, enhancing performance measurement, and aligning incentives, carbon accounting strengthens the overall effectiveness of EMC. This integrated approach ensures that environmental considerations are embedded in strategic and operational decisions.

3.5 Carbon Accounting for "Biodiversity", "Net-Zero", and "Nature Positive" Targets

The integration of carbon accounting into the pursuit of "biodiversity", "net-zero", and "nature positive" targets is a crucial step towards comprehensive environmental management. As organizations increasingly recognize the interconnectedness of carbon emissions, biodiversity conservation, and overall ecological health, carbon accounting becomes an essential tool for measuring, managing, and achieving these ambitious targets. By systematically quantifying GHG emissions, organizations can better align their strategies with broader environmental goals, ensuring that their operations support rather than undermine global sustainability efforts. Carbon accounting, which involves the detailed measurement and reporting of carbon emissions, serves as the foundation for developing and tracking progress towards net-zero and nature positive targets. Net-zero refers to the balance between the amount of GHG emissions produced and the amount removed from the atmosphere, with the ultimate goal of reducing net emissions to zero (IPCC, 2018). Nature positive, on the other hand, aims to halt and reverse biodiversity loss, ensuring that natural ecosystems are preserved and restored (IUCN, 2020).

By integrating carbon accounting into their environmental management systems, organizations can ensure that their activities contribute to these critical objectives. A key aspect of integrating carbon accounting with biodiversity and nature positive targets is the ability to identify and mitigate the ecological impacts of carbon emissions. Carbon-intensive activities often lead to habitat destruction, pollution, and other forms of environmental degradation that threaten biodiversity. By quantifying emissions, organizations can identify the most significant sources of ecological harm and implement targeted mitigation strategies (Schaltegger & Burritt, 2010). For example, reducing emissions from deforestation and land-use change can directly benefit biodiversity by preserving habitats and ecosystems. Furthermore, carbon accounting supports the development of comprehensive strategies to achieve net-zero targets. Organizations can use carbon data to set realistic and science-based emission reduction goals, track their progress, and adjust their strategies as needed (Henri & Journeault, 2010). This involves not only reducing emissions through energy efficiency, renewable energy adoption, and other measures but also investing in carbon offset projects that enhance biodiversity and ecosystem services. For instance, reforestation and afforestation projects can sequester carbon while simultaneously providing habitat for wildlife and supporting ecosystem functions.

The integration of carbon accounting with nature positive targets requires a holistic approach that considers the full spectrum of an organization's environmental impacts. This involves assessing both direct and indirect emissions, as well as the broader ecological consequences of business activities (Epstein & Buhovac, 2014). By expanding the scope of carbon accounting to include impacts on biodiversity and ecosystem health, organizations can develop more comprehensive sustainability strategies that address multiple dimensions of environmental performance. Incorporating carbon accounting into the pursuit of these targets also enhances transparency and accountability. As stakeholders demand greater disclosure of environmental performance, robust carbon accounting practices provide data to meet these expectations (Wiedmann & Minx, 2008). Transparent reporting on carbon emissions and progress towards net-zero and nature positive targets builds trust and credibility, demonstrating an organization's commitment to sustainability and responsible environmental stewardship.

Carbon accounting facilitates the alignment of corporate strategies with global sustainability frameworks and initiatives. Initiatives such as the Science-Based Targets initiative (SBTi) and the TCFD provide guidelines for setting and reporting on climate and biodiversity targets (SBTi, 2021; TCFD, 2017). By adhering to these frameworks, organizations can ensure that their targets are aligned with the latest scientific understanding and best practices, enhancing their contribution to global sustainability efforts. Thus, integrating carbon accounting with biodiversity, net-zero, and nature positive targets is essential for achieving comprehensive and effective environmental management. By providing detailed insights into carbon emissions and their ecological impacts, carbon accounting supports the development and implementation of strategies that address multiple dimensions of sustainability. This approach not only helps organizations achieve their environmental goals but also enhances transparency, accountability, and alignment with global sustainability initiatives.

3.6 *Carbon Accounting for Sustainable Development Goals*

The integration of carbon accounting into the framework of sustainable development goals (SDGs) is pivotal in advancing global sustainability efforts. The 17 SDGs, established by the United Nations in 2015, aim to address a broad range of social, economic, and environmental challenges by 2030 (United Nations, 2015). Carbon accounting, which involves the measurement and reporting of GHG emissions, supports these goals by providing the necessary data to manage and reduce carbon footprints effectively. This integration ensures that efforts to combat climate change are aligned with broader sustainable development objectives, promoting a holistic approach to sustainability.

Carbon accounting contributes directly to several SDGs, particularly Goal 13: Climate Action. By quantifying emissions, organizations can set measurable targets for reducing their carbon footprint, track progress, and implement strategies to mitigate climate impacts (Schaltegger & Burritt, 2010). This process not only supports climate action but also enhances accountability and transparency, as companies can report their emissions and reduction efforts in a standardized and verifiable manner. Transparent reporting builds trust with stakeholders and demonstrates a commitment to sustainability, aligning corporate actions with global climate goals. In addition to supporting Goal 13, carbon accounting indirectly contributes to other SDGs, such as Goal 7: Affordable and Clean Energy, and Goal 12: Responsible Consumption and Production. By identifying high-emission activities, organizations can implement energy efficiency measures and transition to renewable energy sources (Henri & Journeault, 2010). These actions reduce carbon emissions and promote sustainable energy use, advancing Goal 7. Similarly, optimizing resource use and minimizing waste through carbon accounting aligns with Goal 12, fostering more sustainable production and consumption patterns.

The integration of carbon accounting with SDG frameworks also supports Goal 9: Industry, Innovation, and Infrastructure. Carbon accounting data can drive innovation by highlighting areas where technological advancements can reduce emissions and improve efficiency (Epstein & Buhovac, 2014). For example, developing low-carbon technologies and sustainable infrastructure projects can significantly lower an organization's carbon footprint while contributing to economic growth and environmental sustainability. This alignment encourages investments in green technologies and infrastructure, driving progress towards a more sustainable industrial landscape. Moreover, carbon accounting enhances the ability of organizations to contribute to Goal 11: Sustainable Cities and Communities. Urban areas are significant sources of carbon emissions, and effective carbon management is essential for creating sustainable urban environments (Wiedmann & Minx, 2008). By integrating carbon accounting into urban planning and development processes, municipalities and businesses can implement strategies to reduce emissions, such as promoting public transportation, improving energy efficiency in buildings, and enhancing waste management systems. These actions not only lower emissions but also improve the quality of life in urban areas, making cities more sustainable and resilient.

Additionally, carbon accounting supports Goal 17: Partnerships for the Goals, by fostering collaboration among businesses, governments, and civil society. The standardized and transparent reporting of carbon emissions enables various stakeholders to share data, best practices, and technologies, enhancing collective efforts to address climate change (Schaltegger & Wagner, 2006). Collaborative initiatives, such as joint emission reduction projects and cross-sector partnerships, can amplify the impact of individual efforts, driving more significant progress towards achieving the SDGs. The integration of carbon accounting into the pursuit of SDGs also addresses the financial dimensions of sustainability. By providing a clear picture of an organization's carbon footprint, carbon accounting helps attract investment from environmentally conscious investors and access to green financing options (Hopwood et al., 2010).

Investors increasingly prioritize companies that demonstrate strong environmental performance, viewing them as less risky and more sustainable in the long term. This alignment of financial incentives with sustainability goals further drives corporate commitment to achieving the SDGs. To this end, integrating carbon accounting into the framework of SDGs is essential for advancing global sustainability efforts. By providing accurate and comprehensive data on carbon emissions, carbon accounting supports the achievement of multiple SDGs, from climate action and sustainable energy use to responsible production and sustainable urban development. This integration promotes transparency, accountability, and innovation, fostering collaborative efforts and attracting sustainable investments.

4. Mapping Future Research Directions

Future research for carbon accounting could focus on:

4.1 *Theoretical Development*

Exploring Cultural Contexts: Investigating how cultural factors influence perceptions and practices of green accounting in China could provide valuable insights into tailoring accounting principles to cultural norms. This research direction could delve into indigenous philosophies such as Confucianism or Taoism and their implications for environmental stewardship within accounting frameworks.

Integration of Technological Innovations: Examining the integration of emerging technologies, such as blockchain or artificial intelligence, into green accounting practices could enhance efficiency, transparency, and accuracy in environmental reporting. This research direction could explore the potential applications of technology in data collection, verification, and analysis within green accounting systems.

4.2   Legal and Regulatory Framework

Policy Impact Assessment: Conducting comprehensive assessments of the impact of existing environmental policies on green accounting practices could inform policy reform efforts. Research could evaluate the effectiveness, feasibility, and unintended consequences of environmental regulations on corporate reporting standards and sustainability initiatives.

Comparative Analysis: Conducting comparative analyses of green accounting regulatory frameworks across different jurisdictions could identify best practices and lessons learned. Research in this area could explore the strengths and weaknesses of regulatory approaches in promoting environmental accountability and transparency, facilitating cross-border knowledge exchange and collaboration.

4.3   Professional Development

Curriculum Innovation: Investigating innovative pedagogical approaches and curriculum designs for green accounting education could enhance the effectiveness of training programmes. Research could explore interdisciplinary course offerings, experiential learning opportunities, and competency-based assessments to meet the evolving needs of the field.

Industry–Academia Collaboration: Exploring strategies to foster closer collaboration between academia and industry stakeholders could enhance the relevance and applicability of green accounting education. Research in this area could examine best practices for industry partnerships, internship programmes, and knowledge exchange platforms to bridge the gap between theory and practice.

4.4   Impact of Green Accounting on Sustainability

Sector-Specific Analysis of Green Accounting Practices: Future research should focus on conducting a sector-specific analysis to explore the differential impacts of green accounting practices on corporate sustainability across various industrial sectors in China. The current chapter provides an aggregate view of the positive relationship between green accounting and sustainability, but it does not account for the unique environmental challenges and regulatory pressures faced by different industries. By comparing sectors such as manufacturing, mining, technology, and services, researchers can identify sector-specific outcomes and effectiveness of green accounting practices. This analysis would involve gathering detailed data from companies in each sector, applying multiple regression models, and conducting comparative analyses to discern patterns and differences. The insights gained from such research could inform tailored policy recommendations and sector-specific strategies, enhancing the adoption and optimization of green accounting practices to promote sustainability effectively within each industry.

Longitudinal Study on the Evolution of Green Accounting Practices: Another valuable direction for future research is to undertake a longitudinal study that examines the evolution and long-term impacts of green accounting practices on corporate sustainability in Chinese companies. While there are evidence of a positive relationship over a ten-year period, it is crucial to extend the data collection to at least 20 years, utilizing time-series analysis and panel data regression models to evaluate trends, patterns, and temporal dynamics. Additionally, qualitative insights from corporate executives and sustainability officers could complement the quantitative analysis, offering a comprehensive view of the factors driving the changes and sustainability of green accounting practices. Such a study would provide long-term insights into the effectiveness of green accounting, inform policy and corporate strategy, and contribute to the theoretical framework of sustainable accounting and corporate behaviour.

4.5   *Performance Implications of Green Accounting*

Long-Term Financial Implications of Green Accounting: Future research could investigate the long-term financial implications of green accounting practices in various industries. The positive correlation between green accounting and financial performance primarily captures a snapshot over a ten-year period. A longitudinal study extending beyond this timeframe could provide deeper insights into how the financial benefits of green accounting evolve and sustain over time. Such research could involve tracking a cohort of companies that have adopted green accounting practices and comparing their financial performance with those that have not over multiple decades. It is also possible to explore how different phases of economic cycles impact the financial returns from green accounting, offering a more nuanced understanding of its benefits under varying economic conditions. By examining long-term data, researchers can identify persistent trends, fluctuations, and the potential long-lasting advantages of integrating environmental costs into financial reporting, thereby providing robust evidence to support or refine existing theories about the profitability of sustainability initiatives.

Comparative Analysis of Green Accounting in Emerging and developed Markets: Another direction for future research is to conduct a comparative analysis of the impact of green accounting on financial performance between emerging and developed markets. The current study focuses on Chinese companies, providing valuable insights into how green accounting influences financial outcomes in a rapidly growing economy. However, the dynamics might differ significantly in developed markets where regulatory frameworks, market maturity, and stakeholder expectations are distinct. Future research could involve a cross-country comparison, examining companies in developed economies like the United States, Germany, and Japan, alongside those in emerging markets such as India, Brazil, and South Africa. Such a study could analyse how the maturity of financial markets, regulatory environments, and cultural attitudes towards sustainability influence the effectiveness of green accounting practices. By highlighting the differences and similarities across these diverse contexts, the research could offer a more global perspective on the financial viability of green accounting, providing actionable insights for policymakers, investors, and corporate managers in both emerging and developed markets. A comparative analysis would also inform the development of tailored strategies that consider the unique characteristics of different economic environments.

4.6   Carbon Accounting and Management Control Systems

Evaluating the Long-Term Impact of Integrated Carbon Accounting on Financial and Environmental Performance: Future research should focus on evaluating the long-term impacts of integrating carbon accounting into MCS on both financial and environmental performance. While the current body of literature emphasizes the immediate benefits of such integration, including improved sustainability and compliance with regulatory standards, there is a need for longitudinal studies that assess how these practices influence corporate performance over extended periods. Specifically, researchers should investigate how carbon accounting data used in strategic planning, budgeting, and performance measurement affects an organization's ability to achieve long-term sustainability goals and maintain financial health. This could involve case studies of companies that have integrated carbon accounting into their MCS for several years, comparing their performance with those that have not. By analysing trends and outcomes over time, it is possible to provide valuable insights into the sustained efficacy of carbon accounting as a tool for enhancing both environmental and financial performance, thereby offering evidence-based recommendations for businesses and policymakers.

The Role of Technological Advancements in Enhancing Carbon Accounting within MCS: Another promising area for future research is the role of technological advancements in enhancing the effectiveness of carbon accounting within management control systems. With the advent of digital technologies, such as blockchain, artificial intelligence (AI), and Internet of Things (IoT), there are significant opportunities to improve the accuracy, efficiency, and transparency of carbon accounting processes. Research could explore how these technologies can be integrated into existing MCS to provide real-time data on carbon emissions, facilitate more precise measurement and reporting, and enable predictive analytics for better decision-making. For instance, AI can be used to analyse large datasets to identify patterns and predict future emissions, while blockchain technology can ensure the integrity and transparency of carbon data across the supply chain. Future research is suggested to involve theoretical studies and practical implementations, examining the challenges and benefits of adopting these technologies. By identifying best practices and potential pitfalls, such research could guide organizations in leveraging technology to enhance their carbon accounting practices, ultimately supporting more effective environmental management and sustainability efforts.

4.7   Integration of Carbon Accounting and Sustainability Performance Measurement

Impact of Integrated Carbon Accounting and Sustainability Performance Measurement on Organizational Performance: Future research is recommended to focus on conducting longitudinal studies to assess the impact of integrating carbon accounting with SPM on organizational performance over time. While current studies provide insights into the immediate benefits of this integration, researchers could explore how continuous tracking and reporting of carbon emissions influence strategic decisions, operational efficiency, and overall sustainability performance over extended periods. This research could involve case studies of companies that have implemented integrated carbon accounting and SPM systems, analysing data across multiple years to identify trends, challenges, and successes. By examining how sustained efforts in carbon accounting and SPM affect financial outcomes, stakeholder trust, and regulatory compliance, this research would provide valuable insights into the durability and scalability of these practices in order to help organizations refine their sustainability strategies and ensure that environmental considerations remain central to their business operations.

Technological Innovations in Carbon Accounting and Their Integration with SPM: Another area for future research is the exploration of technological innovations in carbon accounting and their integration with SPM systems. With advancements in digital technologies such as blockchain, AI, and the IoT, there are significant opportunities to enhance the accuracy, efficiency, and transparency of carbon accounting practices. Researchers could investigate how these technologies can be utilized to automate data collection, improve real-time monitoring, and provide predictive analytics for better decision-making. AI could be used to analyse large datasets to identify emission patterns and forecast future trends, while blockchain technology could ensure the integrity and traceability of carbon data across the supply chain. Future studies could involve theoretical frameworks and practical implementations, assessing the impact of these technologies on the effectiveness of integrated carbon accounting and SPM systems. Identifying best practices and potential challenges could guide organizations in leveraging technology to enhance their sustainability performance, ultimately supporting more impactful environmental management efforts.

4.8   Carbon Accounting and Reporting for Climate Risk Management

Enhancing Predictive Analytics in Carbon Accounting for Climate Risk Management: While current practices in carbon accounting primarily involve measuring and reporting historical emissions data, integrating advanced predictive analytics can provide a forward-looking approach to managing climate risks. Predictive analytics, powered by AI and machine learning, can analyse vast amounts of data to identify patterns, forecast future emissions, and simulate the potential impacts of different climate scenarios on a company's operations. Researchers could explore the development and application of these technologies to create robust models that predict how various mitigation strategies, such as energy efficiency improvements or shifts to renewable energy, could influence a company's carbon footprint and associated risks over time. By providing a more dynamic and proactive tool for climate risk management, predictive analytics can help organizations anticipate future regulatory changes, market shifts, and physical climate impacts.

Sector-Specific Impacts of Carbon Accounting on Climate Risk Management: Different industries face unique climate risks and regulatory environments, which influence how carbon accounting practices can be effectively integrated into their climate risk management frameworks. For example, the manufacturing sector may focus on reducing emissions through process optimization and energy efficiency, while the financial sector might prioritize the assessment of climate-related risks in investment portfolios and lending practices. Researchers could conduct comparative studies across various sectors to identify best practices, challenges, and opportunities in implementing carbon accounting for climate risk management. This research could involve case studies of companies within different industries, analysing how they utilize carbon accounting data to manage specific climate risks and comply with sector-specific regulations. By highlighting successful strategies and sector-specific adaptations, it is possible to provide valuable insights for companies looking to enhance their climate risk management practices, contributing to more effective and tailored approaches to sustainability and risk mitigation across diverse industries.

4.9   Carbon Accounting and Ecological Management Control

Exploring the Synergistic Effects of Carbon Accounting and Biodiversity Conservation in Ecological Management Control: Future research could investigate the synergistic effects of integrating carbon accounting with biodiversity conservation initiatives within ecological management control frameworks. This research could explore how organizations can simultaneously manage carbon emissions and biodiversity impacts to create more holistic and effective environmental strategies. By conducting case studies of companies that have successfully integrated these practices, researchers can identify best practices for aligning carbon reduction efforts with biodiversity conservation. Future studies could collect and analyse data on carbon emissions and biodiversity indicators, evaluating how combined management approaches influence overall ecological performance. This could provide valuable insights into how organizations can design and implement integrated ecological management control systems that support both carbon neutrality and biodiversity goals, ultimately contributing to more sustainable and resilient ecosystems.

The Role of Digital Technologies in Enhancing Carbon Accounting and Ecological Management Control: Another direction for future research is to explore the role of digital technologies, such as blockchain, IoT, and AI, and IoT in enhancing the integration of carbon accounting within ecological management control systems. This research could focus on how these technologies can improve the accuracy, transparency, and efficiency of carbon data collection, reporting, and analysis. Blockchain could ensure the integrity and traceability of carbon data, AI could be used to analyse large datasets, identify patterns, and predict future ecological outcomes based on different management scenarios, while IoT devices could provide real-time monitoring of emissions and environmental impacts. A suggestion is to develop and test technology-driven solutions in various organizational contexts, assessing their effectiveness in improving carbon accounting and ecological management. The results could offer practical guidance for organizations seeking to leverage digital innovations to enhance their environmental performance and support sustainable development goals.

4.10   *Carbon Accounting for "Biodiversity", "Net-Zero", and "Nature Positive" Targets*

Integrating Carbon Accounting with Biodiversity Metrics to Support Nature Positive Targets: Future research could further explore the integration of carbon accounting with biodiversity metrics to support the achievement of nature positive targets. We encourage studies investigating how organizations can concurrently track carbon emissions and biodiversity impacts to develop comprehensive environmental management strategies. By examining case studies of companies and conservation projects that have successfully merged these two aspects, researchers can identify effective methodologies for integrating carbon and biodiversity data. The development of dual accounting frameworks could include both carbon and biodiversity indicators, and assess their effectiveness in promoting nature positive outcomes. Insights into the synergies and trade-offs between carbon reduction and biodiversity conservation efforts may offer practical guidance for organizations aiming to achieve holistic environmental sustainability.

Evaluating the Long-Term Impacts of Carbon Accounting on Achieving Net-Zero Targets: While current research often focuses on the immediate effects of carbon accounting, a longitudinal study could provide deeper insights into how sustained carbon accounting practices influence an organization's progress towards net-zero over time. This research would preferably involve tracking a cohort of companies that have committed to net-zero targets, analysing their carbon accounting data, emission reduction strategies, and overall environmental performance over an extended period. Identifying patterns, challenges, and success factors, this study could offer a comprehensive understanding of the effectiveness of carbon accounting in driving long-term carbon neutrality, helping organizations refine their strategies and enhance their commitment to net-zero goals.

Exploring the Role of Policy and Regulation in Promoting Carbon Accounting for Net-Zero and Nature Positive Goals: Future research could also examine the role of policy and regulation in promoting the use of carbon accounting to achieve net-zero and nature positive goals. By conducting comparative studies across regions with varying regulatory environments, researchers can assess which policies and incentives are most effective in encouraging organizations to integrate carbon accounting into their sustainability strategies. This could involve evaluating the influence of regulations such as carbon pricing, emission trading systems, and biodiversity offsets on corporate behaviour and environmental outcomes, and may provide policymakers with evidence-based recommendations for designing regulatory frameworks that effectively promote carbon accounting, supporting broader climate and biodiversity objectives.

4.11   Carbon Accounting for Sustainable Development Goals

Assessing the Impact of Carbon Accounting on Achieving Multiple SDGs: Another suggestion for future research is to focus on assessing the impact of carbon accounting on the achievement of multiple SDGs. Such a study could explore how carbon accounting practices influence not only SDG 13 (Climate Action) but also other interconnected goals such as SDG 7 (Affordable and Clean Energy), SDG 12 (Responsible Consumption and Production), and SDG 15 (Life on Land). By conducting cross-sectional analyses of organizations that have implemented robust carbon accounting systems, researchers can identify the broader environmental, social, and economic benefits. Collecting data on carbon emissions, energy usage, production practices, and biodiversity conservation efforts, and evaluating how these factors contribute to achieving multiple SDGs, could provide knowledge of the synergies and trade-offs involved in carbon accounting, offering insights into how organizations can design integrated strategies that maximize their contribution to the SDGs.

The Role of Digital Technologies in Enhancing Carbon Accounting for SDG Achievement: Future research could explore how advancements in blockchain, AI, and the IoT can improve the accuracy, efficiency, and transparency of carbon accounting practices. By implementing pilot projects and case studies, researchers can assess the effectiveness of these technologies in providing real-time data, ensuring data integrity, and enabling predictive analytics for better decision-making. This could involve evaluating the impact of digital technology integration on the ability of organizations to meet their carbon reduction targets and other related SDGs. The results would likely offer practical insights into how technological innovations can be leveraged to enhance sustainability performance and support the global agenda for sustainable development.

Evaluating Policy Frameworks That Support Carbon Accounting for SDGs: Future research could also examine the effectiveness of various policy frameworks in supporting the integration of carbon accounting practices to achieve the SDGs. By conducting comparative analyses of policy environments in different countries, researchers can identify potential gaps in existing frameworks. This research could involve evaluating policies such as carbon pricing, emission trading schemes, and sustainability reporting mandates, and their influence on corporate behaviour and sustainability outcomes, helping policymakers design and implement regulatory frameworks that effectively promote carbon accounting and drive progress towards the SDGs. This may contribute to a deeper understanding of how policy interventions can facilitate the transition to a low-carbon economy and support sustainable development on a global scale.

4.12   Environmental Accounting and Organizational Performance

The Relationship between Environmental Accounting and Firm Performance across Different Industries: A potential direction for future research is to extend the analysis of the relationship between environmental accounting and firm performance across different industries beyond the oil sector. While this chapter focuses on Chinese oil companies, other industries with varying levels of environmental impact and regulatory pressures (such as manufacturing, technology, and finance) may exhibit different dynamics. Comparative studies across these industries could reveal sector-specific insights and help develop tailored environmental accounting frameworks. Additionally, exploring the impact of industry-specific environmental regulations and international sustainability standards could provide a better understanding of how regulatory environments influence corporate environmental practices and financial outcomes. Such research would not only enhance the generalizability of findings but also offer more nuanced guidance for policymakers aiming to integrate sustainability into corporate strategies.

The Role of External Factors, such as Economic Conditions, Corporate Governance Structures, and Cultural Influences: Another promising avenue for future research is to investigate the role of external factors, such as economic conditions, corporate governance structures, and cultural influences, on the relationship between environmental accounting and firm performance. Current studies primarily examines internal corporate practices and direct regulatory impacts; however, external economic conditions, such as economic downturns or booms, could significantly affect how companies prioritize and report environmental costs. Similarly, variations in corporate governance practices, including board composition, ownership structure, and executive incentives, might influence environmental reporting behaviours and their financial implications. Cultural factors, both within China and in a broader international context, could shape corporate attitudes toward environmental responsibility and transparency. By incorporating external variables into empirical analyses, future research can offer a more holistic view of the factors driving environmental accounting practices globally, and their effectiveness in enhancing corporate performance.

4.13   ESG and Organizational Performance

Exploring the Mechanisms through Which ESG Performance Translates into Financial Performance: While this chapter establishes a positive relationship between ESG and financial performance, the specific pathways and processes that drive this relationship remain underexplored. Future studies could employ a mixed-methods approach, combining quantitative analyses with qualitative case studies, to uncover the nuanced ways in which environmental, social, and governance practices contribute to financial outcomes. Researchers could investigate how ESG initiatives influence operational efficiencies, risk management practices, brand reputation, and stakeholder relationships across various industries. By identifying these mechanisms, future research can provide insights into the strategic value of ESG practices and offer more granular guidance for managers seeking to optimize their ESG investments.

The Role of External Institutional Factors in Shaping the Relationship between ESG and Corporate Financial Performance: A final suggestion is to examine the role of external institutional factors in shaping the relationship between ESG performance and corporate financial performance. This chapter highlights the moderating effect of firm ownership, but external factors such as regulatory environments, market maturity, and cultural attitudes towards sustainability may also play significant roles. Comparative studies across different countries and regions could shed light on how these external factors influence the effectiveness of ESG initiatives. Researchers could analyse how varying levels of government support for sustainability, differences in regulatory requirements, and cultural norms impact the ESG–financial performance nexus. Such comparative studies could reveal broader contextual influences on ESG effectiveness, thereby providing more tailored recommendations for policymakers and businesses operating in diverse environments. This line of research could also explore how multinational corporations navigate different institutional landscapes to implement effective ESG strategies globally.

5. Conclusion

This chapter has explored the critical intersections of carbon accounting, management control, performance measurement, and climate risk management, emphasizing their collective importance in fostering sustainable business practices. The integration of carbon accounting into MCS and performance measurement frameworks emerges as a strategic necessity for organizations committed to environmental stewardship and sustainable development.

The chapter highlights the transformative potential of carbon accounting in enhancing corporate transparency, accountability, and sustainability performance. By quantifying GHG emissions, carbon accounting provides essential data that informs decision-making processes aimed at reducing environmental impacts. This quantification is crucial for setting measurable targets, monitoring progress, and implementing effective strategies to mitigate climate risks.

One of the key insights from this chapter is the role of regulatory frameworks in standardizing and institutionalizing carbon accounting practices. Clear and comprehensive regulations, along with adherence to voluntary reporting standards such as the GRI and the CDP, are fundamental in ensuring the consistency and reliability of environmental reporting. Regulatory compliance not only enhances transparency and accountability but also builds stakeholder trust and supports the broader sustainability goals of organizations.

The chapter underscores the importance of professional development in advancing carbon accounting practices. The shortage of skilled professionals in this burgeoning field necessitates targeted training programmes and interdisciplinary collaboration. By integrating carbon accounting principles into academic curricula and professional development initiatives, organizations can cultivate a workforce equipped with the knowledge and skills required to navigate the complexities of green accounting.

Empirical evidence presented in this book demonstrates the positive impact of carbon accounting on corporate sustainability. Companies that adopt robust green accounting practices tend to exhibit superior environmental performance, including improved resource efficiency, reduced emissions, and enhanced stakeholder relationships. These practices not only contribute to environmental sustainability but also offer financial benefits, challenging the traditional notion that environmental initiatives are merely cost centres.

The relationship between carbon accounting and performance management is further explored, revealing that green accounting practices can drive financial value. By incorporating carbon metrics into PMS, organizations can achieve a more comprehensive view of their performance, balancing economic, environmental, and social objectives. This holistic approach to performance management supports the strategic alignment of sustainability goals with broader business objectives, fostering long-term financial and environmental performance.

The chapter also discusses the importance of integrating carbon accounting with climate risk management frameworks. As organizations face increasing pressures from stakeholders and regulatory bodies to address climate risks, the ability to measure and manage carbon emissions becomes paramount. Effective carbon accounting supports compliance with regulatory requirements, enhances resilience to climate-related risks, and positions organizations to capitalize on opportunities arising from the transition to a low-carbon economy.

The chapter concludes by emphasizing the need for ongoing research and innovation in carbon accounting practices. Future research directions include exploring the integration of technological advancements such as blockchain, AI, and the IoT into carbon accounting systems. These technologies have the potential to enhance the accuracy, efficiency, and transparency of carbon data collection, reporting, and analysis, supporting more effective environmental management.

To this end, the integration of carbon accounting into management control, performance measurement, and climate risk management frameworks is essential for advancing corporate sustainability. By providing critical data for decision-making, supporting regulatory compliance, and driving financial value, carbon accounting plays a pivotal role in helping organizations achieve their sustainability goals. This chapter offers a comprehensive perspective on the strategic importance of carbon accounting, providing valuable insights for academics, practitioners, and policymakers committed to fostering sustainable development and environmental stewardship.

References

Bowen, F., & Wittneben, B. (2011). Carbon accounting: Negotiating accuracy, consistency and certainty across organisational fields. *Accounting, Auditing & Accountability Journal*, *24*(8), 1022–1036.

Burritt, R., & Schaltegger, S. (2014). Accounting towards sustainability in production and supply chains. *The British Accounting Review*, *46*(4), 327–343.

Epstein, M. J., & Buhovac, A. R. (2014). *Making sustainability work: Best practices in managing and measuring corporate social, environmental, and economic impacts*. Berrett-Koehler Publishers.

Henri, J. F., & Journeault, M. (2010). Eco-control: The influence of management control systems on environmental and economic performance. *Accounting, Organizations and Society, 35*(1), 63–80.

Hopwood, A. G., Unerman, J., & Fries, J. (2010). *Accounting for sustainability: Practical insights*. Earthscan.

Intergovernmental Panel on Climate Change (IPCC). (2018). *Global warming of 1.5°C*. www.ipcc.ch/sr15/

International Union for Conservation of Nature (IUCN). (2020). *Nature positive by 2030*. www.iucn.org/theme/nature-based-solutions/initiatives/nature-positive

Merchant, K. A., & Van der Stede, W. A. (2017). *Management control systems: Performance measurement, evaluation, and incentives*. Pearson Education.

Schaltegger, S., & Burritt, R. (2010). Sustainability accounting for companies: Catchphrase or decision support for business leaders? *Journal of World Business, 45*(4), 375–384.

Schaltegger, S., & Wagner, M. (2006). Managing and measuring the business case for sustainability: Capturing the relationship between sustainability performance, business competitiveness and economic performance. In S. Schaltegger & M. Wagner (Eds.), *Managing the business case for sustainability: The integration of social, environmental and economic performance*. Greenleaf Publishing.

Science Based Targets initiative (SBTi). (2021). *Science based targets*. https://sciencebased-targets.org/

Task Force on Climate-related Financial Disclosures (TCFD). (2017). *Final report: Recommendations of the task force on climate-related financial disclosures*. www.fsb-tcfd.org/publications/final-recommendations-report/

United Nations. (2015). *Transforming our world: The 2030 agenda for sustainable development*. https://sustainabledevelopment.un.org/post2015/transformingourworld

Wiedmann, T., & Minx, J. (2008). A definition of "carbon footprint". In C. C. Pertsova (Ed.), *Ecological economics research trends*. Nova Science Publishers.

Index

Note: Numbers in **bold** indicate a table. Numbers in *italics* indicate a figure on the corresponding page.

environmental accounting: organizational performance and 352
environmental accounting disclosure quality: chemical manufacturing companies in China and 203–217
environmental accounting information disclosure, quality of (EAIDQ) 18, 26; effects of ownership structure, corporate governance, and external supervision on the quality of 296–324; government supervision and 303–304; monetary (M-EAIDQ) 18, 298–299, 304–305, 307, 309–310, 312, 314, 316–317, 321–322; non-monetary (NM-EAIDQ) 18, 298, 303–304, 307, 309–310, 312, 314, 316–318, 320–323; ownership concentration and 302; proportion of independent directors and NM-EAIDQ 303; supervision by public opinion and M-EAIDQ 304–305; *see also* environmental information disclosure (EID)
environmental accounting information disclosure index (EAIDI) 305–318, **319**, **320**; annual descriptive statistics of **310**; monetary and non-monetary (M-EAIDI and NM-EAIDI) 306–312, **313**, **315**, 316, **317**, 318
Environmental Carrier Information 186
environmental cost of companies (ENVC) 70–73
environmentally conscious investors 224
environmental information disclosure (EID) 300–301; benefit/motivation of 153–154; in chemical manufacturing 213; corporate image and 214; descriptive statistics of **158**; economic performance and 151–152; effectiveness of, measuring of and hypothesis for 149–163; empirical analysis of 151; as essential for companies as communication tool 9, 62; evaluation of 13; factors affecting 152–153; in food industry 14–15, 183–197; GRI and 209; guidelines in China for 149; hypothesis development for 155; institutional development in China for 154–155; lack of 8; methodology for 156–158; result 158–161; as tool for corporate environmental responsibility 12; US norms for 149; variables to measure 150
Environmental Information Disclosure Guidelines for Listed Companies in China 149

environmental management: carbon accounting for 3–20
environmental performance (EP): financial performance (FP) and 106–121
environmental reporting *see* CDP; GRI
environmental, social, and governance (ESG) performance 9–12, 16–17, 84–85, 334; accounting and firm value during COVID-19 237–256; carbon emissions and 88; carbon neutrality and 79; corporate value and 82; COVID crisis in China and 238–239; effect on enterprise value of 98; financial performance and 128–130; firm value and COVID-19 240–241; future stock returns and 27; impact on financial performance of 125–141; organizational performance and 335; score 81, 87; Wind ESG rating 89, 90
EPS *see* earnings per share
Ernstberger, J. 168
Ernst & Young 14, 170
ESG *see* environmental, social, and governance
Eugster, F. 95
European Commission 166
executive compensation **133**, 134
executives 153, 346
external economic conditions 9, 352
externalities: economic 29; negative 24, 34
external stakeholder factors *see* stakeholder factors
external supervision: Pollution Information Transparency Index (PITI) index and media reports used to gauge 18; quality of EAIDQ and 18

FAG *see* firm age
Fangda Steel **233**
Fang, V. 96
FAOSTAT *see* Food and Agriculture Organization of the United Nations Statistical Database
Fatemi, A. 79, 85, 88, 129
FE *see* fixed-effect (FE) model
Fekrat, M. 224
female directors 14, **133**, 172, 174–175, 177, 179; *see also* board gender diversity (BGDIV)
female independent directors (FIDR) 314, **315**, 316, **317**
Feng, J. 304
Fernando, C. S. 85
FIDR *see* female independent directors
financial management risk 30

For Product Safety Concerns and Information please contact our EU
representative GPSR@taylorandfrancis.com
Taylor & Francis Verlag GmbH, Kaufingerstraße 24, 80331 München, Germany

www.ingramcontent.com/pod-product-compliance
Ingram Content Group UK Ltd.
Pitfield, Milton Keynes, MK11 3LW, UK
UKHW022333100726
473146UK00010B/765